AF702493

Sediment Transfer from Shelf to Deep Water—Revisiting the Delivery System

AAPG Studies in Geology 61

Edited By
R. M. Slatt and C. Zavala

MAR 2013
RECEIVED
Colorado School of Mines
Library

Co-published by

The American Association of Petroleum Geologists and SEPM

ARTHUR LAKES LIBRARY
COLORADO SCHOOL OF MINES
GOLDEN, CO 80401

QE 471.15 .T8 S43 2011

Copyright © 2011
By the American Association of Petroleum Geologists
All Rights Reserved

ISBN13: 978-0 89181-068-1

AAPG grants permission for a single photocopy of an item from this publication for personal use. Authorization for additional photocopies of items from this publication in any form (hardcopy, digital/electronic scanning, or other digital transformation into computer-readable and/or transmittable form) for personal or corporate use is granted provided the appropriate fee is paid directly to the Copyright Clearance Center, 222, Rosewood Drive, Danvers, Massachusetts 01923, http://www.copyright.com. For circumstances occurring outside authorization of Copyright Clearance Center, contact permissions@aapg.org.

AAPG Editor: Stephen E. Laubach
AAPG Geoscience Director: James B. Blankenship

COVER: Lower Miocene channel and channel–margin deepwater hyperpycnites at Cabo San Pablo, Tierra del Fuego, Argentina. These deposits are composed of massive medium to coarse grained sandstones laterally grading into laminated sand-silt couplets with plant debris, the last related to lofting processes at flow margin. Photo by C. Zavala, 2006.

This publication is available from:

The AAPG Bookstore
P.O. Box 979
Tulsa, OK U.S.A. 74101-0979
Phone: 1-918-584-2555 or 1-800-364-AAPG (U.S.A. only)
E-mail: bookstore@aapg.org
www.aapg.org

Canadian Society of Petroleum Geologists
600, 640 – 8th Avenue S.W.
Calgary, Alberta T2P 1G7
Canada
Phone: 1-403-264-5610
Fax: 1-403-264-5898
E-mail: reception@cspg.org
www.cspg.org

Geological Society Publishing House
Unit 7, Brassmill Enterprise Centre
Brassmill Lane, Bath BA13JN
United Kingdom
Phone: +44-1225-445046
Fax: +44-1225-442836
E-mail: sales@geolsoc.org.uk
www.geolsoc.org.uk

Affiliated East-West Press Private Ltd.
G-1/16 Ansari Road, Darya Gaaj
New Delhi 110-002
India
Phone: +91-11-23279113
Fax: +91-11-23260538
E-mail: affiliate@vsnl.com

The American Association of Petroleum Geologists (AAPG) does not endorse or recommend products or services that may be cited, used, or discussed in AAPG publications or in presentations at events associated with AAPG.

The American Association of Petroleum Geologists

Elected Editor
Stephen E. Laubach, Austin, Texas

Geosciences Director
James B. Blankenship, Tulsa, Oklahoma

Books Editorial Committee
Chief Books Editor and Publications Committee Chair
Colin North, Aberdeen, Scotland

Association Books Editors
Michael Sweet, Houston, Texas
Timothy Diggs, Houston, Texas

Association Books Advisors
Bret Fossum, Moscow, Russia
Gretchen Gillis, Houston, Texas
Beverly Molyneux, Tulsa, Oklahoma
William A. Morgan, Houston, Texas
Terrilyn Olson, Denver, Colorado

The American Association of Petroleum Geologists Books Refereeing Procedures

The Association makes every effort to ensure that the scientific and production quality of its books matches that of its journals. Since 1937, all book proposals have been refereed by specialist reviewers as well as by the Association's Publications Committee. If the referees identify weaknesses in the proposal, these must be addressed before the proposal is accepted.

Once the book is accepted, the Association Book Editors ensure that the volume editors follow strict guidelines on refereeing and quality control. We insist that individual papers can only be accepted after satisfactory review by two independent referees. The questions on the review forms are similar to those for the AAPG Bulletin. The referees' forms and comments must be available to the Association's Book Editors on request.

Although many of the books result from meetings, the editors are expected to commission papers that were not presented at the meeting to ensure that the book provides a balanced coverage of the subject. Being accepted for presentation at the meeting does not guarantee inclusion in the book.

More information about submitting a proposal and producing a book for The American Association of Petroleum Geologists can be found on its web site: www.aapg.org.

About the Editors

Roger M. Slatt is currently the Gungoll Chair Professor of Petroleum Geology and Geophysics, and Director of the Institute of Reservoir Characterization in the College of Earth and Energy at the University of Oklahoma. Dr. Slatt had a 14-year career in the oil and gas industry with Cities Service Co. and ARCO/ARCO International, focused on various aspects of exploration and reservoir characterization globally. He also has had a 17-year career in academia, including Head of the School of Geology and Geological Engineering at Colorado School of Mines (1992–2000), Director of the School of Geology and Geophysics at University of Oklahoma (2000–2005), and Director of the Rocky Mountain Regional Petroleum Technology Transfer Council (1995–2000).

He received his undergraduate degree in geology at San Jose State (California) University, and his M.S. (1967) and Ph.D. (1970) degrees in geology from the University of Alaska. He has published more than 100 papers and abstracts, and has made numerous presentations on the subjects of petroleum geology, reservoir geology, seismic and sequence stratigraphy, fluvial, shallow marine and turbidite depositional systems, geology of shale, glacial and Pleistocene–Quaternary geology, and geochemical exploration. He sits on various professional society committees, has organized technical conferences for American Association of Petroleum Geologists (AAPG), and teaches short courses for industry organizations on the *"Introduction to the petroleum geology of deep-water (turbidite) depositional systems," "Principles of geologic reservoir characterization," Applied Sequence Stratigraphy, and Geology of Unconventional Resource Shales*. He has authored books on the first two course topics. His courses have been presented in many places in the U.S., as well as in many countries, including Colombia, Mexico, Indonesia, Malaysia, Peru, Australia, Angola, India, New Zealand, Scotland, and England. He also offers a global, web-based course on *"Introduction to geologic reservoir characterization."* In 1996 he received the AAPG Distinguished Service Award. In 1999 he was the Esso Australia Distinguished Lecturer in Petroleum Geology. In 2001–2003 he was both an AAPG and SPE Distinguished Lecturer. In 2003 he received Honorary Membership in the American Association of Petroleum Geologists. In 2006 he received the AAPG Grover Murray Distinguished Educator Award. In 2007 he received the Society of Exploration Geophysicists (SEG) Special Commendation Award.

Carlos Zavala is currently the president of GCS Argentina S.H. (international consulting services). He has been teaching geology at the Universidad Nacional del Sur, Bahia Blanca Argentina since 1989. He is currently working as a consultant on the petroleum industry for different companies in Venezuela, Mexico, Trinidad, Colombia, and Argentina. After receiving his Ph.D. in 1993 from the Universidad Nacional del Sur, he spent two years (1995–1996) in Italy at the Università degli Studi di Parma, performing a postdoctoral fellowship on turbidites under the guidance of Prof. Emiliano Mutti. He was Member of Research Career of the CONICET (Consejo Nacional de Investigaciones Científicas y Técnicas) of Argentina (1996–2009) and guided several Ph.D. Thesis and research studies focused on sedimentology and stratigraphy.

During the last 20 years, he has published more than 130 papers and abstracts in domestic and international journals, concerning different aspects of the sedimentology, stratigraphy, and sequence stratigraphy of aeolian, lacustrine, fluvial, shallow marine, shelfal, and deep-water clastic deposits. He has taught numerous conferences, field seminars, and short courses including *"Predictive Sedimentology"* and *"Hyperpycnal Systems"* in Argentina (2000–2010), Mexico (2004), Venezuela (2005–2010), and Trinidad (2006–2009). In 2008 he organized the AAPG Hedberg Research Conference "Sediment Transfer from Shelf to Deepwater – Revisiting the Delivery Mechanisms" in Ushuaia, Argentina. His research interest has moved along the last 12 years to the understanding and petroleum significance of hyperpycnal flows and their deposits, hyperpycnites.

Preface

This collection of chapters is an outgrowth of the 2008 AAPG Hedberg Conference titled, "Sediment Transfer From Shelf to Deepwater: Revisiting the Delivery Mechanisms." The conference was held in Argentina in part so that participants could visit well-exposed outcrops and in part because many of the presentations dealt with South American sedimentary basins. Emphasis at the conference was placed upon the role of riverine processes in transporting sediments into marine shelf and basinal environments through relatively long-lived quasisteady hyperpycnal and related flows.

The topic of hyperpycnal flows and their deposits, hyperpycnites, has recently emerged as the latest in a long list of hotly debated topics on deep-water sedimentary processes, environments, and deposits. Perhaps the propensity for debate on deep-water topics is because of the relative rarity in which submarine processes can be directly observed and documented. A partial historical list of such debates includes

- The origin of graded beds (Kuenan and Migliorini, 1950);

- Early identification of hydrocarbon-productive Tertiary sands in California basins as turbidites (Natland and Kuenen, 1951);

- Distal versus proximal deepwater mudstones (Walker, 1967; Nilson, 1980)

- Mopping-up the turbidite mess (Walker, 1973)

- Sandy debrites versus turbidites (Shanmugam, 1997, and related articles in the 1997 AAPG Bulletin, v. 81; Pratson et al., 2006)

- Paradigm shift away from basinward decrease in slope gradient, current velocity, grain size, and bed thickness (Kneller, 1995; Kneller and Buckee, 2000)

- Migration of sinuous submarine channels (Peakall et al., 2000; Posamentier and Kolla, 2003).

All of these, and other debates have led to stimulation of additional research and, ultimately, to advancement in our knowledge of the deep-water sedimentary realm. Such is the the intent of this volume, designed specifically to explore the growing interest in hyperpycnal and associated flows and hyperpycnites as significant contributors to the deep-water sedimentary record. Although first suggested by (Bates, 1953) and noted by Walker (1978), it was really the seminal article by Mulder and Syvitski (1995; followed by several more recent publications, listed in Mulder and Chapron, and Zavala et al., this volume) that elevated interest in hyperpycnal and related processes to the forefront. Numerous modern studies and ancient examples of real and potential hyperpycnites are referenced in the individual chapters comprising this volume, so are not repeated here. Hyperpycnites now stand ready for full and open dialog and debate, which hopefully is prompted by this volume. It places into one document (1) recent advances in understanding modern sediment delivery processes, (2) classifications of sedimentary and stratigraphic features that might (or might not) be diagnostic of hyperpycnal and other types of submarine flows, and (3) examples of both hyperpycnite and nonhyperpycnite deposits and how to distinguish them.

The lead chapter (Mulder and Chapron) focuses on the processes of hyperpycnal floods and their deposits using modern environments, outcrop, core, and seismic reflection profile examples. This key chapter, which sets up the remaining chapters on ancient rocks in outcrop and the subsurface, is followed by the proposed genetic facies tract classification for hyperpycnites by Zavala et al.; not only are the genetic facies groups described in detail, but possible hyperpycnal processes are presented to account for these deposits. A companion chapter by Zavala et al. applies this classification to determine proximality and lateral facies relationships within an oil province, with suggestions on possible extension of reservoir sands. Ponce and Carmona apply a similar facies association classification to some Tertiary channel-levee complexes. Next, Paim et al. document the down-dip changes from fluvial incised channel facies through shelf and slope channel deposits and deeper lobe/fringe facies in continuous outcrop exposures, then discuss their origins within the context of short (surgelike)- versus long (sustained)-lived sediment gravity flows. Campos et al. provide a well-documented example of gravelly and sandy sediments and also discuss whether they are the product of surgelike or sustained flows. Henriksen et al. use three studies to document the effect of high-sediment supply systems on sequence stacking patterns and the role of hyperpycnal flows within these sequences; in particular, they stress the importance of along-strike variability and sediment supply in sequence-stratigraphic interpretation. Two key chapters provide paleoecologic input into the sedimentologic mix—one paper that provides ichnologic characterization of a Cretaceous deltaic system dominated by hyperpycnal processes (Buatois et al.), and the other that documents ichnologic trends from proximal to middle deep-marine hyperpycnites (Carmona and Ponce). Both of these chapters relate trace-fossil assemblages to sedimentologic processes and provide a framework for future ichnologic identification and classification of hyperpycnal systems. The final chapter, by Gamero et al., reinterprets a gas field area as shelf/slope hyperpycnal deposits, which substantially modifies sand trends and geometries for additional exploration and field development.

Taken together, this collection of chapters offers important new insights into the sediment delivery system to deep-marine waters. There are some recurrent themes that thread through the volume, including (1) direct and/or secondary linkages of river-derived sediment into deep-water systems, (2) criteria for identifying hyperpycnal processes and deposits in the stratigraphic record, (3) origin(s) of very coarse-grained deep-water facies, and (4) applications to hydrocarbon exploration and development. All of the authors, as well as the editors, hope that this collection of chapters provides the stimulus for more research into the delivery system(s) by which river-derived sediments ultimately reach the deep ocean.

Roger M. Slatt

Bates, C. C., 1953. Rational theory of delta formation: AAPG Bulletin, v. 37, p. 2119–2162.

Kneller, B., 1995. Beyond the turbidite paradigm: Physical models for deposition of turbidites and their implications for reservoir prediction, *in* A. J. Hartley, and D.J. Prosser, eds., Characterization of deep-marine clastic systems: Geological Society (London) Special Publication 94, p. 31–49.

Kneller B. and C. Buckee, 2000. The structure and fluid mechanics of turbidity currents: A review of some recent studies and their geological implications: Sedimentology, v. 47, p. 62–94.

Mulder, T. and Syvitski, J. P. M., 1995. Turbidity currents generated at river mouths during exceptional discharges to the world oceans. Journal of Geology, v. 103, 285–299.

Natland, M. L., and P. H. Kuenen, 1951. Sedimentary history of the Ventura Basin, California, and the action of turbidity currents: SEPM Special Publication 2, p. 76–107.

Nilsen, T. H., 1980. Modern and ancient submarine fans: Discussion of papers by R. G. Walker and W. R. Normark: AAPG Bulletin, v. 64, p. 1094–1101.

Peakall, J., W. D. McCaffrey, B. C. Kneller, C. E. Stelting, T. R. McHargue, and W. J. Schweller, 2000, A process model for the evolution of submarine fan channels: Implications for sedimentary architecture, *in* A. H. Bouma and C. G. Stone, eds., Fine-grained turbidite systems: AAPG Memoir 72; SEPM Special Publication 68, p. 73–88.

Pratson, L. F., J. Imran, G. Parker, J. P. M. Syvitski, and E. Hutton, 2000. Debris flows vs. turbidity currents: A modeling comparison of their dynamics and deposits, *in* A. H. Bouma, and C. G. Stone, eds., Fine-grained turbidite systems: AAPG Memoir 72; SEPM Special Publication 68, p. 57–71.

Posamentier, H. W., and V. Kolla, 2003. Seismic geomorphology and stratigraphy of depositional elements in deepwater settings: Journal of Sedimentary Research, v. 73, p. 367–388.

Shanmugam, G., 1997. The Bouma sequence and the turbidite mind set: Earth-Science Reviews, v. 42, p. 201–229.

Walker, R. G., 1967. Turbidite sedimentary structures and their relationship to proximal and distal depositional environments: Journal of Sedimentary Petrology, v. 37, p. 25–43.

Walker, R. G., 1973. Mopping-up the turbidite mess, *in* R. N. Ginsburg, ed., Evolving concepts in sedimentology, Johns Hopkins Press, Baltimore, Maryland, p. 1–37.

Walker, R. G., 1978. Deep water sandstone facies and ancient submarine fans: Models for exploration for stratigraphic traps: AAPG Bulletin v. 62, p. 932–966.

Table of Contents

1

Mulder, T., and E. Chapron, 2011, Flood deposits in continental and marine environments: Character and significance, *in* R. M. Slatt and C. Zavala, eds., Sediment transfer from shelf to deep water—Revisiting the delivery system: AAPG Studies in Geology 61, p. 1–30.

Flood Deposits in Continental and Marine Environments: Character and Significance

Thierry Mulder

Université Bordeaux 1, Unité Mixte de Recherches (UMR), Environnements et Paléoenvironnements Océaniques (EPOC), Talence, France

Emmanuel Chapron

Institut des Sciences de la Terre d'Orléans (ISTO), Université d'Orléans, Institut National des Sciences de l'Univers (INSU)-Centre National de la Recherche Scientifique (CNRS), Orléans, France

ABSTRACT

This chapter focuses on the processes, character, and significance of hyperpycnal floods and their deposits (hyperpycnites) in the geologic record based on selected examples from modern environments, outcrops, sediment cores, and seismic reflection profiles.

For bed-load- or suspended-load-dominated hyperpycnites, the rating curve of a flood event can be predicted and is represented by a power-law relationship between discharge and load. Because of reconcentration processes in marine waters, rivers other than those considered dirty (high sediment concentration) may form hyperpycnal flows over long distances on the sea floor. Hyperpycnal floods generally have a meteorologic origin, but catastrophic hyperpycnal floods can also occur because of dam outburst in the catchment area of a river. In some cases, catastrophic floods are associated with earthquakes or volcanic activity. Both in marine and lacustrine environments, frequent and powerful hyperpycnal flows develop meandering channel levee systems and more distal basinal fan lobes.

Hyperpycnites are characterized by a coarsening-upward basal sequence formed during the rising limb of the flood (waxing flow) and a fining-upward upper sequence formed during the falling limb of the flood (waning flow). The lower sequence is not always preserved, so hyperpycnites may be difficult to distinguish from classical turbidites. The common occurrence of organic debris of continental origin at the base of deposits is an indicator of hyperpycnites.

Copyright ©2011 by The American Association of Petroleum Geologists.

DOI:10.1306/13271348St613436

In glacial valleys of mountain ranges, a large amount of clastic material can be deposited in large valley lakes or in fjords by frequent hyperpycnal flows. Both in lacustrine and marine environments, large hyperpycnal floods, which are sometimes catastrophic, are documented in association with significant environmental changes such as deglaciation of the continent or volcano-glacier interactions (jökulhlaup).

INTRODUCTION

Submarine turbidity currents can be initiated by ocean currents of various types, including the transformation of a submarine slide into a laminar and then turbulent flow or the direct flux from continental rivers to submarine environments (Normark and Piper, 1991; Piper and Normark, 2009). The latter case is an important issue because rivers represent the main entry point for terrigenous material provided by erosion on the continents toward sedimentation areas such as lakes or marine sedimentary basins.

Forel (1885, 1892) observed direct transfer from the Rhone River to Lake Geneva (Switzerland), but it was a half century before a real improvement in knowledge of sedimentary processes was made, with the recognition of en masse sediment transfer and the discovery of turbidity currents (Heezen and Ewing, 1952; Kuenen, 1952). The cable failures related to the earthquake-triggered 1929 Grand Banks (eastern Canada) turbidity current demonstrated the existence of energetic submarine gravity processes able to play a part in shaping continental margins. Graded deposits were related to turbidity currents (Kuenen and Migliorini, 1950; Kuenen, 1953). Bouma (1962) defined a sequence (turbidite) consisting of a series of fining-upward sedimentary facies (Ta–Te) that could be related to a waning (velocity decreasing with time) flow (Kneller, 1995; Kneller and Branney, 1995), in which bed-load transport and fallout transport act simultaneously. Hydrodynamically, this turbidity current is a water-sediment mixture, in which particles are maintained in suspension mainly by the upward component of the fluid turbulence (Middleton and Hampton, 1973). Such a mechanism occurs only for low-concentration flows (volume concentration, <9%; Bagnold, 1962). Low concentration allows particle fallout and grain-size sorting. Bed-load transport generates sedimentary structures of the sedimentary facies superposed in the Bouma sequence.

Flow transformation toward a turbidity current (e.g., Heezen and Ewing, 1952; Bourrouilh and Offroy, 1983; Bourrouilh, 1987; Piper et al., 1992) is a concept that includes threshold in flow velocity and concentration (see "ignition" in Parker, 1982). Flow transformation includes different phases from the initial mass failure: debris (cohesive) or hyperconcentrated (granular) density flow (Mulder and Alexander, 2001) including the Lowe sequence (Lowe, 1982), in which bed-load transport dominates; concentrated flow leading to the deposition of the Ta facies of the Bouma sequence; and finally suspension fallout (Tb–Te facies of the Bouma sequence).

The importance of intense sediment transport at river mouths is demonstrated, for example, by Heezen et al. (1960) in the Gulf of Corinth (Figure 1), who compared the frequency of submarine cable failures with the locations of the river mouths, which represent a natural depocenter for terrigenous sediment eroded on the continent. No evidence is seen in the Heezen et al. (1960) article that cable failures are directly related to long-duration flows originating from river floods, instead of the transformation of submarine slides triggered by high sedimentation rates. However, this study indicates the relationship between rivers, sediment flux, and gravity processes. A pioneer study of Houbolt and Jonker (1968) on the Rhone delta in Lake Geneva combined seismic reflection profiling and coring to provide a clear identification of sedimentary facies, channels, levees, and fan lobes in a river-supplied deep basin. In a stratified alpine lake (Lake Thun, Switzerland), Sturm and Matter (1978) further documented with cores the influence on the basin fill of river-borne sediment, which developed either hypopycnal flows (overflows), mesopycnal flows (interflows), or hyperpycnal flows (underflows; Figure 2). Lambert and Giovanoli (1988) used moorings (current meters and thermistor chains) in the active channel of the Rhone delta in Lake Geneva to provide the first identification in a lake of both river-borne dense turbidity underflows (i.e., hyperpycnal flows) and turbidity currents triggered by the remobilization of delta deposits. A similar and quantitative study by Bornhold et al. (1994) used current meter measurements at the mouth of the Homathko and Klinaklini rivers (British Columbia, western Canada) to record turbidity currents related to river floods. However, the measurements of very low suspended-matter

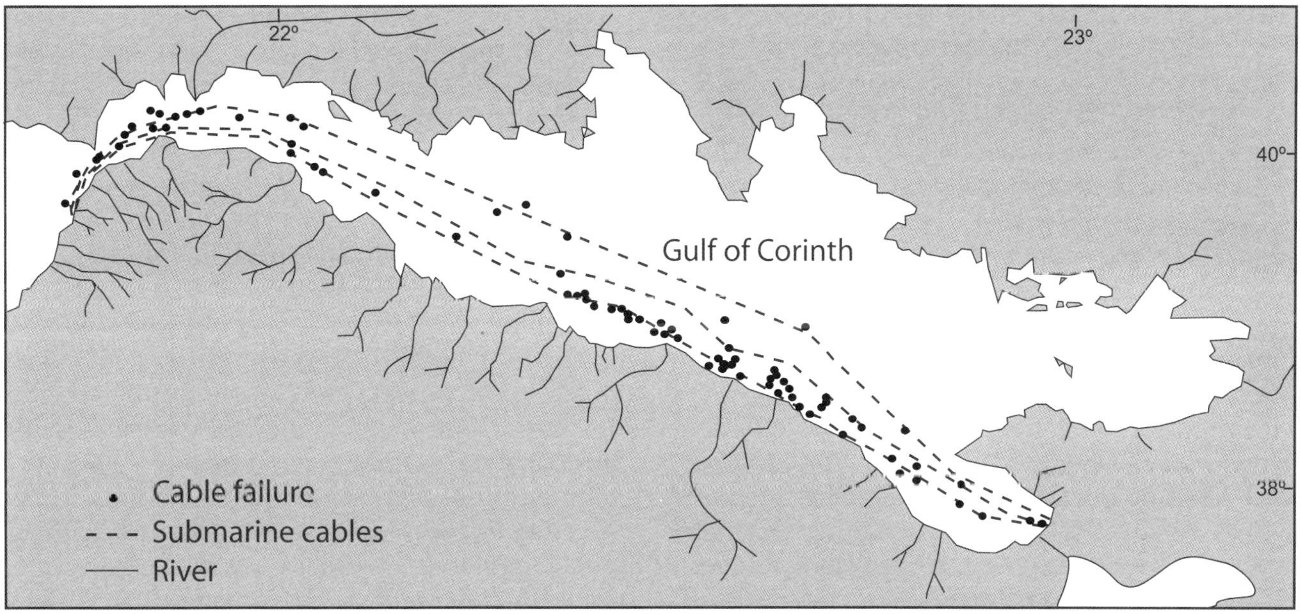

FIGURE 1. Map of the Gulf of Corinth showing the location of cable failures between Corinth and Patras during the period from 1884 to 1939 and the small rivers discharging in the Gulf. Note the clusters of failures located seaward of the river mouth (modified after Heezen et al., 1960).

concentration at these two river mouths showed that these turbidity currents were related to sediment failures generated by mouth-bar progradation instead of the direct transformation of a river flow into a subaquatic sediment flow (hyperpycnal process; see Itirbilung Fjord in Syvitski and Hein, 1991).

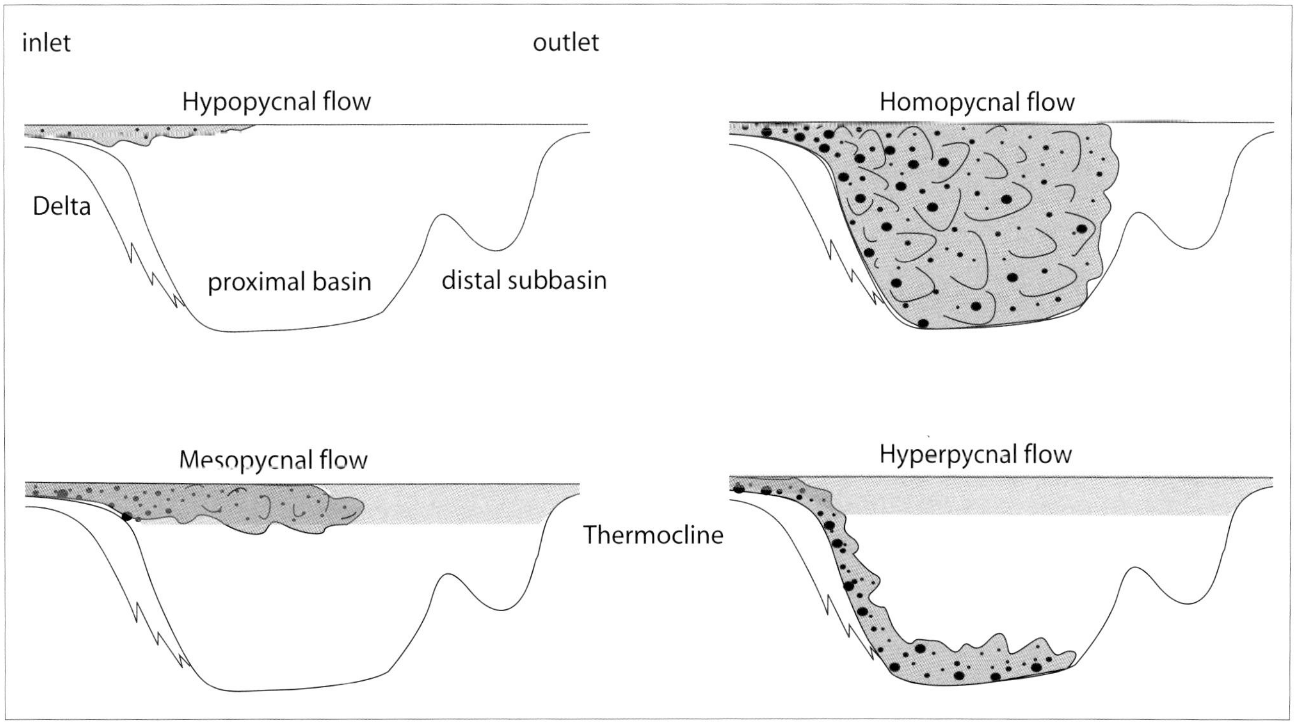

FIGURE 2. Types of flows according to density differences between the effluent and the receiving basin: hypopycnal, homopycnal, mesopycnal, and hyperpycnal flows.

DEFINITION AND TYPES OF RIVER-BORNE FLOWS AND RELATED DEPOSITS

Definition and Types of River-borne Flows

Hyperpycnal flows (Bates, 1953; Mulder and Syvitski, 1995; Nakajima, 2006) are defined in this chapter as sediment-laden subaerial flows entering a subaquatic basin containing either fresh water or salt water. The flow forms a continuum between the subaerial environment and the subaquatic environment. The flow is maintained along the basin floor because of the excess density in the water effluent when compared with the density of the river. Using this definition, hyperpycnal flows differ strongly from turbidity currents resulting from the transformation of sediment failures. Three major differences between hyperpycnal flows and flow resulting from the transformation of a sediment failure can be listed:

1) In a submarine environment, the initial interstitial fluid in a hyperpycnal flow is fresh water. Denser salt water is progressively incorporated through entrainment (e.g., Garcia and Parker, 1993) from the ambient surrounding fluid. In the marine environment, dense fluid entrainment in the moving flow maintains density in excess of the surrounding water and prevents flow lofting. In the submarine environment, seawater entrainment is necessary to maintain the hyperpycnality because initial flow density tends to decrease as a result of sediment deposition, unless ignition occurs. In lakes, entrainment is not necessary. Cold river flows can plunge over a lake floor only because of the difference in temperature between the river water and the lake water. However, lakes frequently are stratified, and a suspended-sediment load is necessary in the river water to plunge through the pycnoclines.
2) Except when related to a catastrophic dam break, hyperpycnal flows are long-duration (range, days to weeks) processes always corresponding to sustained turbidity currents (Lüthi, 1981; Ravenne and Beghin, 1983; Laval et al., 1988). Dam-break hyperpycnal flows and other subaerial flows entering a subaquatic basin, although related to a direct input of fresh water from a river system, can also be considered as turbulent surge.
3) Because of the long duration of hyperpycnal events, bed-load transport can occur over long distances even if the flow is dominated by suspended load and the volume of the bed-load-transported material is small.

Flood deposits of this kind are called "hyperpycnal flows" in the rest of the chapter. They form hyperpycnal flood deposits. The other deposits are related to buoyant (hypopycnal) and partially buoyant (mesopycnal) flows that can form deposits either by simple particle fallout or by reconcentration processes. In the chapter, these are called "hypopycnal flows" or "mesopycnal flows," respectively. Both can also be considered as "flood deposits."

Hyperpycnal flows can be turbulent, and thus correspond to the hyperpycnal turbidity currents of Mulder and Syvitski (1995), or nonturbulent, and thus correspond to the inertia flows of Bates (1953) or the hyperpycnal flows of Mutti et al. (1996).

In terms of origin, floods are dominantly related to climatic events: rainfalls or snowmelt. Rainfall-producing floods are usually seasonal. They can be related to short-duration (range, hours to days) events (thunderstorms, hurricanes) or long-duration (range, weeks to months) events (monsoon, spring snowmelt, or glacier melt). At the geologic scale, global inlandsis melt (combining of a large number of glacier melts) generates larger floods. Geothermal activity can locally and temporarily increase glacier melt. The intensity of discharge during the flood is related to the storage of water by a natural or artificial dam along a stream related to riverbank failure. In the case of a classical climatic event (without damming), the flood intensity is close to steady. The sediment load is mainly suspended load and can be predicted using a rating curve. In the case of a dam break, the flood is unsteady (surgelike or outburst). The discharge and sediment flux (mainly bed load) is not predictable. In both cases (climatic event or dam break), the sediment load can be increased by riverbank and channel-bed erosion.

Suspended-load-dominated Hyperpycnal Flows

The nature of transported particles is fundamental to explain the type of transfer to the river mouth (bed load or suspended load). Syvitski (2003), using a worldwide database, showed that river load represents 89% of the total terrigenous load toward the ocean (Table 1). Seventy-two percent of the transport is represented by suspended load, and 8% is represented by bed load. The remaining 20% is composed of dissolved load that can play an important part in biologic and diagenetic subaquatic processes. Figure 3 illustrates the importance of suspended load in generating surface hypopycnal plumes during flooding at the mouth of a large river draining the Himalayas in Southeast Asia (Ganges-Brahmaputra River; Figure 3A)

Table 1. Transport mechanism for particles attaining the world ocean (modified from Syvitski et al., 2000 and Syvitski, 2003).

Transport Mechanism	*Global Flux Estimate, % (10^{12} kg/yr)*	*Reference*
Rivers, suspended load	64 (18)	Milliman and Syvitski, 1992
Rivers, bed load	7 (2)	Milliman and Syvitski, 1992
Rivers, dissolved load	18 (5)	Summerfield and Hulton, 1994
Glaciers, sea ice, icebergs	7 (2)	Hay, 1994
Wind	2.5 (0.7)	Garrels and Mackenzie, 1971
Coastal erosion	1.5 (0.4)	Milliman and Syvitski, 1992

and a small temperate mountainous river draining the southern Alps (Var River, southeastern France; Figure 3B).

This analysis of the sediment delivery at a river mouth suggests that two types of flood-related flows can occur at a river mouth: Hyperpycnal turbulent flows are mainly dominated by suspended load, and nonturbulent hyperpycnal flows are dominated by bed load.

The flood magnitude depends mainly on two parameters: (1) the climatic event at the origin of the flood and (2) the shape of the drainage basin.

1) If the climatic event is of short duration (e.g., storm, thunderstorm, hurricane), the increase in water discharge is sudden. This is the case for the Var River floods related to the Mediterranean climate (thunderstorms in autumn). This is also the case for floods related to tropical cyclones in Southeast Asia (Warrick and Milliman, 2003; Milliman and Kao, 2005). Erosion on the drainage basin can be intensified, significantly increasing the concentration of suspended matter. In addition, during a flow, erosion of the riverbank increases. Bank failures can occur in places along the river course and transform rapidly into sediment flows that are rapidly diluted and provide soft material mixed with the river water. In that case, the flood hydrograph is sharp, the peak flood condition is attained shortly after the beginning of the flood, and the flood duration is short. The shape of the hydrograph resembles the hydrograph related to a catastrophic event (dam break), but with a peak of lesser magnitude and a less abrupt phase of increasing discharge. This kind of flood hydrograph is observed in arid-dry climates. Sporadic streams called "oueds" (e.g., Djer and Isser rivers in Algeria) can have sudden catastrophic floods after long periods of drought. However, the river experiencing such flood hydrographs tends to produce high-energy floods with high transport competency that favors bed-load instead of suspended-load transport (flow competency represents the maximum grain size transported by a flow and is related to flow velocity). If the climatic event is seasonal (e.g., monsoon), the flood hydrograph is flatter and the flood can last several weeks to several months. This is the case of the Yellow River (Huang He) in China for which long-duration hyperpycnal turbidity currents have been recorded (Wright et al., 1986; Liu et al., 2004). These long seasonal floods lead to very high yearly average concentrations, making Chinese and Taiwanese rivers among the dirtiest rivers in the world (Milliman and Syvitski, 1992; Mulder and Syvitski, 1995).
2) The shape of the drainage basin also has a big influence on the shape of the flood hydrograph (Figure 4A). A long or large drainage basin for rivers flowing on flat continental areas (e.g., Precambrian shields) tends to flatten the flood hydrograph by distributing the water discharge over a long time. In addition, a concentration time ([CT] Figure 4A) can exist between the beginning of the rainfall and the peak flood conditions. For very large rivers, this is emphasized by the fact that the drainage basin might be only partially affected by rainfall during a rainy event. A very large drainage basin can cross different climatic areas, limiting the climatic impact of rainfalls. Conversely, on a rounded drainage basin, the time for a drop of water to reach the river mouth is approximately the same where every point on the drainage basin is where the drop falls. No concentration time before the beginning of the flood is seen, and the flood hydrograph is not flattened. This is typically the case for small drainage basins of rivers in mountainous environments.

FIGURE 3. Subaerial photograph of the hypopycnal plume generated at the Ganges-Brahmaputra River mouth during monsoon floods (A) and a map of the Var River mouth (B). (C) Flooding Choshui River in Taiwan showing the very high turbidity caused by a high suspended-sediment load.

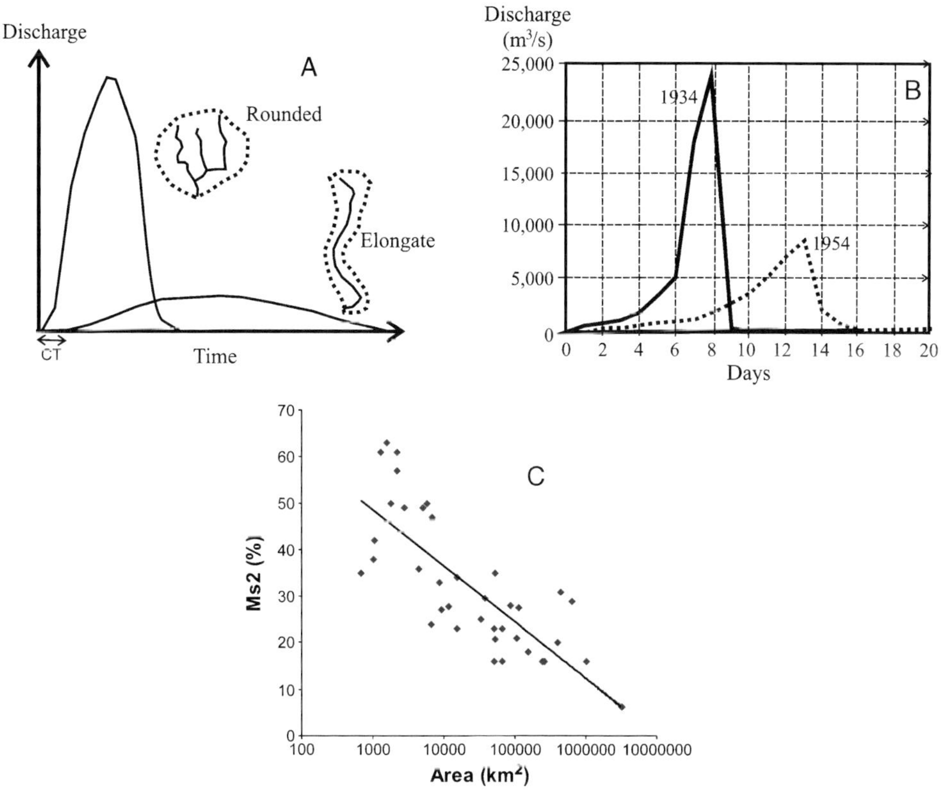

FIGURE 4. Example of flood hydrographs. (A) Importance of the size and shape of the drainage basin for classical climatic-related floods. CT = concentration time (modified from Lefèvre and Schneider, 2002). (B) Hydrograph of 1934 and 1954 jökulhlaups related to Grimstvötn eruptions in Iceland (modified from Björnsson, 1992). (C) Relationship between the total number of suspended-matter fluxes carried more than 2% of the time (Ms2) and the size of the drainage basin (modified from Dang, 2007).

Hypopycnal Flows and Reconcentration Processes

The critical concentration of 35 to 40 kg/m^3 for initiation of hyperpycnal flows, which was initially given by Mulder and Syvitski (1995), can be considerably reduced by convective instability that occurs in a surface plume (Chikita, 1991; Hoyal et al., 1999; Maxworthy, 1999). Hyperpycnal flow with a suspended-sediment concentration at the river mouth as low as 1 kg/m^3 could produce hyperpycnal flows. During the experiments of Parsons et al. (2001), convection took the form of millimeter-scale sediment-laden fingers descending from the base of the surface (hypopycnal) flow. Finally, convective instability can transform most hypopycnal plumes into hyperpycnal flows. Ducassou (2006) and Ducassou et al. (2008) applied this concept to explain very fine-grained ungraded sequences cored on the Nile deep-sea fan. The sequences form contemporaneously with sapropels and during monsoon-related wet (pluvial) periods in the Nile drainage basin. During these periods, the discharge of the Nile River increased, leading to the formation of large low-density hypopycnal plumes and to the intensification of the water stratification in the eastern Mediterranean Basin, where the stratification is normally already larger than in the western Mediterranean Basin. Ducassou (2006) and Ducassou et al. (2008) suggest that several successive convective reconcentration processes occurred at the base of the surface hypopycnal plume and successively at the base of each mesopycnal plume forming along each pycnocline that separates each water mass. The process was repeated until the deepest plume reached the Mediterranean sea floor and formed a very low-density turbidity current.

Flood Events and Lake Stratification

In lakes, the development of river-borne flows, related deposits, and the construction of specific clastic sedimentary environments depend, respectively, on the existence of stratification in the water column during the flood event, the duration and intensity of the event, and the size of the basin.

Four types of flows can form in a lake (hypopycnal, mesopycnal, hyperpycnal, and homopycnal flows; Figure 2) depending on the density contrast between inflowing water and lake waters, density here being chiefly controlled by water temperature and suspended-sediment load. Hypopycnal flows in lakes generally supply very little sediment, and no reconcentration processes of hypopycnal flows have been documented. Thus, only three main types of flood deposits are documented in lakes: mesopycnal, hyperpycnal, and homopycnal flood deposits. In stratified lakes, essentially two types of river-borne flows may develop contrasting deposits: mesopycnal and hyperpycnal flows (Sturm and Matter, 1978; Schröder et al., 1998; Van Rensbergen et al., 1998, 1999; Chapron et al., 2002, 2005). Although mesopycnal flow deposits result from suspended-sediment fallout in the water column, hyperpycnal flows follow the basin-floor morphology. When the basin is large enough, the hyperpycnal flows develop channel levee systems in proximal environments and feed fan lobes in deep distal environments. In nonstratified lakes, river-borne hyperpycnal flows can form, but homopycnal flows are as common, especially in proglacial lakes (Brodzikowski and Van Loon, 1991). During homopycnal flood events, sediment plumes expand in three dimensions from the tributary mouth (Bates, 1953). They can be associated with significant currents and can develop contrasting deposits in proximal and distal basins (cf. Chapron et al., 2006, 2007).

THE RATING CURVE

The formation of flood deposits is directly related to the flux of suspended sediment in the fresh river-water flux. The rating curve is the curve that relates sediment concentration (or the sediment discharge) to the water discharge. Thus, knowledge of this rating curve is a key issue in predicting the frequency of hyperpycnal flows and their magnitude and in quantifying the volume of sediment they transport toward sedimentary basins.

The rating curve is also indispensable in predicting the type of sediment transport at a river mouth. For suspended load, the rating curve is a power law. Its shape induces the major conclusion that during low-discharge periods, sediment transport is very reduced, and most particles (either bed load or suspended load) are transported during high-discharge periods (floods). A quantitative example is provided for the Red River (Hong He) by Dang (2007). A total of 23 to 27% of the annual suspended sediment flux is carried more than 2% of the time (Figure 4C). Only 5% of the annual water discharge occurs during the same period. The analysis shows also that 97% of the annual suspended flux is carried along during precisely half the year. For the Red River, the variation in amount of suspended load carried more than 2% of the time is stable, just as it is for other rivers with their sediment loads dominated by suspended matter, including the Mississippi (United States), Isle, and Adour (France; Meybeck et al., 2003; Coynel et al., 2004; Coynel, 2005). Only 75% of the annual water flux occurs during the same period. Meybeck et al. (2003), Coynel et al. (2004), and Coynel (2005) showed that small mountainous streams in which bed-load transport is important such as in the Mad, Siuslaw, and Iowa rivers (United States) or in the Isère, Nivelle, and Nive rivers (France) have a very high amount of suspended load carried more than 2% of the year, but with large variations (50–70%).

Using the simple rating curve and a worldwide database, Mulder and Syvitski (1995) showed that 71% of the world's rivers discharging into the world's oceans could generate hyperpycnal flows with a frequency of one event every year (dirty rivers) to one event every 100 yr (moderately clean rivers). Parsons et al. (2001) showed that this number could be potentially increased by considering processes such as convective sedimentation, leading to the reconcentration of a surface (hypopycnal) plume. In that case, 84% of the world's rivers could generate hyperpycnal plumes in the submarine environment.

A small number of rivers can have a naturally high concentration of suspended matter (Mulder and Syvitski, 1995, use the term "dirty rivers"). High suspended-matter concentration can also occur exceptionally in clean rivers because of the noncatastrophic erosion along the river's course or because of a temporal natural dam made of fine-grained sediments. This occurs frequently in areas of moderate (Saguenay Fjord in western Quebec, Saint-Onge et al., 2003) to high (Choshui River in Taiwan, Milliman and Kao, 2005; Figure 3C; Golgol River in Chile, Chapron et al., 2006) seismicity, coupled with the presence of fine-grained poorly consolidated sediments (quick clays in Saguenay Fjord, eolian loess in Taiwan, andosols in Chile). An earthquake produces subaerial sediment failures. Some of the slides might dam the river. The dam is later eroded by snowmelt-related floods in Quebec and in Chile, as well as the typhoon-related flood in Taiwan. In the Saguenay Fjord, for example, the formation of a hyperpycnal flow is demonstrated by the superposition of a graded Bouma sequence (contemporaneous with the earthquake) and a hyperpycnite (contemporaneous with the hyperpycnal flows generated by erosion of the natural dam during the spring snowmelt flood). In the world, only 10 rivers out of more than 200 can be considered as dirty (Mulder and Syvitski, 1995). The other 71% (84% in Parsons et al., 2001) of rivers generates hyperpycnal flows solely during floods. This strong relationship between hyperpycnal flows and flooding conditions and the shape of the rating curve suggests that terrigenous sediment flux from continent to sedimentary basins, and particularly oceans, is very restricted both in space (river mouth) and time (flood periods). In general, rating curves are robust laws and can be used for sediment delivery prediction when established using daily data or measurement made with a higher frequency (Mulder and Syvitski, 1995).

Although a classical rating curve can be a simple power law, a significant number of rivers might show a curve with a more complex shape (Syvitski and Alcott, 1993, 1995) that takes the form of an orthograde or retrograde hysteresis. Usually, this particular shape suggests that the erosion rate varies in the drainage basin according to the period of the year. For example, Mulder et al. (1997) showed that the Var River has a rating curve with higher sediment load for floods occurring in autumn when strong thunderstorm-related rainfalls erode very dry soils and sediments (Figure 5A). A similar conclusion is suggested for the Cointzio Lake in Mexico (Figure 5B) (Susperregui, 2008). The rating curve for this freshwater reservoir has a high concentration of suspended matter corresponding to floods that occur at the beginning of the wet season. Erosion is enhanced by strong rainfall onto dry soil after intense soil remolding by agriculture. Conversely, at the end of the wet season, the suspended-matter concentration decreases because of the dilution effect and the consumption of the material available for erosion.

This analysis of the rating curve suggests that monitoring of sediment load at a river's mouth is a key issue for improving the accuracy of quantification of sediment transfer from erosional areas to the sedimentary basin. However, this monitoring is very difficult to perform for three main reasons: (1) monitoring needs a long time series that covers a large number of major floods; (2) monitoring must be concentrated during major floods (this is not an easy task because major floods are commonly related to inhospitable weather conditions); (3) intensification of artificial damming during recent decades has considerably reduced the sediment load (Milliman and Meade, 1983; Milliman and Syvitski, 1992; Farnsworth and Milliman, 2003; Guyard et al., 2007) and suggests that even if a large monitoring is performed, it is presently impossible to measure the natural (= previous to damming) sediment delivery at a river's mouth. This suggests that predicting rating curves is a good alternative to measuring them.

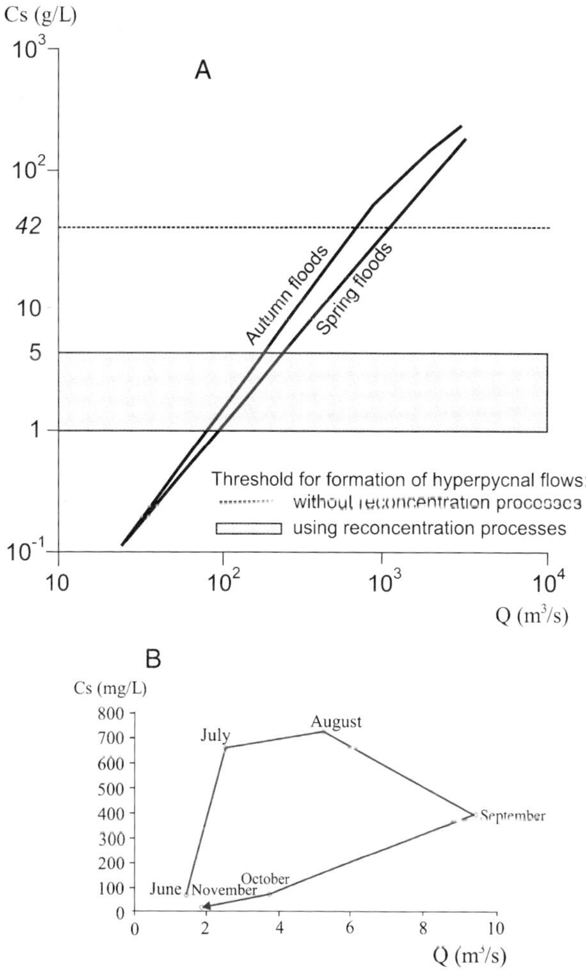

FIGURE 5. Example of rating curves: (A) Var River (modified from Mulder et al., 1997); (B) Retrograde hysteresis, Cointzio Reservoir, Mexico (modified from Susperregui, 2008). June to mid-August: Beginning of wet season and activation of the hydrologic network. Intense rainfall eroded soils too easily after the dry season generated a high sediment load. Mid-August to end of November: Maximum and end of wet season; sediment load decreased because the particles that eroded easily were transported during the preceding period. The load comes from the cohesive particles eroded along the streambed. Q = water discharge; Cs = suspended-sediment concentration.

Morehead et al. (2003) proposed a general relationship to estimate sediment load ($\overline{Q}_S$) using morphometric and climatic parameters:

$$\overline{Q}s = \alpha H^{3/2} A^{1/2} e^{k\overline{T}}$$

where H is river basin relief (m), A is river basin area (km^2), $\overline{T}$ is mean surface temperature of the drainage basin (°C), and α is a dimensionless constant (2 10^{-5}), as is k (0.1331).

This relationship suggests that elevation of the drainage basin is important: sediment load is higher for small mountainous rivers (Milliman and Syvitski, 1992) and rivers with a small drainage basin (Mulder and Syvitski, 1995, 1996). Drainage basins with a cold mean temperature (and lesser vegetal cover) are subjected to more intense erosion, particularly if the small mean temperature is related to relief elevation instead of to a location at a high latitude.

Regional parameters can influence the rating curve such as the geology of the area. In the Var River drainage basin, the Terres Noires (black shale) formation that mostly extends into the western Alps provides a large amount of easily eroded material, producing mudflows during intense rainfalls. For rivers discharging in the Tanganyika Lake in the East African rift (Tanzania), the suspended load is provided by the presence of easily eroded volcanic material such as ashes and pumice (Tiercelin et al., 1987, 1992). In China and Taiwan, the loess cover is intensively eroded during the monsoon rains, generating unusual suspended-sediment concentrations after typhoons, particularly if sediment has been remobilized by earthquakes (Dadson et al., 2004; Milliman and Kao, 2005). The picture of the flooding Choshui River (Figure 3C) illustrates this high sediment load. For the Nile River, the easily eroded material comes essentially from the Blue Nile tributary and the alteration (clay transformation) of basaltic rocks on the Ethiopian highlands. Similarly, in the Ogooué (Gabon) and Zaire basins, the fine particles are coming from the intense alteration of Precambrian shields in tropical climates (Giresse and Kouyoumontzakis, 1973; Séranne et al., 2008).

The geologic and climatic environment and the mean elevation of the drainage basin not only influence the volume of sediment delivery at the river mouth, but also control the nature of the particles that are carried at the river mouth. Using the Reading and Richards (1994) classification of submarine deep-sea fans, rivers located in nonglaciated mountainous environments will have a drainage basin with a small surface area and high relief. The sediment transported at the river mouth will be coarse (pebbles and sands) because the mean slopes on the basin are steep and the river's length is too short to form fine-grained particles. These rivers will preferentially generate bed-load-dominated hyperpycnal flows that develop alluvial fans that preferentially generate suspended load, where the river length is long enough to produce small particles; that is, when the mean slope of the drainage basin is low and the steep slopes where alluvial fans are concentrated are located far away from the river mouth. In glaciated mountainous environments, a significant part of the sediment load supplied to the marine realm through drainage basins also comes from glacier erosion that produces fine-grained particles (glacial flour). These clayey and silty particles are exported from subglacial and proglacial environments by glacier melt water pulses and transported downstream as suspended material in proglacial meltwater streams (Dhal et al., 2003). Glacier meltwater streams can develop large flood events in fluvioglacial outwash plains, proglacial lakes, and fjords (Figure 6).

Lake Type and Basin Size

Lakes in a catchment area can either trap or filter sediment during a flood event, depending on its stratification and size. For example, Lake Saint-Jean in Quebec acts as a trap for particles on the Saguenay River course. Thus, when entering the Saguenay Fjord, the Saguenay River is very clean (Syvitski and Schafer, 1996). As shown in Figure 2, large homopycnal and mesopycnal flows in small lakes can reach the outlet and export downstream fine-grained suspended sediment load, either in stratified or in nonstratified lake basins. Hyperpycnal flows will therefore be deposited in smaller lakes, whereas most of the flood-related sediment plumes will be trapped in large lakes (e.g., lakes Geneva, Houbolt and Jonker, 1968; Tanganyika, Tiercelin et al., 1992; or Baikal, Back et al., 1998). In glaciated catchment areas, for example, glacier melting (related to climate change or increased heat flux by geothermal or volcanic activity) results in the formation of numerous small proglacial lake basins, where large volumes of sediment can be exported with glacial meltwaters by the development of homopycnal flows (Figure 6). In mountain belts, during flood events, a wide range of lakes formed by glacier erosion, frontal moraines, landslides, lava flows, or tectonic basins can occur in valleys and will influence the amount of clastic particles delivered to the ocean.

Continental Shelf

The presence of a narrow continental shelf is important for the longitudinal persistence of a submarine hyperpycnal flow and its ability to reach the continental slope, canyons, and finally, the deep-marine environment. Suspended-load-dominated hyperpycnal flows are low-concentration flows in which particles are mixed with fresh water when entering the sea. If the nonbuoyancy is not rapidly maintained by fluid entrainment, the flow rapidly collapses. Particles are then mixed with pelagic sediment to form classical hemipelagites. The intensity of water entrainment

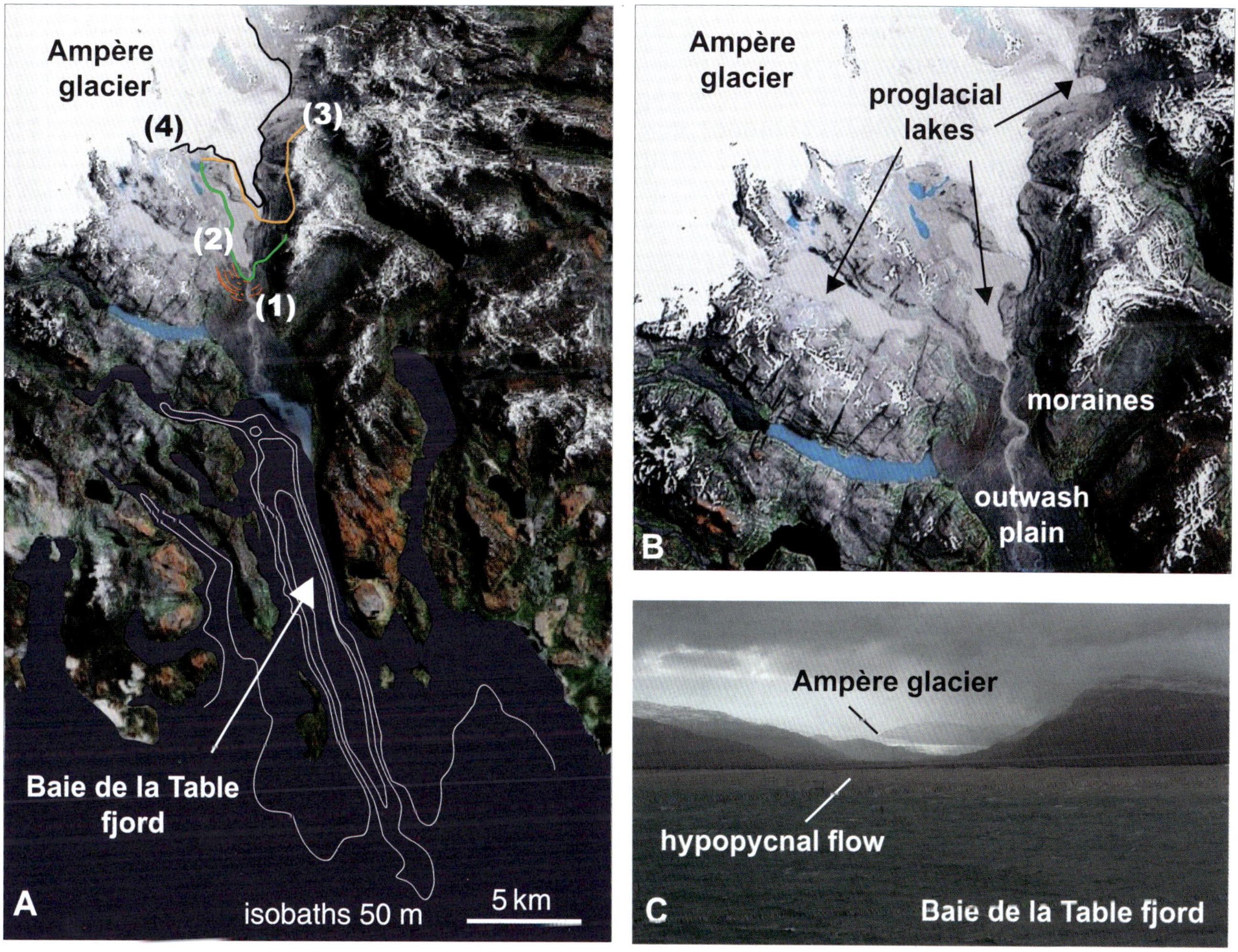

FIGURE 6. Development of hypopycnal flows in Baie de la Table Fjord (49°32′S/69°13′E; Kerguelen Island) in relation with Ampère glacier (Cook ice cap) melting (A). The historical glacier snout positions, (1) Little Ice Age, (2) 1962, (3) 1995, and (4) 2008, were redrawn from the studies of Vallon (1977) and Frenot et al. (1997). Development of homopycnal flows in recent proglacial lakes and in the Ampère outwash plain (B). Picture of a hypopycnal flow in Baie de la Table Fjord in March 2009 (C).

is defined by the Richardson number, *Ri* (inverse of squared Froude number, *Fr*). When *Ri* is small (<0.25, meaning *Fr* is large), the velocity shear is large enough to produce turbulent mixing. When *Ri* is large, no turbulent mixing occurs (Turner, 1973). On steep slopes, such as along canyon floors, hyperpycnal flows rapidly accelerate and become supercritical (*Fr* >1), generating a high entrainment rate. Conversely, on a flat continental shelf, hyperpycnal flows remain subcritical, no entrainment occurs, and the hyperpycnal flow stops and eventually lofts (Sparks et al., 1993; Hesse and Khodabakhsh, 2006).

Bed-load-dominated flows are typically nonturbulent. As in the case of frictional flows, in which the particle support is maintained by the dispersal pressures because of grain-to-grain collisions (Middleton and Hampton, 1973), bed-load-dominated flows require steep slopes to remain in motion. On an even narrower continental shelf, the flow rapidly freezes (Mulder and Alexander, 2001, use the term "frictional freezing").

Bed-load-dominated Hyperpycnal Flows

Bed-load transport in rivers is very difficult to measure and predict (Inbar and Schick, 1979). Bed-load transport in coarse-grained alluvial channels can vary from flood to flood, depending on flood frequency (Reid et al., 1985). Bed-load transport is less important than suspended-load transport and occurs with a lower frequency (Syvitski, 2003). In addition, monitoring is very difficult, particularly in large catchment areas.

Hyperpycnal flows with bed load as the dominant transport mechanism are commonly generated by catastrophic (outsized) events. The examples are essentially related to natural or artificial dam breaks, to the entering of a mass flow in a subaquatic basin, or to flash floods. Even during flood conditions, major bed-load transport in rivers is restricted to small rivers draining mountainous basins and to high-magnitude floods (Inbar and Schick, 1979; Mao and Lenzi, 2007).

The Icelandic jökulhlaup is typical of the sudden drainage of a subglacial lake; it represents the sudden drainage of a subglacial reservoir filled by ice melting during the eruption of a subglacial volcano (Björnsson, 1992). Jökulhlaups occur with a frequency of one event per year to one event every 10 yr. For example, in 1996, the Grimsvötn volcano drained a 3.2-km^3 (0.77-mi^3) lake in 40 hr and produced a jökulhlaup with a peak discharge of 50,000 m^3/s (Björnsson, 1998). Generally, the ice breakage occurs beneath the glacier with the formation of an ice tunnel. A mixture of dark volcanic rock fragments and ice boulders are deposited on a Skeidarársandur (Grönvold and Jóhannesson, 1984; Einarsson et al., 1997; Gudmunsson et al., 1997). Other events produced hyperpycnal flows with concentrations reaching 200 kg/m^3 when entering the Atlantic Ocean (Mulder et al., 2003). Jökulhlaups also occur in Alaska (Baker, 1995). For example, Lake Russell at the front of the Hubbard Glacier was drained by a Jökulhlaups burst in 1986, producing a discharge of 10^5 m^3/s (Krimmel and Trabant, 1992).

At geologic scale, similar triggering can occur at the end of glacial cycles but with a higher magnitude by the drainage of proglacial lakes (moraine-dammed lakes). Brunner et al. (1999) and Zuffa et al. (2000) described deposits related to a hyperpycnal outburst of glacial Lake Missoula (Utah) during the late Pleistocene (17–12 ka). The dam height reached 610 m (2001 ft). The lake that reached a surface area of 7770 km^2 (3000 mi^2) and maximum volume of 2100 km^3 (503 mi^3) was drained 40 to 80 times by the breaking of the moraine front during the melting of the North American ice sheet. Each episode of drainage generated an individual flood with an estimated peak discharge of 10^7 m^3/s. The floods, associated with the Columbia River drainage system, eroded 210 km^3 (50 mi^3) of volcanic rock and sediment and transported the eroded products downstream (Richmond et al., 1965; Mullineaux et al., 1978; Baker and Bunker, 1985; Waitt, 1985).

Lake Bonneville is a pluvial lake that formed 32,000 yr ago when the climate in the western United States was wetter. It occupied a surface of 51,000 km^2 (19,690 mi^2) at the location of the present Great Salt Lake in Utah and had an estimated water volume of 1583 km^3 (379 mi^3) (Gilbert, 1878). The lake probably emptied and filled several times because of alternation of wet-dry cycles. However, its final drainage 14,500 yr ago was catastrophic and produced a flood through the Snake River system, with an estimated peak discharge of 425 10^3 m^3/s (Malde and Powers, 1962; Malde, 1968).

Barber et al. (1999) and Lajeunesse and Saint-Onge (2008) suggested that the outburst flood of the Agassiz-Ojibway subglacial lake, 8.47 ka ago, which resulted from the melting of the Laurentide ice sheet, produced a flux of fresh water into the Atlantic Ocean high enough to perturb its thermohaline circulation and produce the cold climatic event at 8.2 ka (Broecker et al., 1989; Clarke et al., 2004; Broecker, 2006). The flood produced large scours because of iceberg transport by the high energy flow and hectometer-large sand waves in the Hudson Bay.

During the last glacial period, the presence of glacial lake sediments in the area of the Dogger Bank, North Sea (Laban, 1995), suggests that the European ice sheet produced probably a large marginal lake by damming the drainage system of western Europe (Toucanne, 2008). After ice melted, the Manche River drained the lake and the river system before the last Holocene sea level rise. If this lake existed, no evidence for either a quick (outburst) or a steady drainage and the production of hyperpycnites similar to what happened in the North American margin were seen.

Two characteristics of flood hydrographs related to dam breaks or jökulhlaups are the magnitude of the peak flood and the shortness of the increasing discharge period (Figure 4B). The two characteristics show that the flow is unsteady and resembles a surge-like event with a very high sediment concentration instead of a classical flood.

Subglacial meltwater can occur without a lake outburst. Piper et al. (2007) described a large amount of gravel and sand from the Laurentian Rise to the Sohm Abyssal Plain (offshore southern Newfoundland) that seems to be unrelated to North Atlantic Heinrich events and large melting periods of the Wisconsinan ice sheet. Deposition can occur through either rare large bed-load-dominated hyperpycnal flows or hypopycnal flows and lofting hyperpycnal flows. Similarly, for the Orphan Basin (offshore eastern Newfoundland), Tripsanas and Piper (2008) described eight major meltwater events preserved as red layers (plumites). Three of them produced hyperpycnal flows

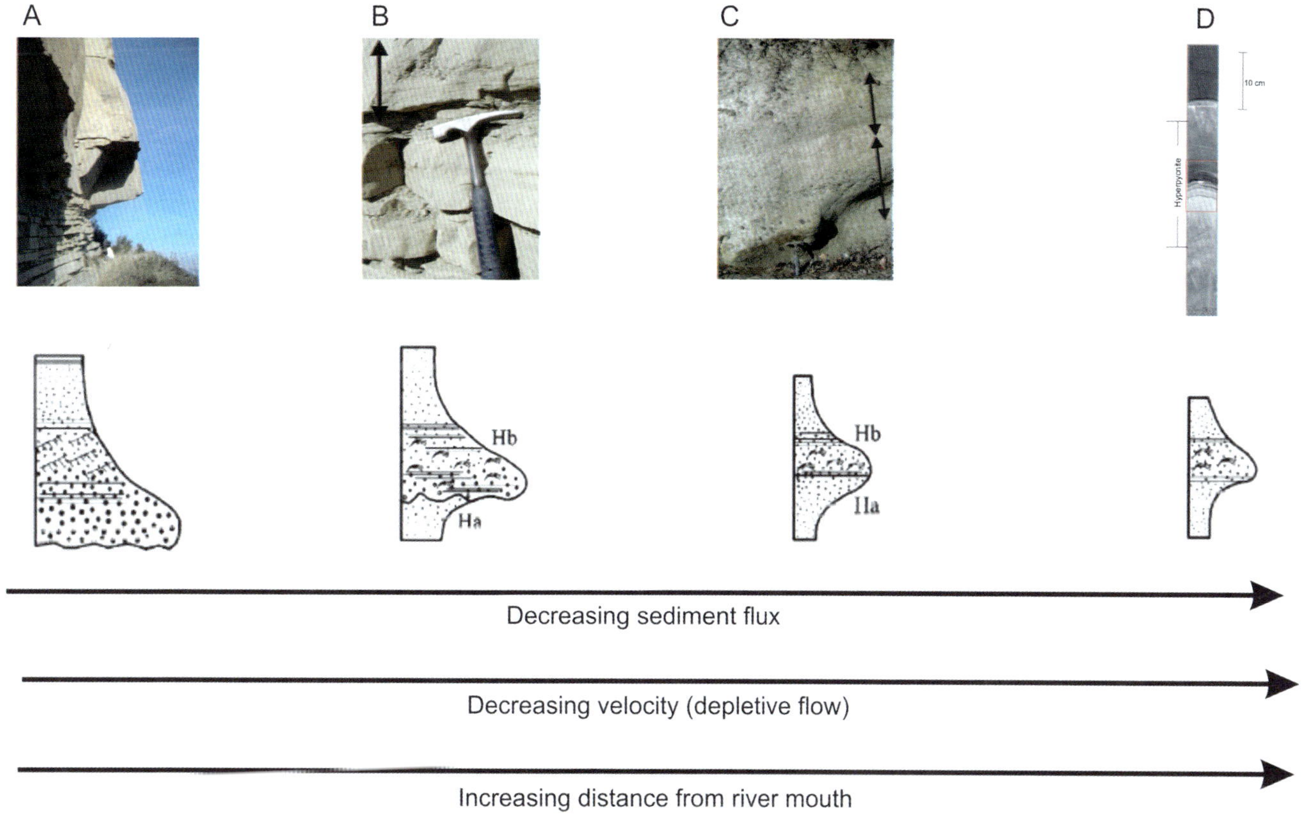

Figure 7. Longitudinal distal-proximal facies evolution of hyperpycnites: from left to right. (A) Base-truncated hyperpycnites similar to Bouma sequence. Example of several stacked sequences in Les Scaffarels outcrops, Annot Sandstone in Annot, southeastern France. (B) Truncated hyperpycnite with partial truncation of the basal unit. Example of one sequence (arrow) in Les Scaffarels outcrops, Annot Sandstone in Annot, southeastern France. (C) Complete coarse-grained hyperpycnite. Example in Saint-Antonin outcrop, southeastern France. (D) Fine-grained hyperpycnite. Example in Oman Sea (Bourget et al., 2010a, b). Ha = basal coarsening-up unit; Hb = top fining-up unit (from Mulder et al., 2003).

that deeply imprinted the channel system and transported sand into the distal part of the system.

Dromart et al. (2007) reported steep clinoforms in Valles Marineris on Mars that could be related to outburst flows formed during Hesperian times (3.5 Ga) when the Martian atmospheric pressure was high enough to temporarily allow ice melting. These deposits can be interpreted as fan deltas formed by the sporadic outbursts of floods related to the temporary melting of the Martian ice and deposition in subaquatic basins, similar to what happens on outwash plains, such as Skeidarársandur in Iceland.

TYPES OF FLOOD DEPOSITS

Establishing links between initiation processes (flood or ice melting, in this chapter) and the resulting deposits is not an easy exercise, as pointed out by Piper and Normark (2009), who noted a lack of simple relationship between both and other sedimentary or hydrodynamic processes, as well as with sea-floor morphology that can interact with the hyperpycnal flow. We thus provide lengthy descriptions for the interpretation of these deposits.

SUSPENDED-LOAD-DOMINATED HYPERPYCNITES

Suspended-load-dominated hyperpycnal flows are turbulent flows (turbidity currents) with velocities that reflect variations of the flood hydrograph at the river mouth. Their deposits are traction-fallout mixed hyperpycnal deposits, and the variations in flow velocity are recorded in the hyperpycnites (Mulder et al., 2002). The typical sequence shows a coarsening-up interval deposited during the increasing discharge period and a fining-up interval deposited during the decreasing discharge period (Figure 7).

Such sequences can be very fine grained and millimeter-to-centimeter thick when deposited in an unconfined environment, for example, on a deep sea fan such as the Var Fan (Mulder et al., 2001a, b) and Toyama (Nakajima, 2006) or be several meters thick when the flow is laterally confined, such as in the Saguenay Fjord (Mulder et al., 1997; Saint-Onge et al., 2003).

For fine-grained hyperpycnites generated by suspended-load-dominated hyperpycnal flows, hyperpycnites with the vertical grain size recording both the waxing and waning phases of the flood hydrograph can be preserved in very deep-sea environments. In ancient systems, such as the Annot Sandstone, the water depth is from shoreface to deeper than storm-wave base (Joseph et al., 2005). In modern environments, hyperpycnites have been recognized in water depths of 2000 m (6560 ft) at the mouth of the Var Canyon and at 3200 m (10,500 ft) in the Arabian Sea (Bourget et al., 2010a, b). Nakajima (2006) described hyperpycnites in the Toyama deep-sea fan, more than 700 km (435 mi) from the river mouth. Ungraded beds related to reconcentration processes during Nile floods have been collected between 1000 m (3280 ft) and greater than 3000 m (9840 ft) water depth (Ducassou, 2006; Ducassou et al., 2008). In these beds, organic particles are observed up to 300 km (186 mi) from the river mouth. Cores collected in central Hudson Bay and western Hudson Strait show a decimeter-thick red silt bed with a reverse to normal grading, which is interpreted as the hyperpycnite related to the final outburst of Lake Agassiz-Ojibway (Lajeunesse and Saint-Onge, 2008).

For high-magnitude floods, sediment forming the basal unit can be eroded during the peak flood conditions, generating a base-absent hyperpycnite resembling a classical Bouma sequence (Figure 7A) (Bouma, 1962; Mulder et al., 2001b; Mulder et al., 2002). The Saint-Antonin outcrops are interpreted as the proximal part of the turbidite system supplying the Annot Sandstone in the main outcrop of Annot Station (du Fornel, 2003; Joseph and Lomas, 2004). It was supplied by small river systems eroding the Corsica-Sardinia massifs during the Eocene and Oligocene. Saint-Antonin outcrops show complete hyperpycnites (Figure 7C), deposited mainly in a shallow-water environment (above the lower limit of wave action). Les Scaffarel outcrops in the Annot Sandstone are located distally of Saint-Antonin outcrops. These outcrops show complete hyperpycnal sequences, including base-truncated hyperpycnites that could be related to shallow conglomerates (bed-load-dominated hyperpycnal deposits) in the Saint-Antonin outcrops.

Sedimentary structures in modern marine hyperpycnites are rare, even when deposits are fine grained. The Saguenay hyperpycnites show only rare horizontal lamination in their basal part corresponding to the waxing phase of the flood hydrograph. Ancient outcrops in the Annot Sandstone (du Fornel, 2003; Joseph and Lomas, 2004; Joseph et al., 2005) or in the Apennines (Mutti et al., 1996; Mavilla, 2000) show classical sedimentary structures (planar lamination, cross-bedding). Soyinka and Slatt (2009) described beds with diffuse lamination and coarsening/fining-up trends interpreted as hyperpycnites in mud-rich prodeltaic deposits from the Cretaceous Seaway in western North America. However, it seems that climbing ripples are more frequent in hyperpycnites than in classical slide-related turbidites (Mutti et al., 1996). In addition, the nature of sedimentary components includes continental species such as Chara oogonia (Bourcart, 1964). The very good preservation of organic matter is also observed in ungraded clastic beds related to reconcentration processes during Nile floods (Ducassou, 2006; Ducassou et al., 2008). Some wood fragments still show the preservation of the woody texture, suggesting a very short period between the death of the vegetation and its final burial.

In clastic lakes, modern suspended-load-dominated hyperpycnites can be millimeter-to-centimeter thick sharp-based deposits with the typical vertical grain-size signature of the waxing and waning phases of the flood hydrograph (Guyard et al., 2007). In-situ measurements in Kluane Lake (Yukon Territory, Canada) (Crookshanks and Gilbert, 2008) show the correlation between increasing/decreasing flow velocity and deposition of coarsening/fining-up units, respectively. High-magnitude flood events also frequently do not preserve the typical coarsening-upward sequence associated with a waxing flow (Guyard et al., 2007); instead, they are characterized by graded silty deposits in proximal environments (Figures 8, 9), and by very fine-grained (clayey-silt) sharp-based deposits in the deeper distal basin (Figure 9) (Chapron et al., 2002). A common feature of these hyperpycnites is the occurrence of organic debris at their base (Chapron et al., 2005).

Catastrophic Floods and Bed-load-dominated Hyperpycnal Flows

Bed-load-dominated hyperpycnal flows are drastically different from suspended-load-dominated hyperpycnal flows. Because of their high energy, the catastrophic flows can produce substantial onland erosion. The Missoula floods transported several meter-size basalt boulders (Malde and Powers, 1962) and the Bonneville flood carried meter- to house-size rounded

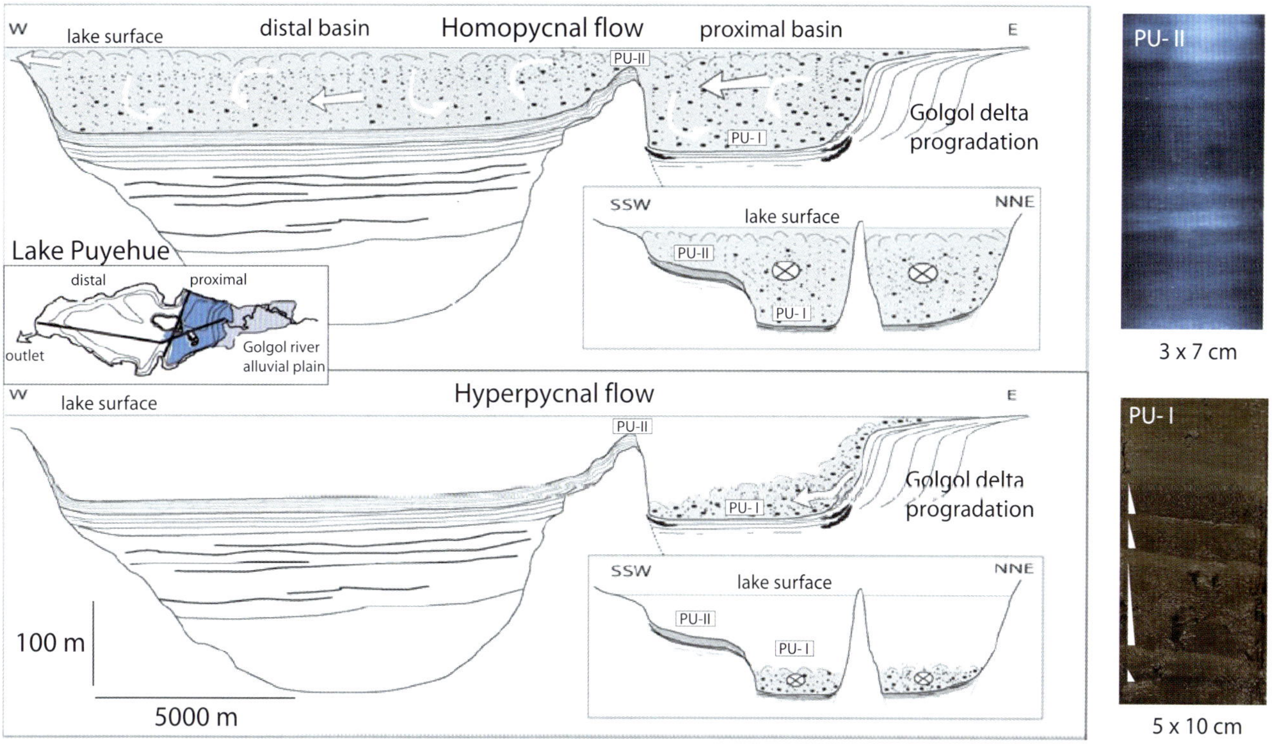

FIGURE 8. **Illustration of homopycnal and hyperpycnal flows in Lake Puyehue (Chile) and related deposits on radiography (PU-II) and photography (PU-I). Line drawings and interpretation of seismic profiles previously published by Moernaut et al. (2006), Chapron et al. (2007), and Charlet et al. (2008) highlight the typical geometry of lacustrine sedimentation resulting from the contribution of authigenic and clastic supply in the upper seismic unit of this basin fill. Although homopycnal flows are affecting the whole lake basin and developing finely laminated sediments at coring site PU-II, sporadic hyperpycnal flows are only developing sharp-based graded beds at site PU-I in the proximal basin facing the Golgol delta. These sandy-silt graded beds were formed by riverbank incision after the deposition of a catastrophic hyperpycnal flood event resulting from the outburst in 1960 of an earthquake-induced landslide-dammed lake further detailed in Chapron et al., 2006, 2007.**

blocks called "melon gravels." Blocks of volcanic rocks with a similar size were transported by the 1996 Grimstvötn jökulhlaup and deposited along the Skeidarársandur.

Brunner et al. (1999) and Zuffa et al. (2000) described submarine deposits related to hyperpycnal outbursts of glacial Lake Missoula (Utah) in the Astoria deep-sea fan (Figure 10). The related deposits were cored during the Integrated Ocean Drilling Program (IODP) leg (Shipboard Scientific Party, 1998a, b, c; Brunner et al., 1999; Zierenberg et al., 2000). They show two types of deposits: (1) several meter-thick massive sand beds with no grading and (2) massive sand beds with a crude grading occurring at the top (Figure 10A). These beds correspond to either (1) hyperconcentrated flow deposits or (2) concentrated flow deposits (Mulder and Alexander, 2001) equivalent to Ta facies of the Bouma sequence (Bouma, 1962).

The deposits related to individual drainage of Lake Missoula are particularly easy to identify on 3.5-kHz seismic records. The top of the massive sand beds (top of the hyperpycnal layer or limit between ungraded and crudely graded sand; Figure 10A) forms a good contrast in acoustic impedance and so provides good seismic reflectors. Deposits of the Malpasset Dam break (located along the Reyran/Argens River, southeastern France) in the Fréjus Bay have been cored and provide a good example of a surgelike hyperpycnal flow (Figure 10B) (Mulder et al., 2009). They also show a few decimeter-thick, organic-rich, fining-up sandy-silt layers, with no grading or with crude grading developing at the top of the deposits. The sand component includes rock fragments and minerals eroded along the drainage basin of the Reyran/Argens River. The material is mixed with broken mollusk and foraminifer shells coming from the beach area and the shoreface. The deposits form a sediment wedge extending from the shore area pinching seaward down to the 30-m (98-ft) isobath. Characteristics of the deposit strongly suggest behavior corresponding to an inertia-dominated flow. This behavior of the flow and the type of deposits are explained by the short

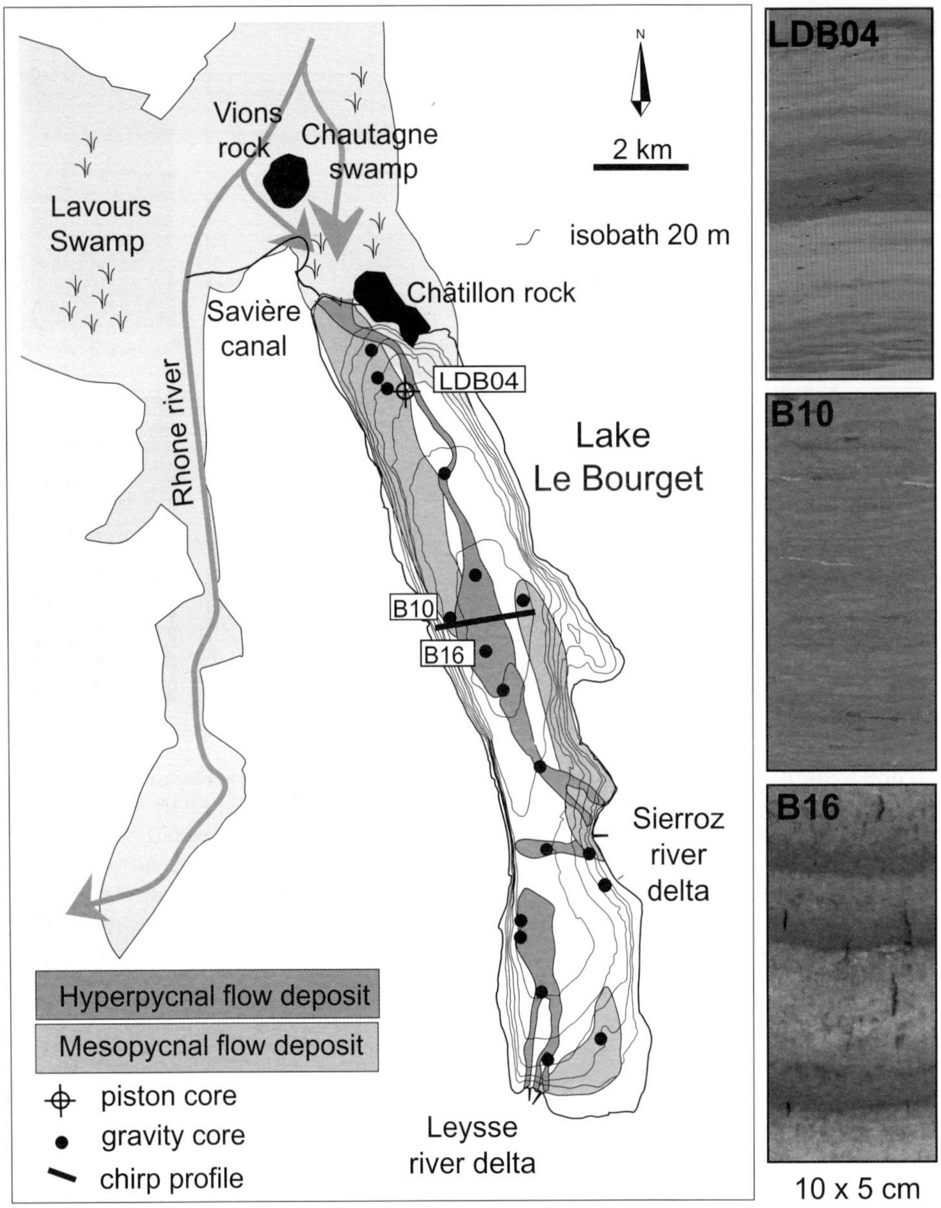

FIGURE 9. Illustration of mesopycnal and hyperpycnal flow deposits in Lake Le Bourget (northwestern Alps, France). Short coring, side scan sonar, subbottom profiling, and long coring in this marginal valley lake (Chapron et al., 1996, 2002, 2005) allow mapping specific sedimentary environments associated with mesopycnal and hyperpycnal flood deposits associated with the different tributaries. Since the early Holocene, only major Rhone River floods are entering the lake and developing either mesopycnal or hyperpycnal flood deposits. In proximal environments (LDB04), Rhone River mesopycnal and hyperpycnal flood deposits are coarser, and the latter is characterized by a sharp base rich in organic debris. In distal environments, Rhone River mesopycnal flood deposits (B10) are producing finely laminated clastic facies, whereas hyperpycnal flood deposits (B16) are producing sharp-based thick clayey-silt layers rich in organic debris. The three flood layers shown in B16 correspond to a historical flood during the Little Ice Age (Chapron et al., 2002). Also shown is the location of a Chirp seismic profile presented in Figure 14.

duration of the unsteady surge flow, dominated by bed-load transport. In addition, Bathurst and Ashiq (1998) showed that a major dam-break flood can increase the bed-load transport in a river several years after the initial outburst flow.

Catastrophic flood events related to modern or Holocene dam outbursts in mountainous valleys can also produce substantial onland erosion and develop unusually thick hyperpycnites in clastic lakes. These catastrophic deposits are characterized by an erosive base, a laminated and coarsening-upward basal sequence, and a massive to graded upper sequence (Schneider et al., 2004; Chapron et al., 2007). The outburst of the Llanz lake (ca. 40 km [25 mi] long, 400 m [1300 ft] deep) dammed by the Flims sturzstrom (a 12-km^3 [3-mi^3] rock avalanche deposit) 9400 yr ago produced two outstanding stacked sandy-silt deposits more than 100 km (62 mi) downstream from the Rhine Valley (Switzerland) in the Rhine lacustrine fan of Lake Constance (Pollet and Schneider, 2004; Schneider et al., 2004). These catastrophic deposits retrieved at two locations in Lake Constance at 160 (525 ft) and 180 m (590 ft) water depths (Figure 11) are in sharp contrast with background sedimentation (black spotted lake marls). In Lake Puyehue (Chile), the main impact of the 9.5 magnitude Valdivia earthquake in 1960 was the outburst of an earthquake-induced landslide dam in the Golgol River valley, several weeks after the main seismic shock (Chapron et al., 2006, 2007). This outburst occurred only 5 km

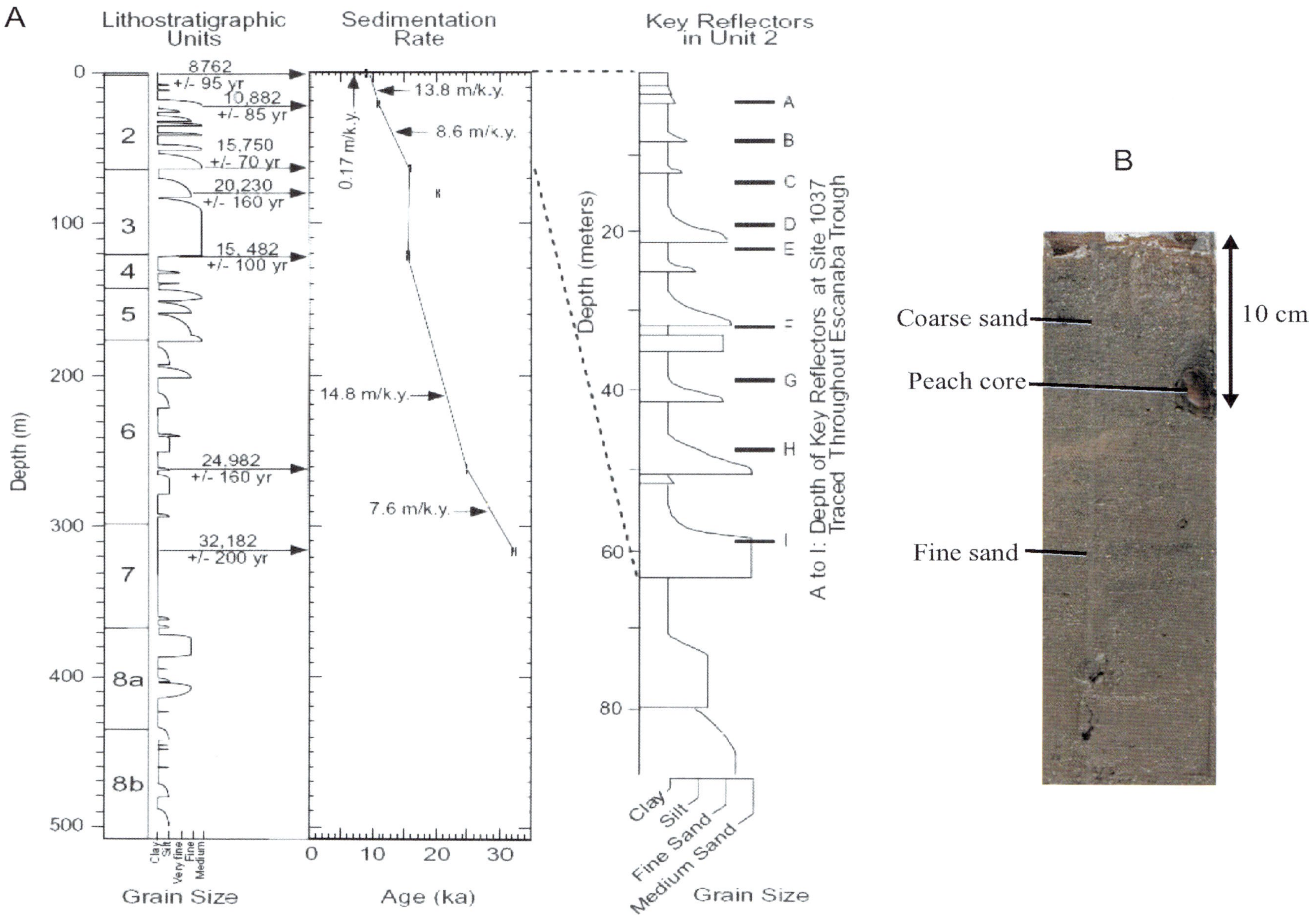

FIGURE 10. Bed-load-dominated hyperpycnal deposits. (A) Log of IODP core 1037B showing succession of hyperconcentrated and concentrated (Ta) flow deposits related to successive floods of Lake Missoula (modified from Brunner et al., 1999). (B) Example of the hyperconcentrated flow deposits (ungraded organic-rich sand) deposited in the Bay of Fréjus (French Mediterranean) by the flood related to the Malpasset Dam break (modified from Mulder et al., 2009). IODP = Integrated Ocean Drilling Program.

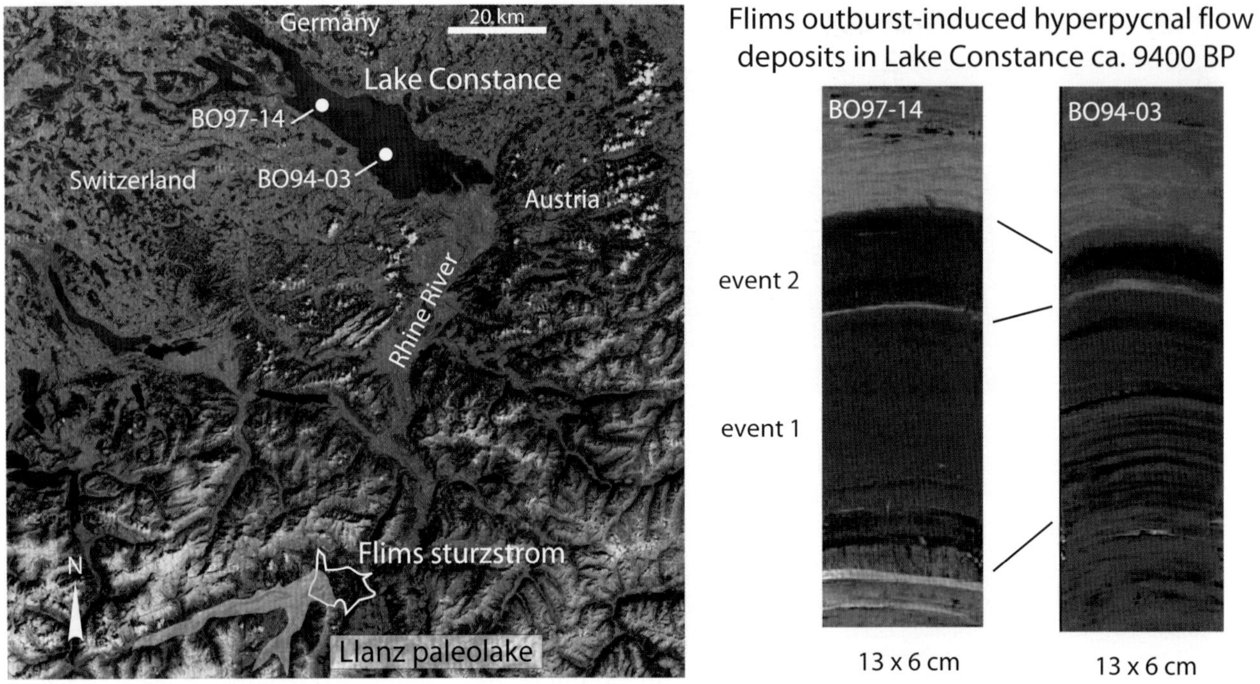

FIGURE 11. Location of Llanz paleolake temporally dammed by the Flims rock avalanche in the Rhine valley (Switzerland). The outburst of the Llanz Lake formed two successive catastrophic hyperpycnal flood deposits in the Rhine lacustrine fan delta of Lake Constance. As detailed in Schneider et al. (2004), these two hyperpycnites are outstanding features dated to ca. 9400 B.P.

(3 mi) upstream from the Golgol delta in Lake Puyehue, and the related megaflood event initiated a new course of the river in the Golgol alluvial plain, as well as a catastrophic hyperpycnite (ca. 3106 m^3) up to 30 cm (11.8 in.) thick in the proximal basin of Lake Puyehue, facing the Golgol delta. This catastrophic event reworked terrestrial, volcanic, and lacustrine material in the deep proximal basin of this piedmont lake.

Low-magnitude Floods and Reconcentration Processes

Ducassou (2006) and Ducassou et al. (2008) described fine-grained beds generated by convective sedimentation related to successive concentration processes (Figure 12). These beds are made of structureless slightly graded clastic mud (grain size, <50 μm) containing lignite fragments, amorphous organic matter patches, and neoformed gypsum. They are almost Azoic and without bioturbation. These beds essentially contain terrigenous particles and reworked microfossils. Rapid sedimentation is indicated by the good preservation of wood fragments, and the lack of planktonic fauna and bioturbation. Forty events are detected during the deposition of sapropel S1 (during ~2000 yr), supporting rapid deposition.

The absence of sedimentary structure in these clastic mud beds is not related to the absence of traction because the grain size of the beds is probably too fine to allow formation of sedimentary structures. The transitional contact between the clastic beds and the interbedded pelagic deposits is more convincing evidence for a short transport distance along the sea floor. The grading of the beds suggests that particle fallout is the most important process in their formation, suggesting that these beds are fine-grained flood-related turbidites.

Wave-modified Hyperpycnites

Wright et al. (1986), Myrow et al. (2002), and Lamb et al. (2008) described deposits from the Pennsylvanian in north-central Colorado showing both reverse and normal grading and sedimentary structures, suggesting oscillatory flow (large-scale hummocky cross-stratification, quasi-parallel lamination, combined flow-ripple lamination) interpreted as storm-influenced hyperpycnal flow deposits. In that case, a storm would be at the origin of both storm waves and flooding rivers. The abundance of sole marks (flute and gutter casts) and ripple cross-stratification suggests that the unidirectional flow was dominant. The paleoenvironmental reconstruction shows that the system was fed by a fan delta. Sediments in hyperpycnites include numerous plant fragments coming from the highlands located in the drainage basin. Wood fragments provided the tool marks, indicating flow direction.

FIGURE 12. Structureless slightly graded clastic mud beds resulting from reconcentration processes by convective sedimentation during floods in the Nile deep-sea fan (modified from Ducassou, 2006 and Ducassou et al., 2008).

SPATIAL EVOLUTION AND TIME DISTRIBUTION OF FLOOD DEPOSITS

Coarse-grained Flood Deposits

Spatial evolution of coarse-grained flood deposits (bed-load-dominated hyperpycnal flow deposits) has been synthesized in a sigmoidal bar model by Mutti et al. (1996). These bars are formed during major floods at the mouth of a channeled system by the transformation of a concentrated flow into a more dilute turbulent flow. Mutti et al. (1996) defined six facies (Figure 13) numbered from 1 to 6 from the proximal (river mouth) to the distal part. The base of the bar lies on erosion surfaces (scours). Facies 1 corresponds to nonstratified pebbles with or without sandy matrix. It results from the frictional freezing (en masse deposition) of the coarse basal part of a hyperconcentrated flow (Mulder and Alexander, 2001). Facies 2 is a coarse sandstone containing pebbles in the proximal part and cross-bedding with stratification dip increasing in the direction of progradation. Stratification planes are enhanced by pebbles and gravels and become convex up distally. This facies is deposited by bed-load transport in a hyperconcentrated flow, in which flow dilution allows individual transport of particles. Facies 3 is made of nonstratified coarse-to-medium sandstone with a crude fining up that would correspond to a concentrated flow deposit. Facies 4 corresponds to medium sandstone with horizontal or slightly undulated planar bedding. Facies 5 corresponds to fine to very fine sandstone with horizontal planar and cross-bedding (ripples). Facies 4 and 5 correspond to deposition by traction plus fallout deposition. Facies 6 corresponds to a clay interval (fallout deposition).

In Saint-Antonin outcrops (du Fornel, 2003; Joseph et al., 2005), the longitudinal facies evolution approximates the Mutti et al. (1996) model. However, beds containing gravel and pebble layers show a clear inverse and then normal grading without an erosion surface and can be interpreted as hyperpycnites (Mulder et al., 2002). The storm-wave-modified hyperpycnites described by Lamb et al. (2008) show that proximal bed forms record mainly an oscillation-dominated motion and distal bed forms record mainly a unidirectional current-dominated motion, consistent with a distal decrease of the wave impact on the hyperpycnal flow as it moves along the shelf floor.

Bed-load/Suspended-load Hyperpycnal Flow Deposits

A synthetic model of longitudinal evolution of hyperpycnal facies can be made (Figure 13). It differentiates bed-load-dominated hyperpycnal and suspended-load-dominated hyperpycnal transport. The resulting deposits depend on the suspended-load/bed-load ratio. This ratio is related to the flood hydrograph. Short and intense climatic-event-related flash floods generate a fast increase of the flow velocity at the river mouth. In this case, hyperpycnal flows are preferentially high-capacity flows dominated by

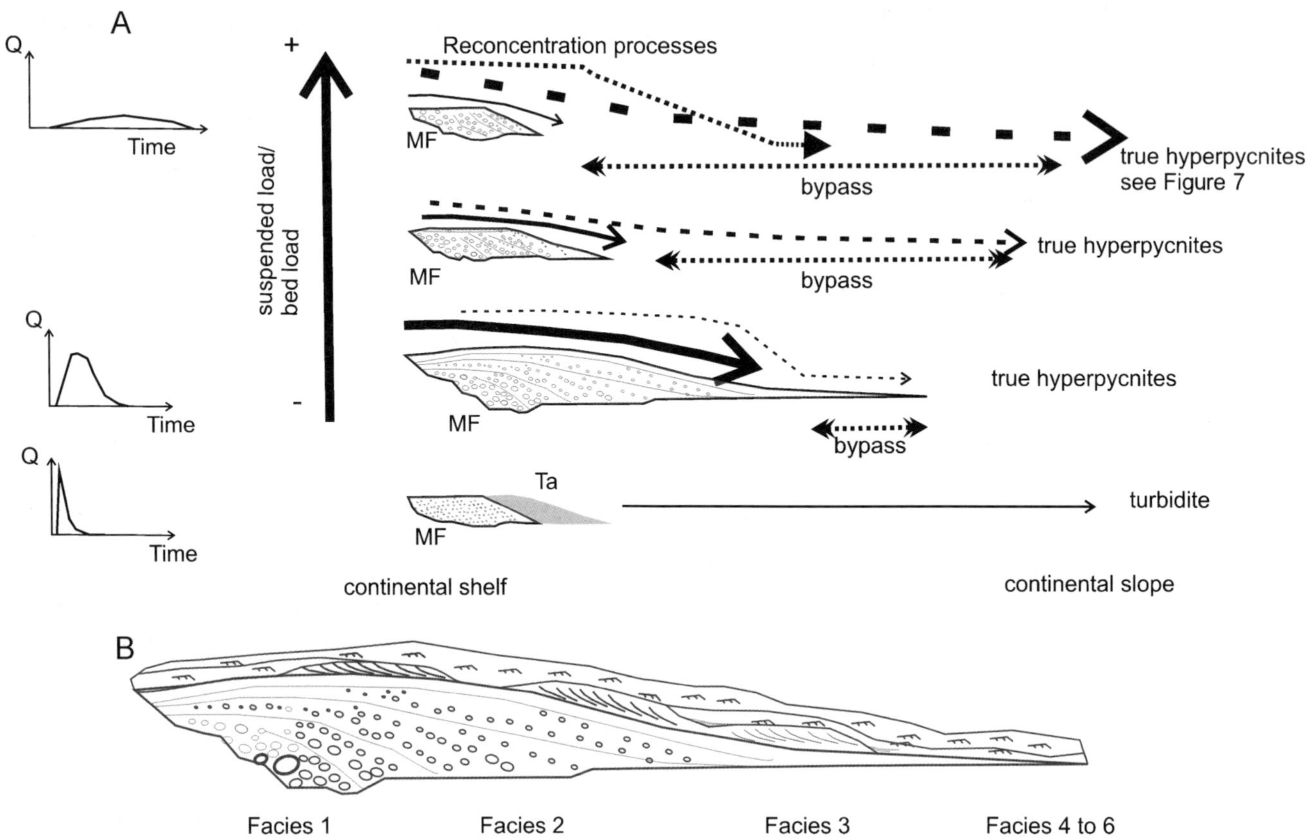

FIGURE 13. **(A) Geometry of a flood-related sigmoidal bar (modified from Mutti et al., 1996 and du Fornel, 2003). (B) Synthetic model of longitudinal evolution of hyperpycnal facies. It differentiates bed-load-dominated hyperpycnal and suspended-load-dominated hyperpycnal transport (vertical path). The resulting deposits depend on the suspended-load/bed-load ratio. This ratio is related to the flood hydrograph. Horizontal path shows main flow evolution from distance to river mouth. This evolution includes the decrease of both flow competency and capacity with distance, the increase of suspended-load versus bed-load transport, the increase of grain sorting, the progressive replacement of en masse deposition by traction, and then particle fallout and the change in grain fabric. Q = water discharge; MF = mass-flow deposit.**

bed-load transport. Long-duration floods have a flat hydrograph generating low-velocity flows dominated by suspended load.

The main differences in deposits between bed-load-dominated and suspended-load-dominated hyperpycnal flows are as follows (vertical path in Figure 13):

- coarse-grained mouth-bar progradation is replaced by fine-grained hyperpycnites
- increase of bypass on the continental shelf and upper slope
- decrease in flow competency
- increase of sediment sorting

Instantaneous catastrophic events or outburst flows (dam breaks and jökulhlaups) can be included in this model; they generate very sudden floods with high-velocity flows favoring bed-load transport. As suggested by Mulder et al. (2009) using the example of the Malpasset Dam break, the hydrodynamics of a catastrophic outburst flow is instead similar to a mass flow or a surgelike turbidity current instead of to a classical flood. The resulting deposit depends on the initial grain size but is first a frictional mass flow deposit in the proximal part (ungraded conglomerate or sandstone). The upper part of the flow can transform into a concentrated flow depositing a Ta Bouma facies with a crude grading. If the exceptional event is of moderate size, such as in the Malpasset example, the deposits stop at this stage and are limited to the continental shelf. If the event is outsized, such as in the case of the Missoula Lake outburst, the flow can reach the continental slope and proceed toward the abyss. The flow becomes accumulative and may transform into a true turbidity current, generating classical bypass on the upper slope and depositing classical turbidites on the deeper part of the continental slope and rise (Piper and Normark, 2009).

Flood Deposits of Climatic Origin

Short-duration floods related to instantaneous climatic events (thunderstorm or hurricane) will generate hyperpycnal flows dominated by bed-load transport. Deposition will occur close to the river mouth by rapid progradation of the mouth bar (Mutti et al., 1996).

The main flow evolution with increasing distance from the river mouth is as follows (horizontal path in Figure 13):

- decrease of both flow competency and capacity.
- increase of suspended-load versus bed-load transport.
- increase of grain sorting.
- en masse deposition progressively replaced by traction and then particle fallout.
- change in grain fabric: no fabric in mass flow, then crude fabric, and then no fabric.
- Bed-load-dominated hyperpycnal flow first moves as mass flows. Deposits include massive pebble to-gravel conglomerates with no internal fabric of clasts and no sedimentary structures (defined as facies 1 in Mutti et al., 1996). They correspond to the frictional flow described by Mulder and Alexander (2001). With distance, clasts begin to be transported individually, generating a crude to clear imbricate fabric in the deposits, corresponding to facies 2 as defined in Mutti et al. (1996). With distance, suspended-load transport becomes larger than bed-load transport. Clast imbrications progressively disappear, and classic hyperpycnites can form. They resemble the hyperpycnites defined by Mulder et al. (2003) in modern marine environments, but with a coarser grain size (Figure 6). They show the classic basal coarsening-up unit and the top fining-up unit. The peak of the flood is marked by the maximum grain size corresponding to a gravel bed or lamina, or coarse sand layer in the deposit (defined as facies 2 in Mutti et al., 1996). When suspended load dominates (defined as facies 3–6 in Mutti et al., 1996), the hyperpycnal flow forms fining-up deposits with facies that mimic the classic Bouma sequence. This deposition is a classic example of a suspended-load depletive waning flow.
- Suspended-load-dominated flows form classical hyperpycnites (Figure 6). In the proximal part, a small gravel bar with facies similar to those formed by bed-load-dominated hyperpycnal flow can form if coarse material is present. If only fine material is present, no deposit occurs in the proximal part, which becomes a bypass zone. Fine-grained material is transported down the continental slope and forms the fine-grained hyperpycnites (Figure 6) (Mulder et al., 2003). In moderately dirty rivers, reconcentration processes dominate, forming fine-grained turbidites (ungraded clastic beds described by Ducassou et al., 2008).

Simultaneously or alternatively, the rapid progradation of mouth bars during floods and particularly when bed-load transport is important can generate sediment failures. The study of bed-load transport along the Yellowstone River showed that floods generate substantial stream-bank erosion and hillslope mass wasting (Holnbeck, 2005). Large quantities of coarse material (sand to cobbles) are entrained by the flows and became flood-bar deposits. When generated close to the river mouth, river-bar failures related to mouth-bar progradation generate submarine flow and classic turbidity currents, as observed by Bornhold et al. (1994) in west Canadian fjords, and these currents deposit classic turbidites (Bouma sequences). This suggests that hyperpycnites and turbidites can be interlayered in sedimentary series (see, e.g., Mulder et al., 2001b, and Saint-Onge et al., 2003, for the Saguenay Fjord; see Mas, 2009, for the Var).

SIGNIFICANCE OF HYPERPYCNAL FLOOD DEPOSITS IN THE GEOLOGIC RECORD

Sedimentary Structure and System Scale in Lacustrine Environments

During the last deglaciation, frequent and powerful clastic hyperpycnal flows in lakes originating from large alpine rivers (e.g., Rhone, Rhine, and Danube rivers) not only allowed the rapid progradation of deltas in proglacial lakes, but also developed migrating, and sometimes meandering, channel levee systems along prodeltas that fed fan lobes in the deep lacustrine basins (Houbolt and Jonker, 1968; Chapron et al., 1996, 2009; Schröder et al., 1998; Van Rensbergen et al., 1998, 1999; Popescu et al., 2001). These lacustrine deposits filled many large alpine valleys, and only fine-grained particles originating from homopycnal or mesopycnal flows could be exported downstream into fluvial systems when the lake basins became smaller. Some of these large valley lakes were completely filled and are today flat alluvial plains. Some large alpine valley lakes are still directly filled by hyperpycnal flows from large alpine rivers; other

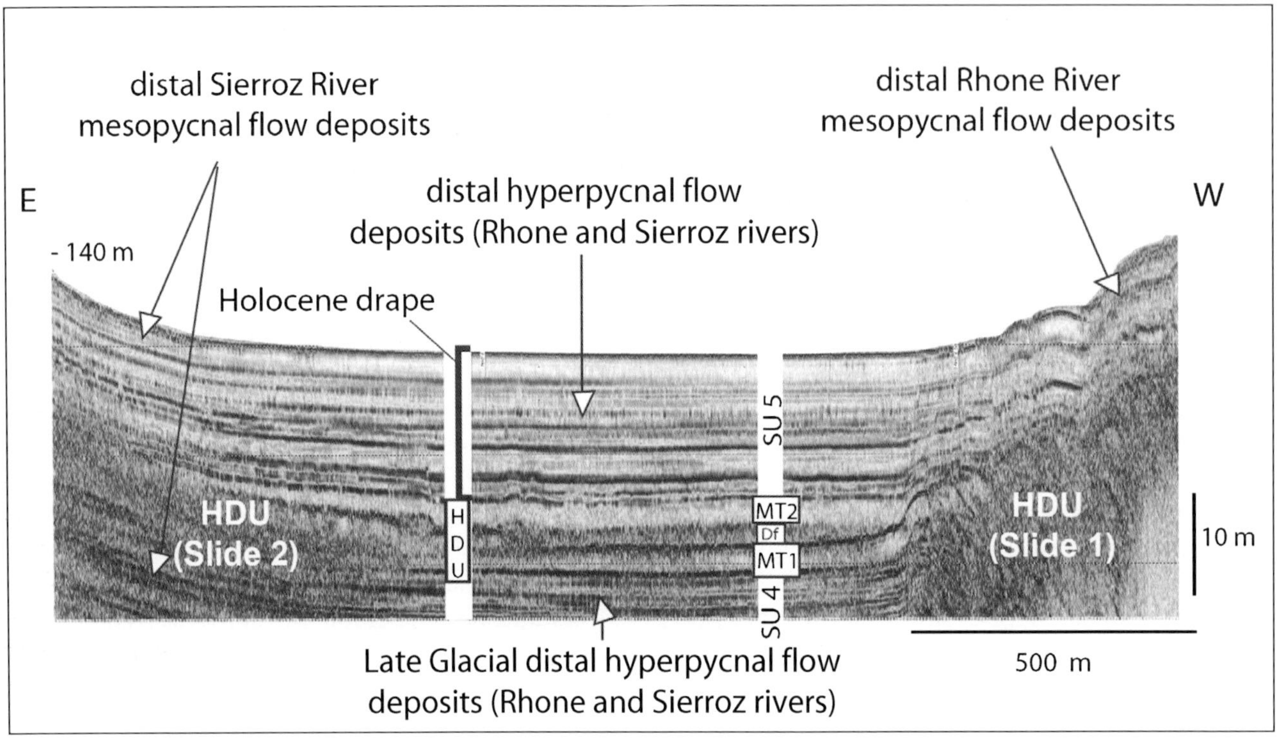

FIGURE 14. Chirp seismic profile across the Lake Le Bourget deep basin (see location in Figure 9) illustrating seismic facies associated with mesopycnal and hyperpycnal flood deposits in the tributaries within the Holocene drape (Seismic Unit 5) and at the end of the Late Glacial period (Seismic Unit 4). High clastic supply during deglaciation enhanced the impact of earthquakes (formation of the Hautecombe Disturbed Unit [HDU]; Chapron et al., 1996, 2005; van Rensbergen et al., 1999). The HDU is here characterized by two almost contemporaneous slides remolding mesopycnal flood deposits and developing one debris-flow deposit (Df) and two megaturbidites in the deep basin (MT1 and MT2).

lakes located in marginal valleys became disconnected from large alpine rivers during deglaciation. Lake Le Bourget (French northwest Alps (Figure 9) is such a marginal valley, where, since the early Holocene, only sporadic major Rhone River floods spread hyperpycnal flows over the alluvial plain and surrounding swamps (Chapron et al., 2005). These sporadic floods do not now develop channel levee systems but have produced centimeter-thick fine-grained hyperpycnites in the deep basin during periods of major environmental change, such as during the Little Ice Age. Similar cold and wet periods associated with the advances of Mont Blanc glaciers enhanced glaciofluvial regimes and high lake levels (Arnaud et al., 2005; Chapron et al., 2005; Debret et al., 2010). These periods of enhanced flooding developed outstanding high-amplitude reflections on Chirp seismic profiles over long distances from the tributary to the deep basin (Figure 14). Today's global warming in mountain ranges favors the melting of glaciers and development of proglacial lakes at the head of large catchment areas, as well as landslides along large valleys because of reduced permafrost. These significant environmental changes, together with earthquake activity, may favor the development of catastrophic floods in large valleys related to the outburst of temporary dammed lakes. Examples presented in this chapter show that these catastrophic floods can be erosive over long distances, both on land and in subaquatic basins.

Depositional System Scale in the Marine Realm

Hyperpycnal flows are involved in the construction of two major structures shaping the continental clastic margins: submarine canyons and fan deltas.

Canyons are steep-sided submarine valleys that are deeply incised into continental slopes and sometimes shelves (Shepard and Dill, 1966). Some canyon heads, such as in the Zaire and the Capbreton canyons, penetrate into the river estuary. This latter canyon was connected to the Adour River until AD 1310, but was anthropically disconnected afterward (Klingebiel and Legigan, 1978). Four hypotheses have been listed for canyon formation (Mulder et al., 2004): (1) the subaerial/shallow-water hypothesis (Mediterranean canyons) as shown by the Messinian fan delta deposits related to the Var deep-sea turbidite system (Clauzon, 1978; Clauzon et al., 1995); (2) retrogressive erosion as suggested by the presence of failure scars on canyon heads and flanks (e.g., Gaudin, 2006; Gaudin et al., 2006); the frequent alignment of pockmarks in the direction of the canyons or a deep-sea channel beyond the head (Le Moigne, 1999) is consistent with this hypothesis; (3) canyons that are bypass areas along prograding margins (Pratson and Haxby, 1996); (4) erosion by downslope-eroding sediment flows (Pratson et al., 1994). This last hypothesis suggests that hyperpycnal flows would be an important process in excavating canyons and preserving their morphology despite margin progradation. Hyperpycnal flows could also be involved in the prograding margin hypothesis (hypothesis 3). The role of hyperpycnal flows in canyon formation and in the preservation of their morphology is supported by the record of continuous activity of particle-laden flows in canyons connected to deep-sea fans, in which hyperpycnites have been detected; for example, the Var system (Gennesseaux et al., 1971) and the Zaire system (Droz et al., 1996; Hiscott et al., 1997; Savoye et al., 2000; Kripounoff et al., 2003). Piper and Normark (2009) suggest that large hyperpycnal flows create broad straight valleys with relatively uniform width. Smaller hyperpycnal flows would be responsible for the maintenance of canyons and slope valleys observed seaward of deltas or ice-covered margins during sea level rise.

The frequent sinuous shape of channels in channel levee complexes also suggests the involvement of hyperpycnal flow in canyon persistence. The frequent sinuous shape of submarine channels is explained by the natural flow behavior in response to the resistance of a hummocky sea floor at its basal interface (Gorycki, 1973). In addition, Leopold and Maddock (1953), Schumm (1981), and Rigaut (1997) suggest that meanders would occur only on moderate-to-low slopes under the action of high-energy low-concentration flows maintained during long periods, that is, hyperpycnal flows. However, recent results (Corney et al., 2006; Keevil et al., 2006) suggest that deposition in the inside bend of meanders results from a particular velocity and density gradient of turbidity currents (Clark and Pickering, 1996). The highest velocity in the basal part of such a current suggests a centrifugal force at the base of the current that is the opposite of what happens in river flows. Meander bars in channels would not be point bars, and long-duration currents would not be necessary to explain the frequent sinuous shape of deep-sea channels. The Amazon example (Pirmez and Inram, 2003) shows that frequent turbidity currents can have the same impact on channel levee construction as more sustained flows.

In addition, the shape of the canyon or the channel is probably strongly correlated to the type of sediment transport, suspended load or bed load. Abbado et al. (2005) show that the anastomosed shape of the Columbia River is related to the frequent avulsion of shallow-channel/small-elevation levees in a river system dominated by bed-load transport. Large sandar (e.g., Skeidarársandur in Iceland) and, in general, river systems in periglacial environments are anastomosed, and bed load is the dominant process of transport (Blum and Törnqvist, 2000; Gomez et al., 2000). Typical examples are the Yukon and Copper Rivers in Alaska and the Waimakariri River in New Zealand. The upper part of the Var Canyon shows an anastomosed network of small channels, with the channel floor covered by coarse sand and gravel (Parize et al., 1989; Savoye and Piper, 1991).

Fan deltas are coarse-grained prograding sedimentary deposits that form in subaquatic environments (marine or lake). The presence of submarine bioturbation or fauna differentiates them from alluvial fans. However, fossil tracks are rare in very coarse-grained facies. These fan deltas form in shelf or slope waters where they might connect to a deep-sea submarine fan (see classification in Westcott and Ethridge, 1990). A Gilbert-type delta is a fan delta composed of coarse sediments (pebbles) with steeply dipping (30–35°) foresets. This type of delta would form from a mountain river depositing sediment into a freshwater lake. All fan deltas are directly connected to a river system and are constructed by the stacking of deposits generated by successive high-energy flash floods (bed-load-dominated hyperpycnal flows). In detail, Gilbert-type deltas are constructed by rapidly decelerating hyperpycnal flows (bed-load-dominated hyperpycnal flows) or flows with a small density contrast between the flow body and the surrounding water, that is, the inertia-dominated homopycnal flow described by Bates (1953) or the jet flow described by Wright (1977). Deep-sea submarine fans are constituted by the stacking of both classic failure-related turbidites (Bouma type) and hyperpycnites. Interbedding of Bouma sequences and hyperpycnites in the Var system (Migeon, 2000; Migeon et al., 2001), the Zaire system (Migeon, 2000), and the Arabian Sea (Bourget et al., 2010a, b) show that the two initiation processes (failure transformation or direct river transfer) can interact in the same deep-sea environment.

CONCLUSIONS

Hyperpycnal floods occurring in modern and late Quaternary lacustrine and marine sedimentary environments are frequent and can extend into deltaic environments. Flood events are generally erosive in shallow waters and contribute to the development of channel levee systems. In deep waters, the deposits (hyperpycnites) can reach fan lobes and be distinguished from classical turbidites by the occurrence of organic debris of continental origin at their base and by the preservation of a coarsening-upward basal sequence associated with the waxing flow of the flood event.

Hyperpycnal flood deposits can result either from suspended-load- or bed-load-dominated hyperpycnal flows triggered by climatic events such as heavy rainfall or snowmelt, or by catastrophic events such as jökulhlaups or dam outbursts in the drainage basin. In addition, reconcentration processes affecting low-magnitude floods along deltaic slopes can develop suspended-load hyperpycnal flows in deep-marine waters. Consequently, most of the world's rivers (>80%) could generate hyperpycnal flows in the submarine environment. Given that hyperpycnal flows are sustained, bed-load transport can be important and clean sandy lag deposits can occur along the floor of hyperpycnal-dominated channels.

Large hyperpycnal floods are documented in both lacustrine and marine environments in association with significant environmental changes, such as fluctuations of continental glaciers. The current global warming in mountain ranges, together with earthquake and volcanic activities, may thus favor the development of temporary dam lakes (proglacial moraine dam lakes caused by melting glaciers; landslide dam lakes caused by reduced permafrost) and expose fluvial valleys, lakes, and coastal regions to catastrophic hyperpycnal flood events in the near future.

ACKNOWLEDGMENTS

This represents UMR Centre National de la Recherche Scientifique 5805 Environnements et Paléoenvironnements Océaniques contribution 1747. We thank David J.W. Piper for his constructive remarks on the manuscript.

REFERENCES CITED

Abbado, D., R. Slingerland, and N. D. Smith, 2005, The origin of anastomosis on the Columbia River, British Columbia, Canada, *in* M. D. Blum, S. B. Marriott, and S. Leclair, eds., Proceedings of the 7th International Conference on Fluvial Sedimentology, IAS Special Publication 35, p. 3–15.

Arnaud, F., M. Revel, E. Chapron, M. Desmet, and N. Tribovillard, 2005, 7200 years of Rhone River flooding activity in Lake Le Bourget: A high-resolution sediment record of NW Alps hydrology: The Holocene, v. 15, p. 420–428.

Back, S., M. de Batist, P. Kirillov, M. Strecker, and P. Vanhauwaert, 1998, The Frolikha fan: A large Pleistocene glaciolacustrine outwash fan in northern Lake Baikal, Siberia: Journal of Sedimentary Research, v. 68, p. 841–849.

Bagnold, R. A., 1962, Auto-suspension of transported sediment; turbidity currents: Proceedings of the Royal Society (London) A, v. 265, p. 315–319.

Baker, V., 1995, Surprise endings to catastrophism and controversy on the Columbia: GSA Today, v. 15, no. 9, p. 169, 171–173.

Baker, V. R., and R. C. Bunker, 1985, Cataclysmic late Pleistocene flooding from glacial Lake Missoula: A review: Quaternary Science Reviews, v. 4, p. 1–41.

Barber, D., et al., 1999, Forcing of the cold event of 8,200 years ago by catastrophic drainage of Laurentide lakes: Nature, v. 400, p. 344–348.

Bates, C. C., 1953, Rational theory of delta formation: AAPG Bulletin, v. 37, no. 9, p. 2119–2162.

Bathurst, J. C., and M. Ashiq, 1998, Dambreak flood impact on mountain stream bed-load transport after 13 years: Earth Surface Processes and Landforms, v. 23, p. 643–649.

Björnsson, H., 1992, Jökulhlaups in Iceland: Predictions, characteristics and simulation: Annals of Glaciology, v. 16, p. 95–106.

Björnsson, H., 1998, Hydrological characteristics of the drainage system beneath a surging glacier: Nature, v. 395, p. 771–774.

Blum, M. D., and T. E. Törnqvist, 2000, Fluvial responses to climate and sea-level change: A review and look forward: Sedimentology, v. 47, Suppl. 1, p. 2–48.

Bornhold, B. D., P. Ren, and D. B. Prior, 1994, High-frequency turbidity currents in British Columbia fjords: Geo-Marine Letters, v. 14, p. 238–243.

Bouma, A. H., 1962, Sedimentology of some flysch deposits: A graphic approach to facies interpretation: Amsterdam, Netherlands, Elsevier, 168 p.

Bourcart, J., 1964, Les sables profonds de la Méditerranée occidentale, *in* A. H. Bouma and A. Brouwer, eds., Turbidites: Developments in sedimentology 3: Amsterdam, Netherlands, Elsevier, p. 148–155.

Bourget, J., N. Mouchot, S. Zaragosi, N. Ellouz-Zimmermann, T. Garlan, V. Lanfumey, A. VanToer, J.-L. Schneider, 2010a, Turbidite system architecture and sedimentary processes along topographically complex slopes: The Makran convergent margin: Sedimentology, doi:10.1111/j.1365-3091.2010.01168.x.

Bourget, J., S. Zaragosi, T. Mulder, T. Garlan, J.-L. Schneider, N. Ellouz-Zimmermann, A. Van Toer, and V. Mas, 2010b,

Hyperpycnal-fed turbidite lobe architecture and recent sedimentary processes: A case study from the Al Batha turbidite system, Oman margin: Sedimentary Geology, v. 229, p. 144–159, doi:10.1016/j.sedgeo.2009.03.009.

Bourrouilh, R., 1987, Evolutionary mass flow-megaturbidites in an interplate basin: Example of the north Pyrenean basin: Geo-Marine Letters, v. 7, p. 69–81.

Bourrouilh, R., and B. Offroy, 1983, Séquence de mass-flow évolutif-mégaturbidite du flysch Sénonien nord-pyrénéen. Traitement statistique et anatomie du bassin sénonien. Colloque sur le Sénonien: Géologie Méditerranéenne, v. 20, p. 345–359.

Brodzikowski, K., and A. Van Loon, 1991, Review of glacigenic sediments: Development in Sedimentology, v. 49, 6888 p.

Broecker, W. S., 2006, Was the Younger Dryas triggered by a flood?: Science, v. 312, p. 1146–1148.

Broecker, W. S., J. P. Kennett, B. P. Flower, J. T. Teller, S. Trumbore, G. Bonami, and W. Wolfli, 1989, Routing of meltwater from the Laurentide ice sheet during the younger Dryas cold episode: Nature, v. 341, p. 318–321.

Brunner, C. A., W. R. Normark, G. G. Zuffa, and F. Serra, 1999, Deep-sea sedimentary record of the late Wisconsin cataclysmic floods from the Columbia River: Geology, v. 27, no. 5, p. 463–466.

Chapron, E., P. Van Rensbergen, C. Beck, M. De Batist, and A. Paillet, 1996, Lacustrine sedimentary records of brutal events in Lake Le Bourget (NW Alps-Southern Jura): Quaternaire, v. 7, p. 155–168.

Chapron, E., M. Desmet, T. De Putter, M. F. Loutre, C. Beck, and J. F. Deconinck, 2002, Climate variability in the NW Alps, France, as evidenced by 600 years of terrigenous sedimentation in Lake Le Bourget: The Holocene, v. 12, p. 59–68.

Chapron, E., F. Arnaud, H. Noel, M. Revel, M. Desmet, and L. Perdereau, 2005, Rhone River flood deposits in Lake Le Bourget: A proxy for Holocene environmental changes in the NW Alps, France: Boreas, v. 34, p. 404–416.

Chapron, E., D. Ariztegui, S. Mulsow, G. Villarosa, M. Pino, V. Outes, E. Juviginié, and E. Crivelli, 2006, Impact of 1960 major subduction earthquake in Northern Patagonia (Chile, Argentina): Quaternary International, v. 158, p. 58–71.

Chapron, E., E. Juvigné, S. Mulsow, D. Ariztegui, O. Magand, S. Bertrand, M. Pino, and O. Chapron, 2007, Recent clastic sedimentation processes in Lake Puyehue (Chilean Lake District, 40.5°S): Sedimentary Geology, v. 201, p. 365–385, doi:10.1016/j.sedgeo.2007.07.006.

Chapron, E., M. Dietrich, C. Beck, P. Van Rensbergen, H. P. Finck, G. Menard, G. Nicoud, F. Lemeille, F. Anselmetti, and M. De Batist, 2009, High-amplitude reflections in proglacial lacustrine basin fills of the NW Alps: Origin and implications, *in* P. Y. Bard, E. Chaljub, C. Cornou, F. Cotton, and P. Gueguen, eds., Proceedings of the Third International Symposium on the Effects of Surface Geology on Seismic Motion: Grenoble, France, Collection Actes des journées scientifiques du LCPC, v. 2, 678 p.

Charlet, F., M. de Batist, E. Chapron, S. Bertrand, M. Pino, and R. Urrutia, 2008, Seismic stratigraphy of Lago Puyehue (Chilean Lake District): New views on its deglacial and Holocene evolution: Journal of Paleolimnology, v. 39, p. 163–177.

Chikita, K., 1991, Dynamic processes of sedimentation by river induced turbidity currents: II. Application of a two-dimensional, advective diffusion model: Transactions, Japanese Geomorphological Union, v. 13, no. 1, p. 1–18.

Clark, J., and K. Pickering, 1996, Quantitative analysis of the geometry of submarine channels: Implications for the classification of submarine fans: Geology, v. 20, p. 633–636.

Clarke, G. K. C., D. W. Leverington, J. T. Teller, and A. S. Dyke, 2004, Paleohydraulics of the last outburst flood from glacial Lake Agassiz and the 8200 BP cold event: Quaternary Science Reviews, v. 23, p. 389–407.

Clauzon, G., 1978, The Messinian Var Canyon (Provence, southern France). Paleogeographic implications: Marine Geology, v. 27, no. 3/4, p. 231–246.

Clauzon, G., J.-L. Rubino, and B. Savoye, 1995, Marine Pliocene Gilbert-type fan deltas along the French Mediterranean coast. A typical infill feature of preexisting subaerial Messinian canyons: IAS Regional Meeting of Sedimentology Field Trip Guide Book: Publication ASF, v. 23, p. 145–222.

Corney, R. K. T., J. Peakall, D. R. Parsons, L. Elliot, K. J. Amos, J. L. Best, G. M. Keevil, and D. B. Ingham, 2006, The orientation of helical flow in curved channels: Sedimentology, v. 53, p. 249–257.

Coynel, A., 2005, Erosion mécanique des sols et transferts géochimiques dans le bassin Adour-Garonne: Ph.D. thesis, Université Bordeaux 1, Bordeaux, France, 572 p.

Coynel, A., J. Schäfer, J. E. Hurtrez, J. Dumas, H. Etcheber, and G. Blanc, 2004, Sampling frequency and accuracy of SPM flux estimates in two contrasted drainage basins: Science of the Total Environment, v. 330, p. 233–247.

Crookshanks, S., and R. Gilbert, 2008, Continuous, diurnally fluctuating turbidity currents in Kluane Lake, Yukon Territory: Canadian Journal of Earth Science, v. 45, p. 1123–1138.

Dadson, S. J., et al., 2004, Earthquake triggered increase in sediment delivery from an active mountain belt: Geology, v. 32, p. 733–736.

Dang, T.-H., 2007, Dynamique des transports de matières dans les eaux du Fleuve Rouge au Vietnam: Master's thesis, Université Bordeaux 1, Bordeaux, France, 30 p.

Debret, M., E. Chapron, F. Arnaud, M. Desmet, M. Rolland-Revel, O. Magand, A. Trentesaux, V. Bout-Roumazeille, and J. Nomade, 2010, Northwestern Alps Holocene paleohydrology recorded by flooding activity in Lake Le Bourget, France, and possible relations with Mont-Blanc glaciers fluctuations: Quaternary Science Reviews, doi:10.1016/j.quascirev.2010.05.016 (published online).

Dhal, S., J. Bakke, O. Lie, and A. Nesje, 2003, Reconstruction of former glacier equilibrium-line altitudes based on proglacial sites: An evaluation of approaches and selection of sites: Quarternary Science Reviews, v. 22, p. 275–287.

Dromart, G., C. Quantin, and O. Broucke, 2007, Stratigraphic architectures spotted in southern Melas Chasma, Valles Marineris: Marine Geology, v. 35, no. 4, p. 363–366.

Droz, L., F. Rigaut, P. Cochonat, and R. Tofani, 1996, Morphology and recent evolution of the Zaire turbidite system (Gulf of Guinea): Geological Society of America Bulletin, v. 108, p. 253–269.

Ducassou, E., 2006, Evolution du système turbiditique profond du Nil au cours du Quaternaire récent: Ph.D. thesis, Université Bordeaux 1, Bordeaux, France, no. 3263, 336 p.

Ducassou, E., T. Mulder, S. Migeon, E. Gonthier, A. Murat, L. Capotondi, S. Bernasconi, and J. Mascle, 2008, Nile floods recorded in deep Mediterranean sediments: Quaternary Research, v. 70, p. 382–391.

du Fornel, E., 2003, Reconstitution sédimentologique tridimensionnelle et simulation stratigraphique du système turbiditique éocène-oligocène des Grès d'Annot (Alpes méririonales): Ph.D. thesis, Université de Rennes 1, Rennes, France, 243 p.

Einarsson, P., B. Brandsdóttir, M. Tumi Gudmunsson, H. Björnsson, K. Grönvold, and F. Sigmundsson, 1997, Center of the Iceland hotspot experiences volcanic unrest: Eos Transactions, American Geophysical Union, v. 78, no. 35, p. 369–375.

Farnsworth, K. L., and J. D. Milliman, 2003, Long-term fluvial sediment delivery to the ocean: Effect of climatic and anthropogenic change: Global Planet Change, v. 39, no. 1–2, p. 53–64.

Forel, F. A., 1885, Les ravins sous-lacustres des fleuves glaciaires: Comptes Rendus de l'Académie des Sciences, Sciences de la Terre, v. 101, no. 16, p. 725–728.

Forel, F. A., 1892, Le Léman: Monographie Limnologique 1, F. Rouge, Lausanne, 543 p.

Frenot, Y., J. C. Gloaguen, B. Van der Vijver, and L. Beyens, 1997, Datation de quelques sédiments tourbeux holocènes et oscillations glaciaires aux iles Kerguelen: Comptes Rendus de l'Académie des Sciences – Series III – Sciences de la Vie, v. 320, p. 567–573.

Garcia, M. H., and G. Parker, 1993, Experiments on the entrainment of sediment into suspension by a dense bottom current: Journal of Geophysical Research, v. 98, no. 3, p. 4793–4807.

Garrels, R. M., and F. T. Mackenzie, 1971, Evolution of sedimentary rocks: New York, New York, W. W. Norton and Co., 397 p.

Gaudin, M., 2006, Processus et enregistrements sédimentaires dans les canyons sous-marins Bourcart et de Capbreton durant le dernier cycle climatique: Ph.D. thesis, Université Bordeaux 1, Bordeaux, France, no. 3322, 296 p.

Gaudin, M., T. Mulder, P. Cirac, S. Berné, and P. Imbert, 2006, Past and present sedimentary activity in the Capbreton Canyon, southern Bay of Biscay, *in* T. Mulder, ed., Special Issue on Deep-Sea Turbidite Systems on French Margins: Geo-Marine Letters, v. 26, no. 6, p. 331–346.

Gennesseaux, M., M. Guibout, and H. Lacombe, 1971, Enregistrement de courants de turbidité dans la vallée sous-marine du Var (Alpes-Maritimes): Comptes Rendus de l'Académie des Sciences, Paris, Série D, v. 273, p. 2456–2459.

Gilbert, G. K., 1878, The ancient outlet of Great Salt Lake: American Journal of Science, 3rd Series, v. 15, p. 256–259.

Giresse, P., and G. Kouyoumontzakis, 1973, Cartographie sédimentologique des plateaux continentaux du sud du Gabon, du Congo, du Cabinda et du Zaïre: Cahiers de l'ORSTOM, Série Géologie, v. 5, no. 2, p. 235–257.

Gomez, B., L. C. Smith, F. J. Magilligan, L. A. K. Mertes, and N. D. Smith, 2000, Glacier outburst floods and outwash plain development: Skeidarársandur, Iceland: Terra Nova, v. 12, no. 3, p. 126–131, doi:10.1046/j.1365-3121.2000.00277.x.

Gorycki, M. A., 1973, Hydraulic drag: A meander initiating mechanism: Geological Society of America Bulletin, v. 84, p. 175–186, doi:10.1130/0016-7606(1973)84<175:HDAMM>2.0.CO;2.

Grönvold, K., and H. Jóhannesson, 1984, Eruption in Grímsvötn 1983, course of events and chemical studies of the tephra: Jökull, v. 34, p. 1–11.

Gudmunsson, M. T., F. Sigmundsson, and H. Björnsson, 1997, Ice-volcano interaction of the 1996 Gjálp subglacial eruption, Vatnajökull, Iceland: Nature, v. 389, p. 954–957, doi:10.1038/40122.

Guyard, H., G. St Onge, E. Chapron, F. Anselmetti, and P. Francus, 2007, The AD 1881 earthquake-triggered slump and late Holocene flood-induced turbidites from Proglacial lake Bramant, western French Alps, *in* V. D. Lykousis, D. Sakellariou, and J. Locat, eds., Submarine mass movements and their consequences: Dordrecht, Netherlands, Springer, p. 279–286.

Hay, W. H., 1994, Pleistocene-Holocene fluxes are not the earth's norm, *in* W. W. Hay, ed., Global sedimentary geofluxes: Washington, DC, National Academy of Sciences Press, p. 15–27.

Heezen, B. C., and M. Ewing, 1952, Turbidity currents and submarine slumps and the 1929 Grand Banks earthquake: American Journal of Science, v. 250, no. 12, p. 849–873.

Heezen, B. C., M. Ewing, and G. L. Johnson, 1960, Cable failures in the Gulf of Corinth: A case history, report of the Lamont Geological Observatory: Palisades, New York, Columbia University, 100 p.

Hesse, R., and S. Khodabakhsh, 2006, Significance of fine-grained sediment lofting from melt-water generated turbidity currents for the timing of glaciomarine sediment transport into the deep sea: Sedimentary Geology, v. 186, p. 1–11, doi:10.1016/j.sedgeo.2005.10.006.

Hiscott, R. N., C. Pirmez, and R. D. Flood, 1997, Amazon submarine fan drilling. A big step forward for deep-sea fan models: Geoscience Canada, v. 24, p. 13–24.

Holnbeck, S. R., 2005, Sediment-transport investigations of the upper Yellowstone River, Montana, 1999 through 2001: Data collection, analysis, and simulation of sediment transport: U.S. Geological Survey Scientific Investigations Report 2005-5234, 69 p.

Houbolt, J., and J. Jonker, 1968, Recent sediments in the Eastern part of the Lake of Geneva (Lac Léman): Geologie en Mijnbouw, v. 47, p. 131–148.

Hoyal, D. C. J. D., M. I. Bursik, and J. F. Atkinson, 1999, The influence of diffusive convection on sedimentation from buoyant plumes: Marine Geology, v. 159, p. 205–220, doi:10.1016/S0025-3227(99)00005-5.

Inbar, M., and A. P. Schick, 1979, Bedload transport associated with high stream power, Jordan River, Israel: Proceedings of the National Academy of Sciences of the United States of America, v. 76, no. 6, p. 2515–2517, doi:10.1073/pnas.76.6.2515.

Joseph, P., and S. A. Lomas, eds., 2004, Deep-water sedimentation in the Alpine basin of SE France. New perspectives on the Grès d'Annot and related systems: Geological Society (London) Special Publication 221, 456 p.

Joseph, P., Y. Callec, F. Guillocheau, and C. Robin, 2005, Sédimentologie, stratigraphie séquentielle et architecture réservoir des séries Éocène-Oligocène du Sud-Est de la France (Grès d'Annot et systèmes associés). 10ème Congrès Français de Sédimentologie, Excursion Guidebook and ASF Publication, Paris, no. 54, 152 p.

Keevil, G. M., J. Peakall, J. L. Best, and K. J. Amos, 2006, Flow structure in sinuous submarine channels: Velocity and turbulence of an experimental submarine channel: Marine Geology, v. 229, p. 241–257, doi:10.1016/j.margeo.2006.03.010.

Klingebiel, A., and P. Legigan, 1978, Histoire géologique des divagations de l'Adour. Actes du Congrès de Bayonne des 28 et 29 Octobre 1978 Organisé pour le IVème Centenaire du Détournement de l'Adour, p. 23–34.

Kneller, B., 1995, Beyond the turbidite paradigm: Physical models for deposition of turbidites and their implications for reservoir prediction, *in* A. J. Hartley and D. J. Prosser, eds., Characterization of deep marine clastic systems: Geological Society (London) Special Publication 94, p. 31–49.

Kneller, B. C., and M. J. Branney, 1995, Sustained high-density turbidity currents and the deposition of thick massive beds: Sedimentology, v. 42, p. 607–616, doi:10.1111/j.1365-3091.1995.tb00395.x.

Krimmel, R. M., and D. C. Trabant, 1992, The terminus of Hubbard Glacier, Alaska: Annals of Glaciology, v. 16, p. 151–157.

Kripounoff, A., A. Vangrieshem, N. Babonneau, P. Crassous, B. Dennielou, and B. Savoye, 2003, Direct observation of intense turbidity current activity in the Zaire submarine valley at 4000 m water depth: Marine Geology, v. 194, p. 151–158, doi:10.1016/S0025-3227(02)00677-1.

Kuenen, P. H., 1952, Estimated size of the Grand Banks turbidity current: American Journal of Science, v. 250, p. 874–884.

Kuenen, P. H., 1953, Significant features of graded bedding: AAPG Bulletin, v. 37, p. 1044–1066.

Kuenen, P. H., and C. I. Migliorini, 1950, Turbidity currents as a cause of graded bedding: Journal of Geology, v. 58, p. 91–127, doi:10.1086/625710.

Laban, C., 1995, The Pleistocene glaciations in the Dutch of the North Sea. A synthesis of sedimentary and seismic data: Ph.D. thesis, University of Amsterdam, Amsterdam, Netherlands, 194 p.

Lajeunesse, P., and G. Saint-Onge, 2008, The subglacial origin of the Lake Agassiz-Ojibway final outburst flood: Lajeunesse and St-Onge, Nature Geoscience, v. 3, p. 184–188, doi:10.1038/ngeo130.

Lamb, M. P., P. M. Myrow, C. Lukens, K. Houck, and J. Strauss, 2008, Deposits from wave-influenced turbidity currents: Pennsylvanian Minturn Formation, Colorado. Journal of Sedimentary Research, v. 78, p. 480–498, doi:10.2110/jsr.2008.052.

Lambert, A., and F. Giovanoli, 1988, Records of riverborne turbidity currents and indications of slope failures in the Rhone delta of Lake Geneva: Limnology and Oceanography, v. 33, p. 458–468, doi:10.4319/lo.1988.33.3.0458.

Laval, A., M. Cremer, P. Beghin, and C. Ravenne, 1988, Density surges: Two-dimensional experiments: Sedimentology, v. 35, p. 73–84, doi:10.1111/j.1365-3091.1988.tb00905.x.

Lefèvre, C., and J.-L. Schneider, 2002, Les risques naturels majeurs: Paris, Editions scientifiques GB, 324 p.

Le Moigne, M., 1999, Compréhension des mécanismes de formation des pockmarks sur la pente du Golfe de Guinée: Master's thesis, Université des Sciences et Techniques de Lille, Lille, France, 62 p.

Leopold, L. B., and T. Maddock, Jr., 1953, The hydraulic geometry of stream channels and some physiographic implications: U.S. Geological Survey Professional Paper 252, 53 p.

Liu, J. P., J. D. Milliman, S. Gao, and P. Cheng, 2004, Holocene development of the Yellow River's subaqueous delta, North Yellow Sea: Marine Geology, v. 209, p. 45–67, doi:10.1016/j.margeo.2004.06.009.

Lowe, D. R., 1982, Sediment gravity flows: II. Depositional models with special reference to the deposits of high-density turbidity currents: Journal of Sedimentary Petrology, v. 52, p. 279–297.

Lüthi, S., 1981, Experiments on nonchannelized turbidity currents and their deposits: Marine Geology, v. 40, p. M59–M68, doi:10.1016/0025-3227(81)90139-0.

Malde, H. E., 1968, The catastrophic late Pleistocene Bonneville flood in the Snake River plain, Idaho: U.S. Geological Survey Professional Paper 596, 69 p.

Malde, H. E., and H. A. Powers, 1962, Upper Cenozoic stratigraphy of the western Snake River plain, Idaho. Geological Society of America Bulletin, v. 73, no. 10, p. 1197–1219, doi:10.1130/0016-7606(1962)73[1197:UCSOWS]2.0.CO;2.

Mao, L., and M. A. Lenzi, 2007, Sediment mobility and bedload transport conditions in an alpine stream:

Hydrological Processes, v. 21, p. 1882–1891, doi:10.1002/hyp.6372.

Mas, V., 2009, Caractérisation de l'activité hydrosédimentaire dans le Système Turbiditique du Var (SW Méditerranée) et de son enregistrement dans l'archive sédimentaire: Ph.D. thesis, Université Bordeaux 1, Bordeaux, France, 187 p.

Mavilla, N., 2000, Stratigraphie et analyse de faciès de la succession d'âge oligo-miocène du bassin tertiaire piémontais (Italie nord occidentale): Ph.D. thesis, Université Bordeaux 1, Bordeaux, France, 225 p.

Maxworthy, T., 1999, The dynamics of sedimenting surface gravity currents: Journal of Fluid Mechanics, v. 392, p. 27–44, doi:10.1017/S002211209900556X.

Meybeck, M., L. Laroche, H. H. Durr, and J. P. M. Syvitski, 2003, Global variability of daily total suspended solids and their fluxes in rivers: Global and Planetary Change, v. 39, no. 1–2, p. 65–93, doi:10.1016/S0921-8181(03)00018-3.

Middleton, G. V., and M. A. Hampton, 1973, Sediment gravity flows: Mechanics of flow and deposition, *in* G. V. Middleton and A. H. Bouma, eds., Turbidites and deep-water sedimentation: SEPM Pacific Section Short Course Lecture Notes, Anaheim, California, p. 1–38.

Migeon, S., 2000, Levées sédimentaires en domaine profond: processus de mise en place et enregistrement par les faciès et la géométrie des dépôts: Ph.D. thesis, Université Bordeaux 1, Bordeaux, France, 288 p.

Migeon, S., B. Savoye, E. Zanella, T. Mulder, J.-C. Faugères, and O. Weber, 2001, Detailed seismic and sedimentary study of turbidite sediment waves on the Var sedimentary ridge (SE France): Significance for sediment transport and deposition and for the mechanism of sediment wave construction: Marine and Petroleum Geology, v. 18, p. 179–208, doi:10.1016/S0264-8172(00)00060-X.

Milliman, J. D., and J. S. Kao, 2005, Hyperpycnal discharge of fluvial sediment to the ocean: Impact of Super-Typhoon Herb, 1996) on Taiwanese rivers: Journal of Geology, v. 113, p. 503–516, doi:10.1086/431906.

Milliman, J. D., and R. H. Meade, 1983, World-wide delivery of river sediment to the oceans: Journal of Geology, v. 91, p. 1–21, doi:10.1086/628741.

Milliman, J. D., and J. P. M. Syvitski, 1992, Geomorphic/tectonic control of sediment discharge to the ocean: The importance of small mountainous rivers: Journal of Geology, v. 100, p. 525–544, doi:10.1086/629606.

Moernaut, J., M. De Batist, F. Charlet, K. Heirman, E. Chapron, M. Pino, and R. Urrutia, 2006, Giant earthquakes in south-central Chile revealed by Holocene mass-wasting events in Lake Puyehue: Sedimentary Geology, doi:10.1016/j.sedgeo.2006.08.005.

Morehead, M. D., J. P. M. Syvitski, E. W. H. Hutton, and S. D. Peckham, 2003, Modeling the inter-annual and intra-annual variability in the flux of sediment in ungauged river basins: Global and Planetary Change, v. 39, p. 95–110.

Mulder, T., and J. Alexander, 2001, The physical character of subaqueous sedimentary density currents and their deposits: Sedimentology, v. 48, p. 269–299, doi:10.1046/j.1365-3091.2001.00360.x.

Mulder, T., and J. P. M. Syvitski, 1995, Turbidity currents generated at river mouths during exceptional discharges to the world oceans: Journal of Geology, v. 103, p. 285–299, doi:10.1086/629747.

Mulder, T., and J. P. M. Syvitski, 1996, Climatic and morphologic relationships of rivers: Implications of sea-level fluctuations on river loads: Journal of Geology, v. 104, p. 509–523, doi:10.1086/629849.

Mulder, T., B. Savoye, J. P. M. Syvitski, and O. Parize, 1997, Des courants de turbidité hyperpycnaux dans la tête du canyon du Var? Données hydrologiques et observations de terrain: Oceanologica Acta, v. 20, p. 607–626.

Mulder, T., S. Migeon, B. Savoye, and J.-C. Faugères, 2001a, Inversely graded turbidite sequences in the deep Mediterranean. A record of deposits from flood-generated turbidity currents?: Geo-Marine Letters, v. 21, p. 86–93.

Mulder, T., S. Migeon, B. Savoye, and J.-M. Jouanneau, 2001b, Twentieth century floods recorded in the deep Mediterranean sediments: Geology, v. 29, no. 1, p. 1011–1014, doi:10.1130/0091-7613(2001)029<1011:TCFRIT>2.0.CO;2.

Mulder, T., S. Migeon, B. Savoye, and J.-C. Faugères, 2002, Inversely graded turbidite sequences in the deep Mediterranean, a record of deposits from flood-generated turbidity currents? A reply: Geo-Marine Letters, v. 22, no. 2, p. 112–120.

Mulder, T., J. P. M. Syvitski, S. Migeon, J.-C. Faugères, and B. Savoye, 2003, Hyperpycnal turbidity currents: Initiation, behavior and related deposits: A review, *in* E. Mutti, G. S. Steffens, C. Pirmez, M. Orlando, and D. Roberts, eds., Turbidites: Models and problems: Marine and Petroleum Geology, Special Issue 20, no. 6–8, p. 861–882.

Mulder, T., et al., 2004, Understanding continent-ocean sediment transfer: Eos Transactions, American Geophysical Union, v. 85, no. 27, p. 257 and 261–262.

Mulder, T., S. Zaragosi, J.-M. Jouanneau, G. Bellaiche, S. Guérinaud, and J. Querneau, 2009, Deposits related to the failure of Malpasset Dam in 1959, an analogue for hyperpycnal deposits from jökulhlaups: Marine Geology, v. 260, p. 81–89, doi:10.1016/j.margeo.2009.02.002.

Mullineaux, D. R., R. E. Wilcox, W. F. Ebauch, R. Fryxell, and M. Rubin, 1978, Age of the last major Scalbland flood of the Columbia Plateau in eastern Washington: Quaternary Research, v. 10, p. 171–180.

Mutti, E., G. Davoli, R. Tinterri, and C. Zavala, 1996, The importance of ancient fluvio-deltaic systems dominated by catastrophic flooding in tectonically active basins: Memorie di Scienze Geologiche, v. 48, p. 233–291.

Myrow, P. M., W. Fisher, and J. W. Goodge, 2002, Wave-modified turbidites: Combined-flow shoreline and shelf deposits, Cambrian, Antarctica: Journal of Sedimentary Research, v. 72, p. 641–656.

Nakajima, T., 2006, Hyperpycnites deposited 700 km away from river mouths in the central Japan Sea: Journal of Sedimentary Research, v. 76, no. 1, p. 60–73.

Normark, W.R., and D. J. W. Piper, 1991, Initiation processes and flow evolution of turbidity currents: Implications for the depositional record, *in* R. H. Osborne, ed., From shoreline to abyss: Contribution in marine geology in honor of Francis Parker Shepard: SEPM Special Publication 46, p. 207–230.

Parize, O., B. Savoye, M. Sahabi, and P. Cochonat, 1989, Observation d'un réseau d'origine sous-marine de chenaux en tresses sur le fond de la tête du canyon du Var (Nice, France). Origine et implications: 2ème Congrès Français de Sédimentologie, Livre des résumés, Publications ASF 10, Paris, France, p. 227–228.

Parker, G., 1982, Conditions for the ignition of catastrophically erosive turbidity currents: Marine Geology, v. 46, p. 307–327.

Parsons, J. D., J. W. M. Bush, and J. P. M. Syvitski, 2001, Hyperpycnal plume formation from riverine outflows with small sediment concentrations: Sedimentology, v. 48, p. 465–478.

Piper, D. J. W., and W. R. Normark, 2009, Processes that initiate turbidity currents and their influence on turbidites: A marine geology perspective: Journal of Sedimentary Research, v. 79, no. 5, p. 347–362.

Piper, D. J. W., P. Cochonat, G. Ollier, E. Le Dreezen, M. Morrison, and A. Baltzer, 1992, Evolution progressive d'un glissement rotationnel en un courant de turbidité: Cas du séisme de 1929 des Grands Bancs (Terre Neuve): Comptes Rendus de l'Académie des Sciences, Paris, v. 314, Series II, p. 1057–1064.

Piper, D. J. W., J. Shaw, and K. I. Skene, 2007, Stratigraphic and sedimentological evidence for late Wisconsinan sub-glacial outburst floods to Laurentian Fan: Palaeogeography, Palaeoclimatology, Palaeoecology, v. 246, p. 101–119.

Pirmez, C., and J. Inram, 2003, Reconstruction of turbidity currents in Amazon Channel: Marine and Petroleum Geology, v. 20, p. 823–849.

Pollet, N., and J. L. Schneider, 2004, Dynamic disintegration processes accompanying transport of the Holocene Flims sturzstrom (Swiss Alps): Earth and Planetary Science Letters, v. 221, p. 433–448.

Popescu, I., G. Lericolais, N. Panin, H. K. Wong, and L. Droz, 2001, Late Quaternary channel avulsions on the Danide deep-sea fan, Black Sea: Marine Geology, v. 179, p. 25–37.

Pratson, L. F., and W. F. Haxby, 1996, What is the slope of the U.S. continental slope?: Geology, v. 24, p. 3–6.

Pratson, L. F., W. B. F. Ryan, G. S. Mountain, and D. C. Twitchell, 1994, Submarine canyon initiation by downslope-eroding sediment flows: Evidence in late Cenozoic strata on the New Jersey continental slope: Geological Society of America Bulletin, v. 106, p. 395–412.

Ravenne, C., and P. Beghin, 1983, Apport des expériences en canal à l'interprétation sédimentologique des dépôts de cônes détritiques sous-marins: Revue de'Institut Français du Pétrole, v. 38, no. 3, p. 279–297.

Reading, H. G., and M. T. Richards, 1994, The classification of deep-water siliciclastic depositional systems by grain size and feeder systems: AAPG Bulletin, v. 78, p. 792–822.

Reid, I., L. E. Frostick, and J. T. Layman, 1985, The incidence and nature of bedload transport during flood flows in coarse-grained alluvial channels: Earth Surface Processes and Landforms, v. 10, p. 33–44.

Richmond, G. M., R. Fryxell, G. E. Neff, and D. E. Trimble, 1965, The Cordilleran ice sheet of the northern Rocky Mountains, and related history of the Columbia Plateau, *in* H. B. Wright and D. G. Frey, eds., The Quaternary of the United States: Princeton, New Jersey, Princeton University Press, p. 231–242.

Rigaut, F., 1997, Analyse et évolution récente d/un système turbiditique méandriforme: L'éventail profond du Zaïre: Ph.D. thesis, Université de Bretagne Occidentale, Brittany, France, 209 p.

Saint-Onge, G., T. Mulder, D. J. W. Piper, C. Hillaire-Marcel, and J. Stoner, 2003, Earthquake and flood-induced turbidites in the Saguenay Fjord (Québec): A Holocene paleoseismicity record: Quaternary Science Reviews, v. 23, p. 283–294.

Savoye, B., and D. J. W. Piper, 1991, The Messinian event on the margin of the Mediterranean Sea in the Nice area, southern France. Marine Geology, v. 97, p. 279–304.

Savoye, B., et al., 2000, Structure et évolution récente de l'éventail turbiditique du Zaïre: Premiers résultats scientifiques des missions d'exploration ZAÏANGO: Comptes Rendus de l'Académie des Sciences, Paris, v. 331, p. 211–220.

Schneider, J. L., N. Pollet, E. Chapron, M. Wessels, and P. Wassmer, 2004, Signature of Rhine Valley sturzstrom dam failures in Holocene sediments of Lake Constance, Germany: Sedimentary Geology, v. 169, p. 75–91.

Schröder, H. G., M. Wessels, and F. Niessen, 1998, Acoustic facies and depositional structures of Lake Constance: Archiv Für Hydrobiologie, Special Issues on Advances in Limnology, v. 53, p. 351–368.

Schumm, S. A., 1981, Evolution and response of the fluvial system, sedimentologic implications, *in* F.G Ethridge and R.M. Flores, eds., Recent and ancient nonmarine depositional environments: SEPM Special Publication 31, p. 19–29.

Séranne, M., et al., 2008, U-Pb single zircon grain dating of present fluvial and Cenozoic eolian sediments from Gabon: Consequences on sediment provenance, reworking, and erosion processes on the equatorial West African margin: Bulletin de la Société Géologique de France, v. 179, no. 1, p. 29–40.

Shepard, F. P., and R. F. Dill, 1966, Submarine canyons and other sea-valleys: Chicago, Illinois, Rand McNally and Co., 381 p.

Shipboard Scientific Party, 1998a, Introduction, *in* Y. Fouquet et al., eds., Proceedings of the Ocean Drilling Program, Initial Reports, 169: College Station, Texas, Ocean Drilling Program, p. 7–16.

Shipboard Scientific Party, 1998b, Escanaba Trough: reference site (Site 1037), *in* Y. Fouquet et al., eds.,

Proceedings of the Ocean Drilling Program, Initial Reports, 169: College Station, Texas, Ocean Drilling Program, p. 205–251.

Shipboard Scientific Party, 1998c, Escanaba Trough: Central Hill (Site 1038), *in* Y. Fouquet et al., eds., Proceedings of the Ocean Drilling Program, Initial Reports, 169: College Station, Texas, Ocean Drilling Program, p. 253–298.

Soyinka, O. A., and R. M. Slatt, 2009, Identification and micro-stratigraphy of hyperpycnites and turbidites in Cretaceous Lewis Shale, Wyoming: Sedimentology, v. 55, no. 5, p. 1117–1133.

Sparks, R. S. J., R. T. Bonnecaze, H. E. Huppert, J. R. Lister, M. A. Hallworth, H. Mader, and J. Phillips, 1993, Sediment-laden gravity currents with reversing buoyancy: Earth and Planetary Science Letters, v. 114, p. 243–257.

Sturm, M., and A. Matter, 1978, Turbidites and varves in Lake Brienz (Switzerland): Deposition of clastic detritus by density currents: International Association of Sedimentologists Special Publication 2, p. 174–168.

Summerfield, M. A., and N. J. Hulton, 1994, Controls of fluvial denudation rates in major world drainage basins: Journal of Geophysical Research, v. 99, no. B7, p. 13,871–13,883.

Susperregui, A.-S., 2008, Caractérisation hydro-sédimentaire des retenues de Cointzio et d'Umécuaro (Michoacan, Mexique) comme indicateur du fonctionnement érosif du bassin versant: Ph.D. thesis, Université de Grenoble, Grenoble, France, 289 p.

Syvitski, J. P. M., 2003, Sediment fluxes and rates of sedimentation, *in* G. V. Middleton, ed., Encyclopedia of sediments and sedimentary rocks: Dordrecht, Netherlands, Kluwer Academic Publishers, p. 600–606.

Syvitski, J. P. M., and J. M. Alcott, 1993, Grain2: Prediction of particle size seaward of river mouths: Computers and Geosciences, v. 19, p. 399–446.

Syvitski, J. P. M., and J. M. Alcott, 1995, River3: Simulation of river discharge and sediment transport: Computers and Geosciences, v. 21, no. 1, p. 89–151.

Syvitski, J. P. M., and F. J. Hein, 1991, Sedimentology of an arctic basin: Itirbilung Fjord, Baffin Island, Northwest Territories: Geological Survey of Canada Paper 9111, 66 p.

Syvitski, J. P. M., and C. T. Schafer, 1996, Evidence for an earthquake-triggered basin collapse in Saguenay Fjord, Canada: Sedimentary Geology, v. 104, p. 127–153.

Syvitski, J. P. M., M. D. Morehead, D.B. Bahr, and T. Mulder, 2000, Estimating fluvial sediment transport: The rating parameters: Water Resources Research, v. 36, no. 9, p. 2747–2760.

Tiercelin, J.-J., et al., 1987, Le demi-graben de Baringo-Bogoria, Rift Gregory, Kenya 30000 ans d'histoire hydrologique et sédimentaire: Bulletin des Centres de Recherche Exploration-Production Elf-Aquitane, v. 11, no. 2, p. 249–540.

Tiercelin, J.-J., A. S. Cohen, M. Soreghan, K. E. Lezzar, and J.-L. Bouroullec, 1992, Sedimentation in large rift lakes: Example from the Middle Pleistocene-Modern deposits of the Tanganyika Trough, East African rift system: Bulletin des Centres de Recherche Exploration-Production Elf-Aquitane, v. 16, p. 83–111.

Toucanne, S., 2008, Reconstructions des transferts sédimentaires en provenance du Paléofleuve Manche et du système glaciaire de Mer d/Irlande au cours des derniers cycles climatiques: Ph.D. thesis, Université Bordeaux 1, Bordeaux, France, 362 p.

Tripsanas, E. K., and D. J. W. Piper, 2008, Glaciogenic debris-flow deposits of Orphan Basin, offshore Eastern Canada: Sedimentological and rheological properties, origin and relationship to meltwater discharge: Journal of Sedimentary Research, v. 78, p. 724–744.

Turner, J. S., 1973, Buoyancy effects in fluids: Cambridge, United Kingdom, Cambridge University Press, 367 p.

Vallon, M., 1977, Bilan de masse et fluctuations récentes du Glacier Ampère (Iles Kerguelen, TAAF): Zeitschrift für Gletscherkunde und Glazialgeologie, v. 13, p. 55–85.

Van Rensbergen, P., M. de Batist, C. Beck, and F. Manalt, 1998, High-resolution seismic stratigraphy of late Quaternary fill of Lake Annecy (Northwestern Alps): Evolution from glacial to interglacial sedimentary processes: Sedimentary Geology, v. 117, p. 71–96.

Van Rensbergen, P., M. de Batist, C. Beck, and E. Chapron, 1999, High-resolution seismic stratigraphy of glacial to interglacial fill of a deep glacigenic lake: Lake Le Bourget, northwestern Alps, France: Sedimentary Geology, v. 128, p. 99–129.

Waitt, R. B., 1985, Case for periodic, colossal jökulhlaups from Pleistocene glacial Lake Missoula: Geological Society of America Bulletin, v. 96, p. 1271–1286.

Warrick, J. A., and J. D. Milliman, 2003, Hyperpycnal sediment discharge from semi-arid southern California rivers: Implications for coastal sediment budgets: Geology, v. 31, p. 781–784.

Westcott, W. A., and F. G. Ethridge, 1990, Fan deltas—Alluvial fans in coastal settings, *in* A. H. Rachocki and M. Church, eds., Alluvial fans: A field approach: Chichester, United Kingdom, Wiley, p. 195–211.

Wright, L. D., 1977, Sediment transport and deposition at river mouths: A synthesis: Geological Society of America Bulletin, v. 88, p. 857–868.

Wright, L. D., Z.-S. Yang, B. D. Bornhold, G. H. Keller, D. B. Prior, and W. J. Wisenam Jr., 1986, Hyperpycnal flows and flow fronts over the Huanghe (Yellow River) delta front: Geo-Marine Letters, v. 6, p. 97–105.

Zierenberg, R. A., Y. Fouquet, D. J. Miller, and W. R. Normark, eds., 2000, Proceedings of the Ocean Drilling Program, Initial Reports, 169: College Station, Texas, Ocean Drilling Program.

Zuffa, G. G., W. R. Normark, F. Serra, and C. A. Brunner, 2000, Turbidite megabeds in an oceanic rift valley recording jökulhlaups of late Pleistocene glacial lakes of the western United States: Journal of Geology, v. 108, p. 253–274.

2

Zavala, C., M. Arcuri, M. Di Meglio, H. Gamero Diaz, and C. Contreras, 2011, A genetic facies tract for the analysis of sustained hyperpycnal flow deposits, *in* R. M. Slatt and C. Zavala, eds., Sediment transfer from shelf to deep water—Revisiting the delivery system: AAPG Studies in Geology 61, p. 31–51.

A Genetic Facies Tract for the Analysis of Sustained Hyperpycnal Flow Deposits

Carlos Zavala and Mariano Arcuri
GCS Argentina, Universidad Nacional del Sur, Bahía Blanca, Argentina

Mariano Di Meglio
GCS Argentina, Bahía Blanca, Argentina

Helena Gamero Diaz and Carmen Contreras
Schlumberger Technology Corp., Addison, Texas, U.S.A.

ABSTRACT

Facies analysis performed for more than 10 yr in several lacustrine and marine basins allowed the distillation of a genetic and predictive facies tract of general application to the recognition of sustained turbulent (noninertial) hyperpycnal flow deposits. The facies tract is composed of three main genetic facies groups related to bed load (B), suspended load (S), and lofting (L) transport processes.

Facies B (bed load) is the coarsest grained and relates to shear and frictional drag forces provided by the overpassing long-lived turbulent (hyperpycnal) flow. Three main subcategories are recognized, termed B1 (massive or crudely bedded conglomerates), B2 (pebbly sandstones with low-angle asymptotic cross-stratification), and B3 (pebbly sandstones with diffuse planar lamination and aligned clasts).

Facies S is finer grained and relates to the gravitational collapse of suspended load transported by the hyperpycnal flow. Four subfacies types are recognized, termed S1 (massive sandstones), S2 (parallel laminated sandstones), S3 (sandstones with climbing ripples), and S4 (massive siltstones and mudstones).

Facies L (lofting) relates to the buoyancy reversal of the hyperpycnal flow provoked by lift of a less dense fluid (in this case, fresh water) typically found in marine and other saline-receiving basins. The finest materials suspended in the

Copyright ©2011 by The American Association of Petroleum Geologists.
DOI:10.1306/13271349St613438

flow (very fine–grained sand, silt, plant debris, and mica) are lifted from the substrate and settle down forming silt/sand couplets of great lateral extent. Facies L develops only in marine/saline environments, whereas facies S3 and S4 are more common in lacustrine environments.

Hyperpycnites are often very complex, showing internal erosional surfaces and gradual facies recurrences related to deposition from long-lived and highly dynamic (fluctuating) flows. This complex behavior results in the accumulation of composite beds having an internal facies arrangement that strongly departs from conventional facies models developed for surgelike flows. Facies B characterizes transfer zones, and its occurrence allows the prediction of sandstone deposits (facies S) basinward. Facies L is mostly developed in flow-margin areas. In marine settings, the reversing buoyancy effect (lofting) at the hyperpycnal flow margin will result in less lateral continuity of sandstone bodies compared with those related to resedimentation processes, with important consequences for hydrocarbon reservoirs.

INTRODUCTION

Recent advances in the understanding of a new category of clastic depositional system, termed "hyperpycnal system," have provided a new frontier for the understanding of clastic facies in marine and lacustrine environments. A hyperpycnal system (Zavala et al., 2006a) is the subaqueous extension of the fluvial system (in the sense of Schumm, 1977, 1981). It originates when a river in flood directly discharges a sustained and relatively more dense turbulent mixture of fresh water and sediments (hyperpycnal flows; Bates, 1953) into a receiving standing body of water. To overcome the density contrast with marine waters, the plunging of a river flow will require a suspended-load concentration varying between 35 and 45 kg/m^3 (Mulder and Syvitski, 1995) according to latitudinal variations. As a consequence of its excess in density, the incoming flow plunges and moves basinward as a land-generated underflow (Figure 1). Increasing evidence shows that hyperpycnal systems can extend hundreds of kilometers basinward (Collins, 1986; Johnson et al., 2001; Nakajima, 2006; Zavala et al., 2006a; Gamero et al., 2007; Cheng-Shing and Ho-Shing, 2008; Bourget et al., 2010) and can explain the accumulation of very thick clastic successions in shelfal and deep-marine settings even during highstands of sea level (Mutti et al., 1996, 2003).

Observations made in 230 contemporaneous rivers (Mulder and Syvitski, 1995) revealed that more than 66% of the rivers produce periodic hyperpycnal discharges, allowing the basinward transfer of huge volumes of sediments. As an example, during a single hyperpycnal discharge (lasting 18 hr in November 1995) the Var river transferred a volume of sediments equivalent to that transported in 20 yr in nonflood conditions (Mulder et al., 2003). Although hyperpycnal discharge seems to be very common in contemporaneous rivers, only a few examples of deposits interpreted as fossil hyperpycnites can be found in the literature (Plink-Björklund and Steel, 2004; Pattison, 2005, 2008; Zavala et al., 2006a; Myrow et al., 2008; Soyinka and Slatt, 2008). Consequently, it is possible that many ancient hyperpycnal deposits were misinterpreted in the past as deposited in more conventional depositional systems (Mutti et al., 1996, 2003). Although present hyperpycnal discharges transport mostly very fine-grained sediments, increasing evidence from ancient strata (Mutti et al., 2003; Plink-Björklund and Steel, 2004; Saitoh, 2004; Gamero et al., 2005, 2007, 2008; Pattison, 2005, 2008; Hesse and Khodabakhsh, 2006; Olariu and Bhattacharya, 2006; Petter and Steel, 2006; Zavala et al., 2006a, b, c, 2007a, b, 2008b; Lamb et al., 2008; Myrow et al., 2008; Ponce et al., 2008a, b; Soyinka and Slatt, 2008; Zavala, 2008) strongly supports the idea that coarse-grained hyperpycnites could be much more common than expected.

Extensive field and core studies performed during more than 10 yr in several different sedimentary basins (Table 1) show that sedimentary facies related to hyperpycnal systems are very distinctive and could easily be differentiated from surgelike (classical) turbidites, tempestites, and other conventional facies types.

A hyperpycnite (Mulder et al., 2003) is a particular type of turbidite having distinctive and at present poorly known facies characteristics (Mulder and Alexander, 2001). Its origin is closely related to a direct fluvial discharge and results in facies types and depositional features that commonly resemble those

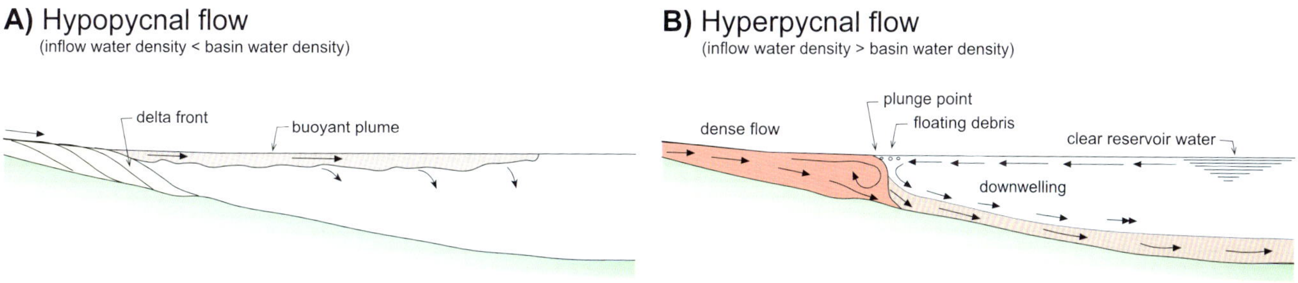

FIGURE 1. Comparison between hypopycnal (A, inflow density < sea-water density) and hyperpycnal (B, inflow density > sea-water density) flows (original concept by Bates, 1953). Note that in the case of the hyperpycnal flow, the fluvial discharge sinks below the sea-water body, continuing its travel basinward as a quasi-steady underflow. Panel B was redrawn from the pioneer work of Knapp (1943). After Zavala et al. (2008a).

considered as typical of fluvial environments (bed load, meandering, etc.) but are commonly associated with clear marine and lacustrine indicators. The transfer and accumulation of a huge volume of continental sediments during a single long-lived flood frequently results in confusing facies types from a conventional point of view. The resulting clastic sedimentary bodies could be biostratigraphically sterile or can display a wide range of water-depth indicators. Very thick hyperpycnites commonly lack or show rare and very specific trace fossils favoring the confusion of these strata with estuarine deposits. As for every sediment gravity flow, the location and thickness of coarse-grained hyperpycnal sedimentary bodies are very sensitive to the contemporaneous subaqueous topography, commonly resulting in the accumulation of very thick laminated or structureless sandstone bodies of difficult interpretation using nonhyperpycnal facies models.

The objective of this chapter is to introduce and to discuss a new approach to a genetic facies model constructed for the analysis and understanding of long-lived and fully turbulent hyperpycnal flows and their coarse-grained deposits illustrated with selected examples from different South American basins.

A GENETIC FACIES MODEL FOR THE ANALYSIS OF HYPERPYCNAL DEPOSITS

Descriptive facies models focus on the careful description and classification of different types of

Table 1. Main units and sedimentary basins considered in this study.

Unit	*Age*	*Basin*	*Country*	*References*
La Lola Fm	Ordovician	Ventania	Argentina	Zavala et al., 2000
Los Molles Fm	Early Jurassic	Neuquén	Argentina	Zavala and Gonzalez, 2001
Lajas Fm	Middle Jurassic	Neuquén	Argentina	Zavala and Gonzalez, 2001
Lotena Fm	Middle Jurassic (Callovian)	Neuquén	Argentina	Zavala, 2005
Mulichinco Fm	Early Cretaceous (Valanginian)	Neuquén	Argentina	Zavala, 2000a
Rayoso Fm	Barremian–Aptian	Neuquén	Argentina	Zavala et al., 2006a
Magallanes Fm	Late Cretaceous	Austral	Argentina	Saccavino et al., 2008
Misoa Fm	Eocene	Maracaibo	Venezuela	Gamero et al., 2005
Pampatar Fm	Eocene	Ensenada	Venezuela	Guzmán and Campos, 2008
Merecure Fm	Oligocene	Maturin	Venezuela	Zavala et al., 2007b
Carapita Fm	Miocene	Maturin	Venezuela	Gamero et al., 2007
Herrera Sst	Miocene	Columbus	Trinidad	Gamero et al., 2007
Mayaro Fm	Pliocene	Columbus	Trinidad	Zavala et al., 2008b
Pliocene sandstones	Pliocene	Columbus	Trinidad	Gamero et al., 2008
Catarozzo Gp	Late Pliocene	Sant'Arcangelo	Italy	Zavala, 2000b
Aliano Gp	Late Pliocene–Early Pleistocene	Sant'Arcangelo	Italy	Zavala, 2000b
Tursi Gp	Late Pliocene–Early Pleistocene	Sant'Arcangelo	Italy	Zavala, 2000b

Fm = Formation; Sst = Sandstone; Gp = Group.

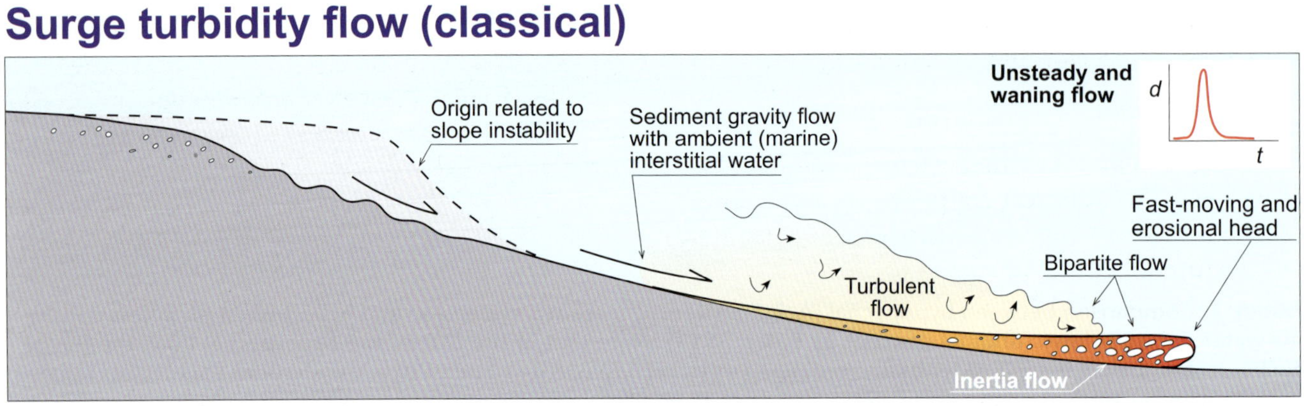

FIGURE 2. **Main characteristics of surgelike turbidites originated by slope instability. Flow bipartition results in a fast-moving inertial flow followed by a slow-moving turbulent (gravitative) flow. Inspired by Mutti et al. (1999). *d* = discharge; *t* = time.**

sedimentary rocks according to their macroscopic characteristics such as sedimentary structures, grain size, color, geometry, and types of bounding surfaces, among others (Walker, 1984). These descriptive facies are then commonly grouped into different categories according to some common characteristics (i.e., coarse-grained facies, cross-bedded facies, massive facies, etc). A descriptive facies analysis is an objective procedure because it is based on observable macroscopic physical characteristics. Therefore, different observers can arrive at similar conclusions. The main objective of a descriptive facies analysis is to determine the paleoenvironment of accumulation.

Genetic facies models and facies tracts represent a step ahead in facies analysis because they require a conceptual-based classification. A genetically oriented facies approach distinguishes facies according to the main sedimentary processes involved during their transport and deposition. As in fossil deposits sedimentary processes can only be inferred, the genetic approach strongly depends on the experience and the paradigm used by the observer. The use of a genetic approach requires an early understanding of the depositional environment, which should be determined using the descriptive facies approach previously described. The main goal of a genetic facies analysis is to determine the conceptual facies tract and to predict the expected facies types outside of the reference section under study. At present, genetic facies tracts are only available for turbidites related to ignitive surgelike processes, mainly associated with slope failures. The use of a genetic approach for the analysis of surgelike turbidites was introduced by Bouma (1962) in his milestone contribution. Bouma described a key facies sequence from the Oligocene Peira Cava flysch that allowed, for the first time, the analysis of the position along the depositional profile of a certain sandstone body and the prediction of landward and basinward directions and processes based on the possible characteristics of the genetically related sedimentary bodies.

A significant improvement to the genetic facies tract applied to the analysis of turbidite deposits was later made by Mutti and Ricci Lucchi (1972), which also incorporated very coarse-grained sedimentary populations. Lowe (1979, 1982) carefully described and discussed the origin of ancient fine- and coarse-grained turbidites based on their flow rheology and inferred particle support mechanisms. Lowe (1982) also emphasized the importance of bed load in coarse-grained turbidites.

In a series of more recent contributions, Mutti (1992) and Mutti et al. (1999, 2003) increased the predictive power of genetic facies tracts by adding a discussion on the importance of hydraulic jumps and flow transformation in controlling the characteristics of the resulting sedimentary facies. Mutti et al. (1999) improved the genetic interpretation by applying the concept of flow bipartition introduced by Sanders (1965). According to Mutti et al. (1999), a turbidity flow will therefore be composed of a fast-moving inertial (granular) flow followed by a slow-moving turbulent flow. Deposition of different facies types (F2–F9) occurs as a consequence of a series of hydraulic jumps as the flow wanes and dilutes basinward (Mutti et al., 1999). As the flow is driven by inertial forces related to the fast-moving granular flow, this kind of flow will therefore require an acceleration on a steep slope and will be very sensitive to slope breaks. In addition, if this turbidity flow originates from slope instability (ignitive classical turbidites), the density of interstitial flow waters will be similar to that of ambient waters, thus preventing the effect of lofting (Figure 2).

In recent years, the origin of turbidites has been also related to the direct supply of relatively dense mixtures of water and sediments by rivers in flood or hyperpycnal flow (Normark and Piper, 1991; Mutti et al., 1996; Bruhn and Walker, 1997; Mutti et al., 1999, 2009; Klaucke et al., 2000; Stow and Mayall, 2000; Mulder et al., 2001a, b; Deptuck et al., 2003; Mutti et al., 2003; Posamentier and Kolla, 2003; Joseph and Lomas, 2004; Migeon et al., 2004; Plink-Björklund and Steel, 2004; Pattison, 2005; Antobrch and Krastel, 2006; Gervais et al., 2006; Mansurbeg et al., 2006; Migeon et al., 2006; Ponce et al., 2008a, b). Mutti et al. (1996, 2003, 2009) emphasized the importance of catastrophic flooding in the accumulation of very thick clastic successions in shallow- to deep-marine settings, mainly associated with small and medium-size mountainous rivers characterized by relatively short- and high-gradient transfer zones. Mutti et al. (2003) recognized the flood-generated delta-front sandstone lobes as the main depositional element associated with hyperpycnal flows. This provides a broad spectrum of relatively small intergradational and coarse-grained fluvial-deltaic systems ranging between two end members, (1) fan deltas and (2) river deltas. According to Mutti et al. (2003), fan-delta systems accumulate in roughly tabular lobes extending from alluvial conglomerates to shelfal siltstone and mudstone. In these systems, flood-generated dense flows enter seawaters as catastrophic and inertia-dominated relatively unconfined flows. Coarser materials tend to collect at the front of the flow, giving way to a horizontally negative grain-size gradient. River-delta systems, however, could be dominated by two intergradational types of flow corresponding to sediment-laden streamflows (long-lived turbulent flows) and composite sediment-laden streamflows (inertial flows, with horizontally negative grain-size gradient). In river-delta systems, sedimentation occurs mainly in a mouth bar (characterized by sigmoidal bedding) and in associated flood-generated delta-front sandstone lobes. Mutti et al. (2003, p. 739) indicate that "in the case of higher outflow efficiency, coarse sediment collapsed at river mouths is transported as tractive bed load by sustained turbulent flows exiting river mouths as hyperpycnal flows. This process sets up the formation of tractional bed forms that are primarily controlled by water depth and the velocity and steadiness of the bypassing hyperpycnal flows."

The understanding of facies types related to sustained turbulent (noninertial) hyperpycnal flows represents a deep challenge for sedimentologists because long-lasting river discharges and their associated facies types could be very different from those expected in surgelike (inertia-dominated) turbidity flows (Mulder and Alexander, 2001; Zavala et al., 2006a; Zavala, 2008). To travel hundreds of kilometers (Nakajima, 2006), the hyperpycnal discharges have to be maintained for weeks or months and consequently they will be mostly associated with low-gradient medium- to large-size rivers.

In this chapter, we focus our analysis on the facies types related to sustained fully turbulent hyperpycnal flows with associated bed load. A sustained hyperpycnal flow has a distinctive behavior that will result in the accumulation of nonconventional beds from a classical point of view. Its origin related to a direct fluvial discharge results in a subaqueous flow having characteristics commonly considered as typical of alluvial sedimentation. A sustained hyperpycnal flow is a land-derived, relatively slow-moving, and fully turbulent sediment gravity flow (Mulder et al., 2003) having the ability to carry basinward interstitial fresh water (Hesse and Khodabakhsh, 1998; Johnson et al., 2001; Hesse et al., 2004; Mansurbeg et al., 2006; Warrick et al., 2008; Zavala, 2008). The diameter of the clastic suspended particles commonly ranges from silt to very fine sand (0.004–0.25 mm; Myrow et al., 2006). The particles are transported together with plant debris and micas (Zavala et al., 2006c, 2008b; Zavala, 2008). In contrast to surgelike (ignitive) turbidites, sustained hyperpycnal flows have a slow-moving and more diluted leading head (Kassem and Imran, 2001) that will be very sensitive to the pre-existing subaqueous topography (Figure 3).

The moving of a sustained hyperpycnal flow will not necessarily require steep slopes (Zavala et al., 2006a) because the flow could be maintained as long as the high-density fluvial discharge continues (Prior et al., 1987). Therefore, the distance reached by a sustained hyperpycnal flow traveling along a gently dipping or near-flat sea bottom will depend mostly on the duration of the related flood event (Zavala et al., 2006a). In contrast with surgelike flows where deposition is dominated by the head, in sustained hyperpycnal flows, deposition is dominated by the body (De Rooij and Dalziel, 2001; Peakall et al., 2001). These characteristics allow the preservation in the hyperpycnal deposit of evidences of flow fluctuations that occurred during the overpassing discharge, resulting in the accumulation of composite

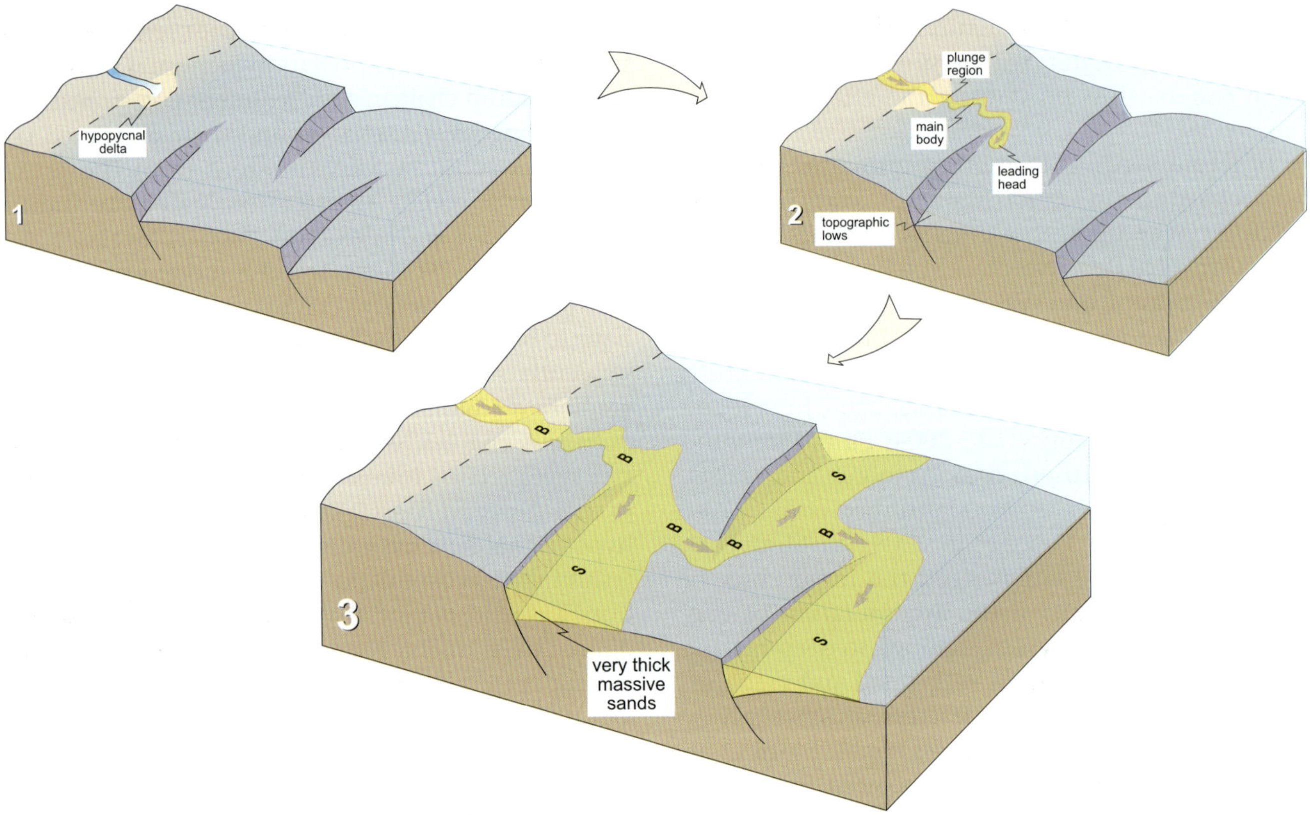

FIGURE 3. **Main steps in the triggering and evolution of a sustained hyperpycnal flow. (1) Normal hypopycnal delta. (2) The increase in the overall density of the fluvial discharge forces the flow to plunge and to produce a sustained hyperpycnal flow, with a leading head followed by the main body. The leading head will advance, trying to achieve the deepest part of the submarine landscape. (3) The submarine lows could be filled with very thick massive sands. Horizontal scale is tens of kilometers.**

beds (Zavala et al., 2007a). Contrary to classical models of turbidity sedimentation, coarse-grained materials are not transported at the flow head but are dragged at the flow bottom as bed load (Figure 4) related to shear forces provided by the overpassing long-lived turbulent flow (Plink-Björklund and Steel, 2004; Zavala et al., 2006b, 2008b). Bed load in sustained hyperpycnal flows is probably inherited from the original subaerial discharge that enters the receiving basin. Recognition of bed-load facies in hyperpycnal deposits is crucial for facies prediction because these facies give fundamental information about (1) the position in the facies tract with respect to the whole hyperpycnal system, (2) flood event magnitude and duration, and (3) expected related facies types in more proximal and distal areas.

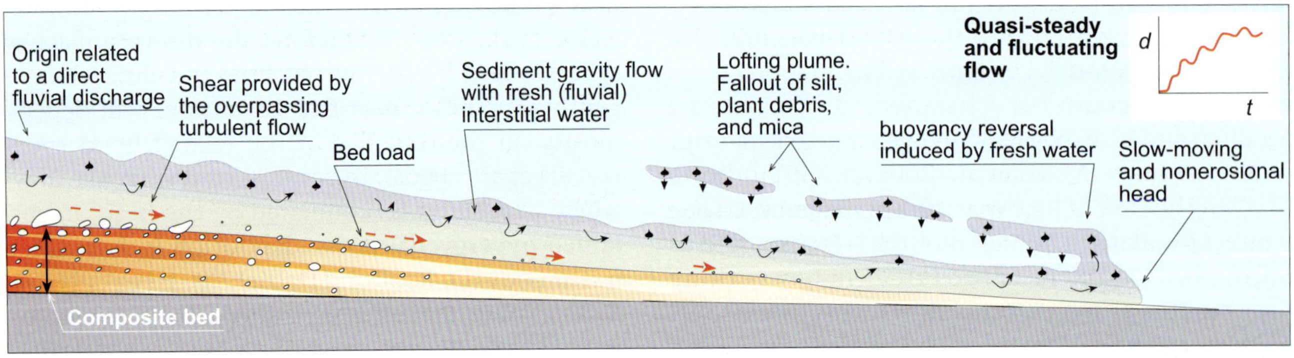

FIGURE 4. **Main characteristics of long-lived hyperpycnal flows and their typical deposits. The complexity of these flows results in the accumulation of composite beds (Zavala et al., 2007a).** ***d*** **= discharge;** ***t*** **= time.**

A fundamental difference between bed load related to the movement of fresh water (fluvial) with respect to that associated with sustained hyperpycnal flows is that in the latter, suspended materials transported by the overpassing turbulent flow will be trapped, filling the interstices between the coarser grained materials. For this reason, open packing in bed load associated with sustained hyperpycnal sedimentation is almost impossible. Although similar facies can be temporarily generated during floods in alluvial channels, the later washout (reworking) by fresh waters commonly results in better sorting and open packing in gravels. Consequently, the recognition of open packing in fossil gravels (commonly filled by chemical precipitates) can be used as a field criterion to distinguish bed-load deposits related to subaerial fluvial streams from those associated with hyperpycnal flows. Other typical characteristics of sustained hyperpycnal flows in marine settings are related to the buoyancy flow reversal induced by the lift-up effect provided by the interstitial fluid once the sediment concentration falls below a certain threshold in a process known as "lofting" (Sparks et al., 1993; Hesse et al., 2004).

The characteristics previously discussed result in the accumulation of three main facies categories related to the three main processes that characterize all sustained hyperpycnal discharges in marine settings: bed load, turbulent suspension, and lofting. These facies categories are here termed as B (bed-load-related sedimentary facies), S (suspended-load-related sedimentary facies), and L (lofting-related sedimentary facies) (Figure 5).

Facies Related to Bed-load Processes (Facies B)

Type B facies includes different coarse-grained deposits related to shear/drag forces exerted by the overpassing long-lived turbulent (hyperpycnal) flow over coarse-grained materials lying on the flow bottom. Three main categories of bed-load facies are recognized, here termed B1, B2, and B3 (Figure 6).

Facies B1

Facies B1 is composed of massive and crudely stratified conglomerates with abundant coarse- to fine-grained sandstone matrix. The largest clasts are embedded in the matrix (facies B1; Figure 7A, C, D) or imbricated following diffuse subhorizontal alignments. The overall texture is matrix supported, although some varieties of clast-supported conglomerates are recognized (facies B1c) (Figure 7A). Unlike the deposits related to laminar dense flows (hyperconcentrated flows), large clasts move freely at the base of the overpassing turbulent flow and commonly appear imbricated. Large clasts are probably transported by sliding and rolling, whereas the finest materials composing the matrix would correspond to suspended load transported by turbulence, which became trapped at the low-velocity and relatively high-concentration basal zone (Manville and White, 2003). Field relationships suggest that a vertical evolution between facies B1c and B1 (Figure 7A) could be related to a progressive decrease of drag forces exerted by the overpassing turbulent flow. In systems that lack coarse-grained clastics, B1 facies could be entirely composed of matrix- or clast-supported clay clasts (facies B1s) (Figures 6, 8A).

Facies B2

Facies B2 is composed of fine-grained conglomerates and pebbly sandstones with low-angle asymptotic cross-stratification (Figures 6, 7B). Large clasts in this facies appear floating in a medium- to coarse-grained sandstone matrix. Individual sets of cross-bedding commonly show thicknesses between 0.3 (1 ft) and 1 m (3.2 ft) and asymptotic relationships with top and base (Figure 8C). The foreset inclination in general does not exceed 20°. Depending on the rate of sediment fallout from the overpassing turbulent flow, bedset-bounding surfaces can be erosional (like an anisotropic hummocky cross-stratification) or transitional (Figure 6), conforming with climbing dunes (in the sense of Mutti et al., 1996). Field observations suggest a close association of this facies with channel-fill deposits. In gravel-free systems, facies B2 is almost entirely composed of medium- to coarse-grained sandstones (Figure 8C, D), commonly displaying abundant clay clasts in the lower foreset. This sandstone variety is termed facies B2s. If clay clasts are numerous and of small size, this structure can easily be confused with a tidal bundle (Figure 8C). The key for a correct interpretation resides in differentiating the small clay clasts from a true mud couplet. Like facies B2, facies B2s relates to the migration of straight or sinuous bed forms at the base of a long-lived turbulent flow carrying high suspended load.

Facies B3

Facies B3 is characterized by coarse-grained to pebbly sandstones with diffuse horizontal to subhorizontal stratification and levels of small aligned pebbles (Figures 6; 7C, D). Pebbles in general do not exceed 10 mm in diameter and seem dispersed in a coarse-grained sandstone matrix. The sandy variety

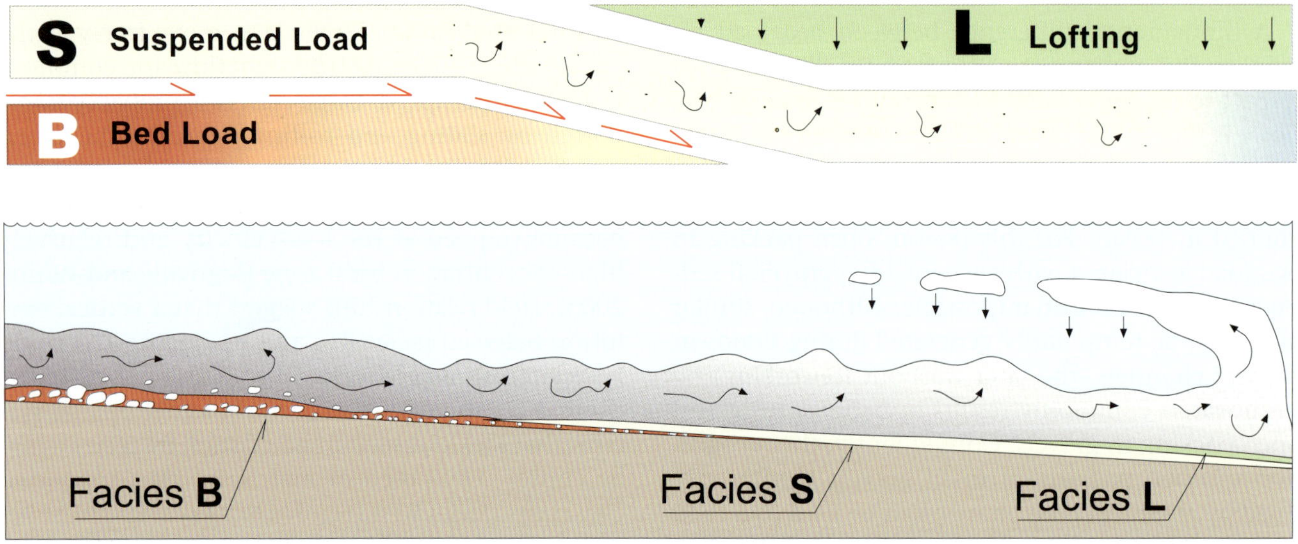

FIGURE 5. **Basic conceptual diagram for the origin of the genetic facies deposited from sustained hyperpycnal flows in a marine environment. Facies B include all those facies related to bed-load processes developed at the base of an overpassing long-lived turbulent flow. Facies S correspond to the gravitative collapse of sand-size materials transported in suspension by the turbulent flow (suspended-load facies). Facies L are related to the fallout of fine-grained materials lifted up in the interstitial fresh water contained in the flow once it has lost part of the sand-size suspended load. After Zavala et al. (2006b); Zavala, 2008; and Zavala et al., 2011.**

of this facies (facies B3s) is characterized by aligned clay clasts and plant fragments (Figure 8B). Facies B3 comprises tabular to lenticular bodies commonly filling erosive depressions. It is interpreted that these facies accumulated by the combined effect of bed load and the gravitational segregation of sandstone materials transported in the overpassing hyperpycnal turbulent flow. In shallow-water settings, B3 facies may display low-angle diverging and truncated laminae that closely resemble gravelly hummocky cross-stratification (facies B3h in Figure 6). Mutti et al. (1994) suggested that subaerially derived gravity flows may develop an internal oscillatory component during their downslope motion through the setting in motion of standing bodies of water in the shallower parts of the receiving basin.

Facies Related to the Collapse of Suspended Load (Facies S)

Facies S are mostly fine grained and composed of sediments transported as suspended load, forming thick and commonly complex intervals that can be massive or display traction plus fallout sedimentary structures.

Facies S1

Facies S1 is one of the most common within the facies tract of hyperpycnal systems. This facies is composed of tabular fine- to medium-grained massive sandstone beds (Figure 6). These sandstones commonly integrate monotonous and very thick successions, internally showing subtle and gradual grain-size variations (Figure 9A). Small floating clay chips are common and could seem dispersed within the sandstone body or grouped toward the top at the boundary with facies related to more diluted flows (facies S2). Carbonaceous remains and wood fragments are also common within massive sands, commonly displaying leaf fragments with exceptional preservation. Leaves in massive sands can be very abundant and have been proposed as the main source of hydrocarbons in the Kutei Basin, Indonesia (Saller et al., 2006). The S1 facies mostly lack any bioturbation. Nevertheless, some massive very thick intervals show isolated *Ophiomorpha* and *Thalassinoides*, which can be related to crustaceans (doomed pioneers), bulked and transported by the turbulent flow from shallower areas (Grimm and Föllmi, 1994).

The origin of this facies would be related to the progressive aggradation from the bottom by long-lived flows having high suspended load (Sanders, 1965; Kneller and Branney, 1995; Camacho et al., 2002). This progressive aggradation has been proposed as a mechanism that inhibits the formation of primary sedimentary structures. Massive deposits could therefore be caused by the absence of a sharp boundary between the moving flow and the deposit, probably related to a zone of aggrading transition

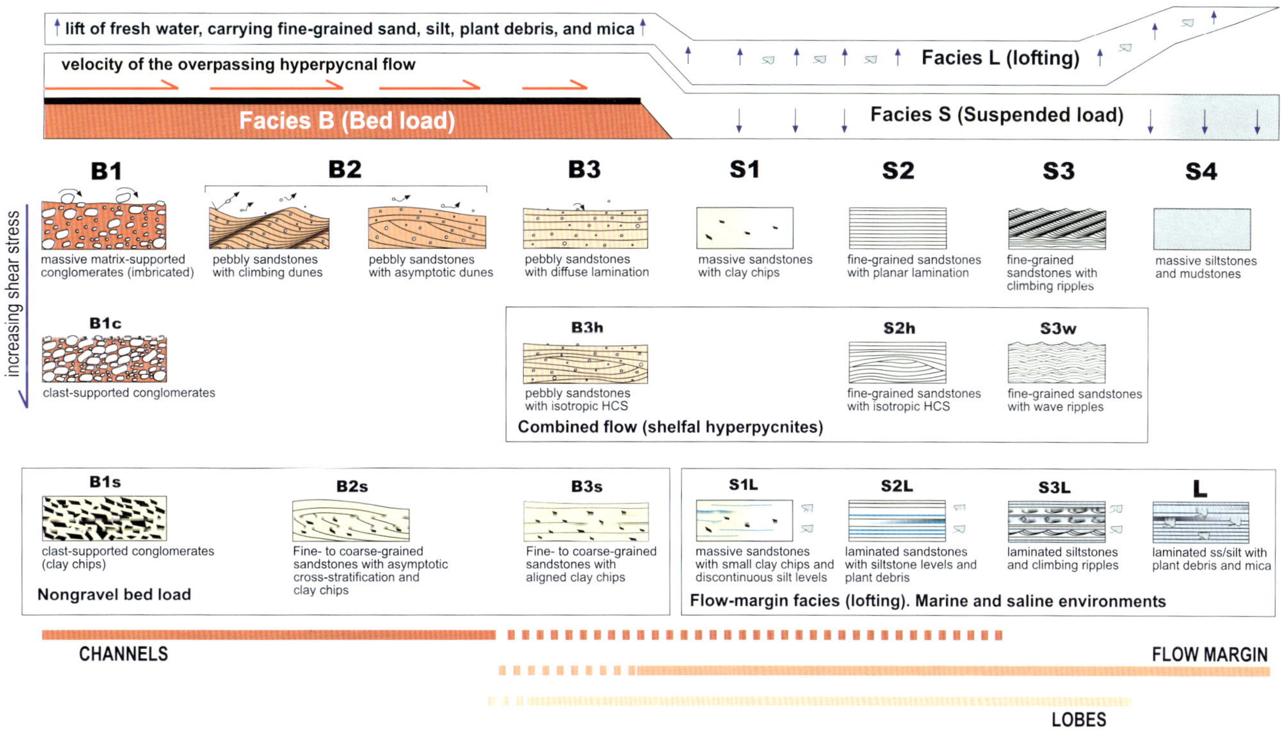

FIGURE 6. Conceptual schema for the genetic classification of clastic facies associated with hyperpycnal systems. Type B facies relate to bed-load processes at the base of an overpassing long-lived turbulent flow. Type S facies originate from the gravitative collapse of sand-size suspended materials carried in the turbulent flow. Facies L are composed of very fine-grained sandstones interbedded with laminated silts with abundant plant debris and mica, accumulated by fallout from lofting plumes. Facies L are exclusive of marine/saline environments. HCS = hummocky cross-stratification; ss/silt = sandstone/silt. After Zavala et al., 2006b; Zavala, 2008; and Zavala et al., 2011.

characterized by high sediment concentration and water escape. Experimental studies (Arnott and Hand, 1989; Sumner et al., 2008) indicate that this facies originates from a turbulent flow with fallout rates greater than 0.44 mm/s. For smaller fallout rates at similar flow velocities, the result will be laminated sands similar to facies S2 (Figure 9B). In structurally controlled depocenters, the individual thickness of single massive sandstone bodies can be very impressive (Amy et al., 2007), in some cases exceeding 45 m (148 ft) (Arcuri and Zavala, 2007, 2008).

Facies S2

Facies S2 is composed of tabular fine-grained sandstone bodies having subhorizontal parallel lamination (facies S2) or internally low-angle diverging (hummocky-like) lamination (facies S2h) disposed over a sharp or transitional boundary. Individual laminae are millimeter thick and in some cases are bounded by thin levels with micas and carbonaceous material.

Facies S2 is commonly associated with parting and other current lineation structures such as those associated with heavy minerals. The origin of this facies has been related to dilute unidirectional flows at the upper flow regime (Simons et al., 1965). However, Sanders (1965) noted that parallel lamination commonly grades laterally into climbing ripples, suggesting a common origin of traction plus fallout processes. This last conclusion is also consistent with the experimental results of Arnott and Hand (1989) and Sumner et al. (2008), and it is also supported by the common association of facies S2 with massive intervals (facies S1) (Figure 9B) commonly constituting very thick rhythmical successions. The low-angle variety (facies S2h) is more common in shallow hyperpycnal systems and is similar to the isotropic hummocky cross-stratification related to combined flow (Harms et al., 1975, 1982; Southard, 1991; Mutti et al., 1994).

Facies S3

Facies S3 is composed of tabular to irregular fine-grained sandstone bodies with climbing ripples (Figures 6, 9C). This facies has been related to traction plus fallout processes from waning turbulent flows with high suspended load (Jopling and Walker, 1968; Mulder and Alexander, 2001; Sumner et al.,

FIGURE 7. **Main characteristics of coarse-grained deposits related to bed-load processes associated with sustained hyperpycnal flows. (A) Clast-supported conglomerates (facies B1c) followed by matrix-supported conglomerates (facies B1), suggesting a decrease in the shear stress. Outer-shelf deposits (early Jurassic), Neuquén Basin, Argentina. (B) Large-scale cross-bedding associated with climbing dunes (facies B2). Deep-marine deposits, Miocene of the Austral Basin, Argentina. (C) Gradual passage between facies B1, B3, and S1, suggesting variations in flow conditions associated with long-lived hyperpycnal flows. Flames at the top of facies S1 indicate almost continuous deposition of facies B1. Deep-marine deposits, Miocene of the Austral Basin, Argentina. (D) Detail showing the internal complexity of a coarse-grained hyperpycnite (composite bed). Gradual changes in texture and sedimentary structures suggest long-lived and fluctuating flows. Shelfal deposits (middle Jurassic), Neuquén Basin, Argentina.**

2008). Facies S3 commonly grades horizontally and vertically with facies S2 (Sanders, 1965; Zavala et al., 2006a), thus evidencing a common origin for both facies, controlled by fluctuations in the velocity of the overpassing turbulent flow. Gradual fluctuations in the flow velocity (in this case indicated by the gradual passage between facies S2 and S3) and in the rate of sediment fallout (passage between facies S1 and S2) are diagnostic elements of long-lived turbulent flow with fluctuating energy. The location of these elements in a subaqueous setting suggests the existence of a hyperpycnal system (Zavala et al., 2006a). In shallow-water environments affected by combined flows, ripples can show aggrading wave structures (facies S3w) (Figure 6), suggesting sediment fallout associated with unidirectional to oscillatory flows.

Facies S4

Facies S4 is characterized by massive to laminated siltstones and mudstones. It is composed of the finest materials transported by the hyperpycnal flow, which accumulated by normal settling when the flow completely stopped. Consequently, facies S4 is useful to identify boundaries between different hyperpycnal events. The differentiation of facies S4 from prodelta/shelfal mudstones may be difficult and sometimes requires micropaleontological studies because facies S4 commonly contains continental/shallow-water species. In the absence of microfossils, facies S4 is characteristic of nonmarine hyperpycnal deposits because, in marine settings, fine-grained materials are lofted up by the lighter interstitial fluid and accumulate as facies L.

Facies Related to Flow Lofting (Facies L)

Facies L

Facies L is characterized by thin couplets of siltstones and sandstones having great lateral extension and commonly showing abundant intercalations of plant debris (Figures 6, 9D) and micas. These couplets

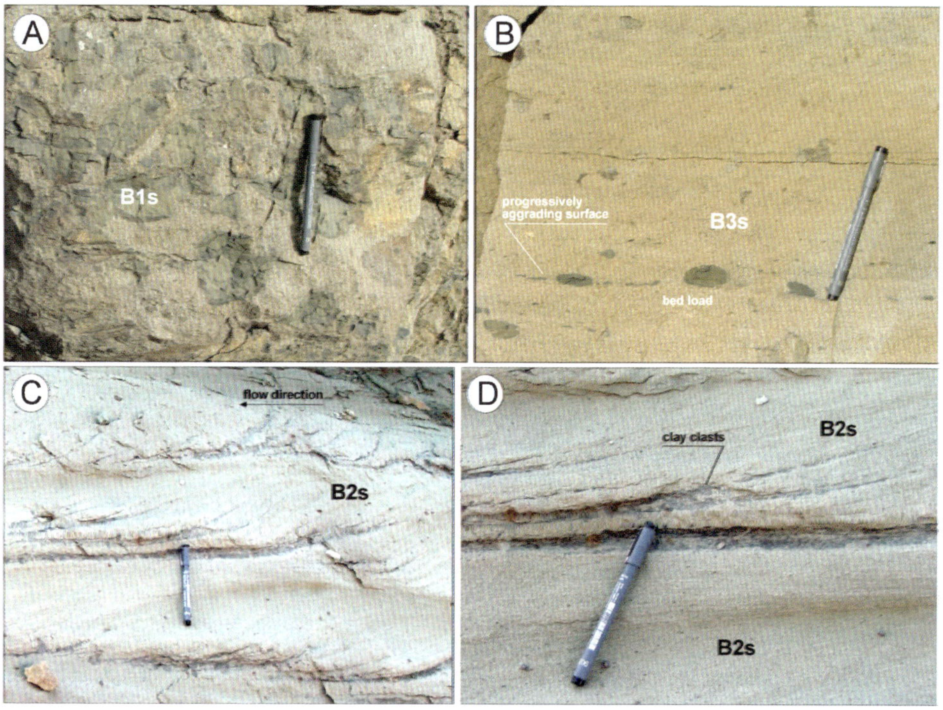

FIGURE 8. Bed-load facies in sand-bearing hyperpycnal systems. (A) Massive conglomerates composed of clay clasts (facies B1s). Early Jurassic outer-shelf deposits, Neuquén Basin, Argentina. (B) Coarse-grained sandstones with aligned clay clasts (facies B3s). Early Jurassic outer-shelf deposits, Neuquén Basin, Argentina. (C) Cross-bedded facies related to migrating sand dunes at the base of an overpassing long-lived hyperpycnal flow (facies B2s). Cross-bedding is low angle and asymptotic and can be easily confused with tidal bundles. The main difference resides in the fact that in these dunes, dark horizons are small clay clasts and not mud couplets such as in tidal deposits. (D) Close-up of picture C, showing a detail of clay clasts levels. Lower Jurassic shallow lacustrine strata, Neuquén Basin, Argentina.

display abundant small-scale load-cast structures, commonly associated with syneresis cracks and siderite nodules. Trace fossils are scarce and mostly limited to some forms of *Palaeophycus*. These laminated levels can form isolated bedsets up to 2 m (6.6 ft) thick or be located alternating at the top of facies B or S, but they are more commonly associated with massive sandstones (facies S1) (Figure 9D). The individual levels commonly display a variable thickness from a few millimeters up to 1 cm (0.39 in.) and are separated

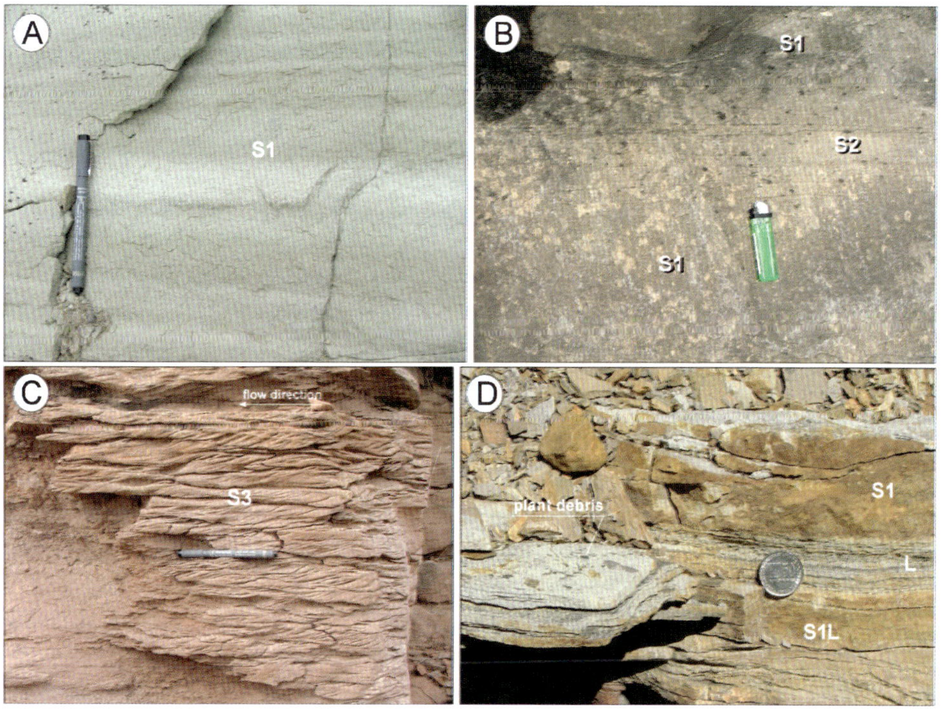

FIGURE 9. Main facies types associated with fallout from sustained turbulent suspensions. (A) Massive sandstones (facies S1). The diffuse banding marks the different positions of the rising depositional surface. Middle Jurassic, Neuquén Basin, Argentina. (B) Laminated sandstones (facies S2). The gradual passage between massive and laminated sandstones probably relates to variations in the fallout rate. Shelfal deposits, Late Cretaceous of Burgos Basin, Mexico. (C) Climbing ripples (facies S3). Shallow lacustrine deposits. Early Cretaceous of the Neuquén Basin, Argentina. (D) Lofting rhythmites (facies L) interbedded within massive sandstones. Shelfal deposits. Late Oligocene of the eastern Venezuela Basin, Venezuela.

by thin layers with abundant plant debris (Figure 9D). Plant debris provides definitive evidence of a direct connection between the fluvial system and the marine-related basin (Petter and Steel, 2005; Lamb et al., 2008). The absence of tractive structures in these sandstones suggests an accumulation by normal settling from a suspension cloud elevated over the depositional surface. Based on the contributions of Sparks et al. (1993), Mutti et al. (2003), and Hesse et al. (2004), these materials are interpreted to be related to the accumulation of the finest materials transported by the hyperpycnal flow, which settled from a lofting plume (Zavala et al., 2008b). Facies L constitutes a diagnostic and characteristic element of hyperpycnal sedimentation in a marine environment because it suggests the existence of a less dense interstitial fluid (fresh water) with respect to that of the environment (marine water), derived from direct fluvial discharge (Zavala et al., 2006c). Vertical and lateral relationships between L and S (S1, S2, and S3) facies are not sharp and result in transitional categories (Figure 6) termed S1L (massive sandstones with discontinuous siltstone levels) (Figure 9D), S2L (laminated sandstones with abundant plant debris and micas), and S3L (siltstone levels interbedded with sandstones having small climbing ripples).

THE LOFTING POINT

The lofting point (LP) is the position in which the hyperpycnal flow reverses its negative buoyancy as a consequence of the progressive loss of suspended load by deposition out of the flow. The velocity threshold for the LP could be quite variable because the hyperpycnal flow is a heterogeneous mixture composed of buoyant and load components (Figure 10).

The buoyant component is provided by the density difference between fresh water and other light materials (e.g., plant debris) with respect to the more dense ambient (marine) water. The load component, however, is related to the heaviest materials transported in suspension by the hyperpycnal flow (fine-grained sand, silt, and mud). If the load component is greater than the buoyant component, the flow will follow along the basin floor. Nevertheless, the load component at the flow head will decay because the amount of sand material in turbulent suspension will decrease by deposition as the flow progressively wanes. The velocity threshold for the flow lift-up (LP) will therefore depend on the amount of clay in suspension because clay content will remain constant until the flow stops completely. Consequently, in flows having a low clay content, buoyancy reversal will occur early at faster velocities, preventing in some cases the development of plane beds and climbing ripples at the top of massive intervals (Mutti et al., 2003). The buoyancy reversal of hyperpycnal flows (lofting) is very sensitive to velocity changes. Velocity changes in a hyperpycnal flow occur not only in the flow direction, but also laterally as the flow spreads across the sea floor (Zavala et al., 2006a), favoring deposition of lofting facies (or lofting rhythmites; Zavala et al., 2006c) at flow margins (Figure 11).

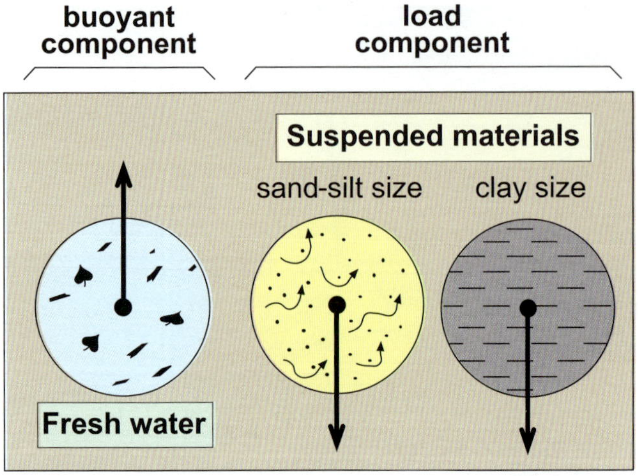

FIGURE 10. **Buoyant and load components as the basic constituents of a turbulent hyperpycnal flow. The buoyant component results from the density difference between fresh water and ambient marine waters. If the load component is greater than the buoyant component, the flow will move along the basin floor. Nevertheless, the load component will decay in time because the amount of sand in turbulent suspension will decrease as the flow wanes, favoring buoyancy reversal (lofting).**

The position of the LP is not fixed. The LP will move basinward or landward depending on the magnitude and concentration of the sustained hyperpycnal discharge. Almost all sustained hyperpycnal flows are triggered from an initial hypopycnal ("normal") delta stage (1 in Figure 12). As the fluvial discharge increases, the bulk concentration of the incoming flow will rise exponentially (Mulder et al., 2003). When the density of the fluvial discharge equals that of seawater, a transitional homopycnal flow will develop (2 in Figure 12). In stages 3 and 4 (Figure 12), a fluvial discharge with a progressively increasing density exceeds that of seawater. The flow therefore has plunged below marine waters and has started its travel basinward as a sustained hyperpycnal flow. The sinking of fluvial waters in the sustained hyperpycnal flow has two important consequences.

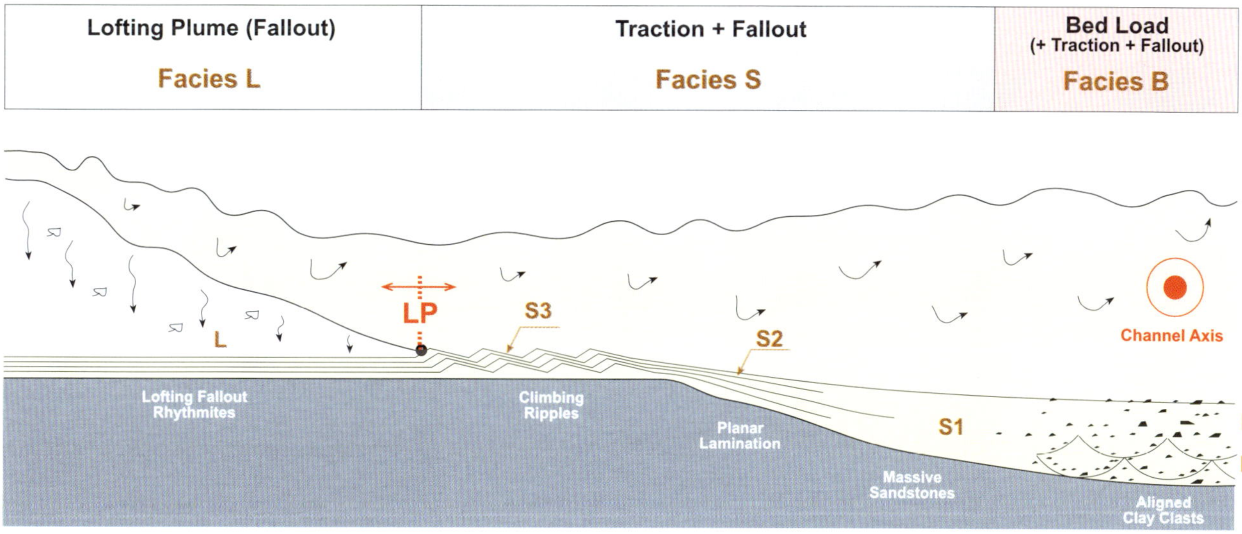

FIGURE 11. Diagram showing the lateral relationships between channel and levee deposits. Note that as a consequence of drastic velocity variations recognized between the channel axis and adjacent levees, facies changes tend to show the entire facies association within a short distance. This facies association is the same as that expected to occur for hundreds of kilometers in the longitudinal direction. Note that lofting rhythmites are typical of flow-margin areas. LP = lofting point.

The first consequence is related to the downwelling effect with the ensuing suppression of the surficial buoyant plume previously developed contemporaneously with the hypopycnal delta. As a result, fine-grained light materials contained in the flow, including plant debris suspended in the previous buoyant plume, will be suctioned and dragged basinward with the sustained hyperpycnal flow. Once the surficial plume disappears, the dragging effect provided by the sinking hyperpycnal flow continues, thus generating a marked contrast between the dirty (by sediment load) fluvial discharge and the clear waters of the basin (Mulder et al., 2003).

The second consequence is related to the less dense interstitial fluid contained in the hyperpycnal flow, which will cause a reversal of the flow density at the LP. In the flow head, the LP is characterized by a lift-up effect associated with a sudden flow deceleration. Consequently, the advancing density flow tends to move beneath the lofting plume that contains the finest and lightest materials transported by the hyperpycnal flow (including plant debris and micas). The lofting plume will have a great lateral dispersion with respect to the more confined hyperpycnal flow. Sediments will be deposited from fallout, forming characteristic sand/silt couplets (facies L) or lofting rhythmites (Zavala et al., 2006c, 2008b). The reversing buoyancy effect (lofting) at the hyperpycnal flow margin will result in less lateral continuity of sandstone bodies compared with those related to resedimentation processes, with important consequences for the case of sandstone reservoirs.

Once the density of the incoming flow begins to decay, the LP will move progressively landward (5–6 in Figure 12). This retraction can be simple and continuous or it can be a series of forward-backward pulses related to a more dynamic long-lived hyperpycnal flow with fluctuating energy. Finally, when the flow density returns to the normal low-density values, the fluvial-deltaic system will evolve into homopycnal (7 in Figure 12) and then hypopycnal (8 in Figure 12) conditions (littoral delta).

COMPOSITE BEDS AND HYPERPYCNAL DEPOSITS

The complex nature of the fluvial drainage network associated with medium to large rivers results in long-lived discharges (floods) having very complex hydrographs. This complexity is commonly reflected in the associated hyperpycnal deposits and results in the accumulation of internally highly complex sedimentary bodies. Hyperpycnal flows related to long-lasting floods have short- and long-term variations not only in the flow speed, but also in the concentration of the overpassing flow. These variations can be clearly read from the fossil record in single beds that internally show gradual and sharp facies transitions without having a definite and predictable internal arrangement (Figures 13, 14). The resulting

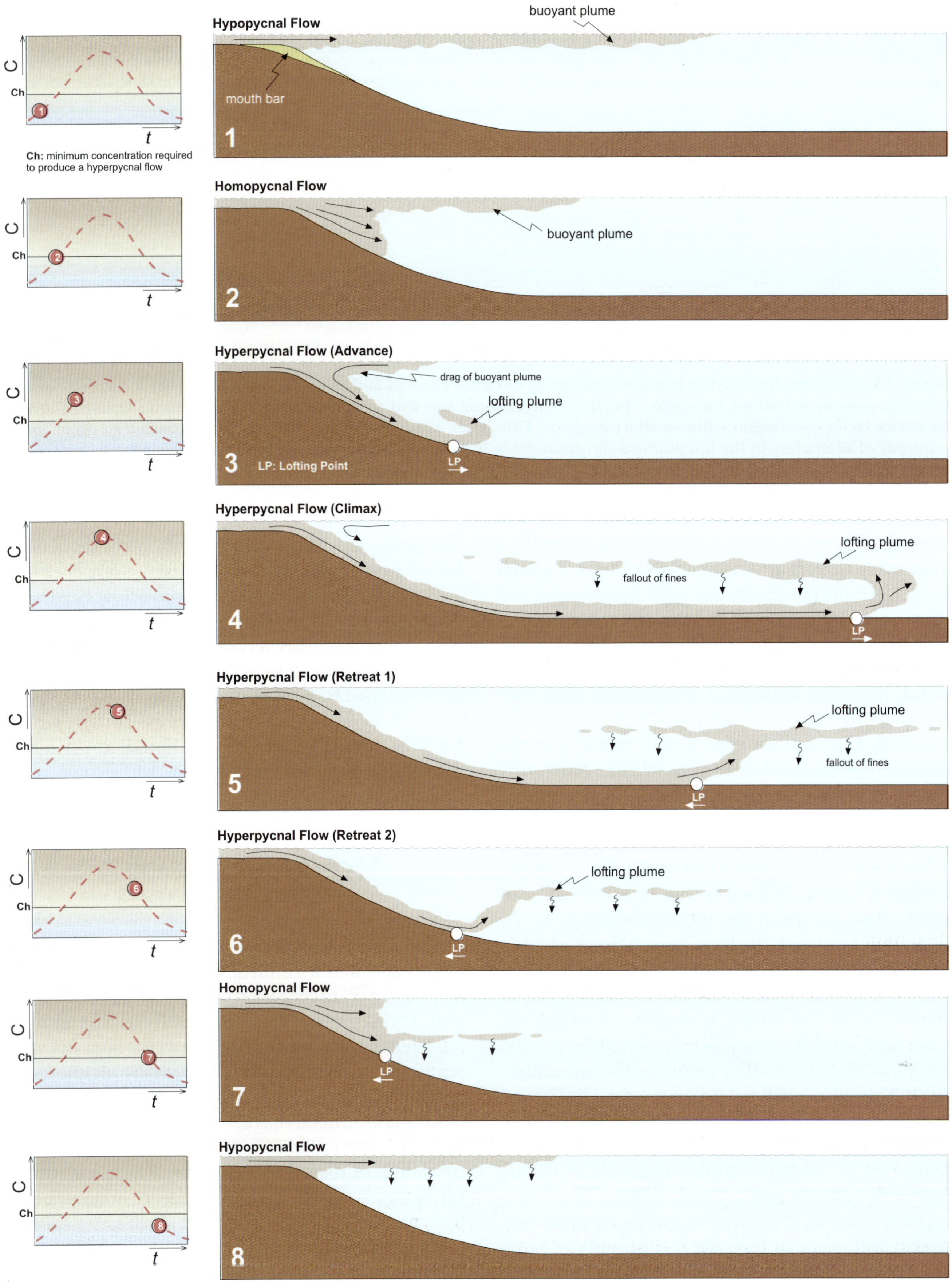
C
Ch
t
Ch: minimum concentration required to produce a hyperpycnal flow
Hypopycnal Flow
buoyant plume
mouth bar
1
Homopycnal Flow
buoyant plume
2
Hyperpycnal Flow (Advance)
drag of buoyant plume
lofting plume
3
LP: Lofting Point
LP
Hyperpycnal Flow (Climax)
lofting plume
fallout of fines
4
LP
Hyperpycnal Flow (Retreat 1)
lofting plume
fallout of fines
5
LP
Hyperpycnal Flow (Retreat 2)
lofting plume
6
LP
Homopycnal Flow
7
LP
Hypopycnal Flow
8

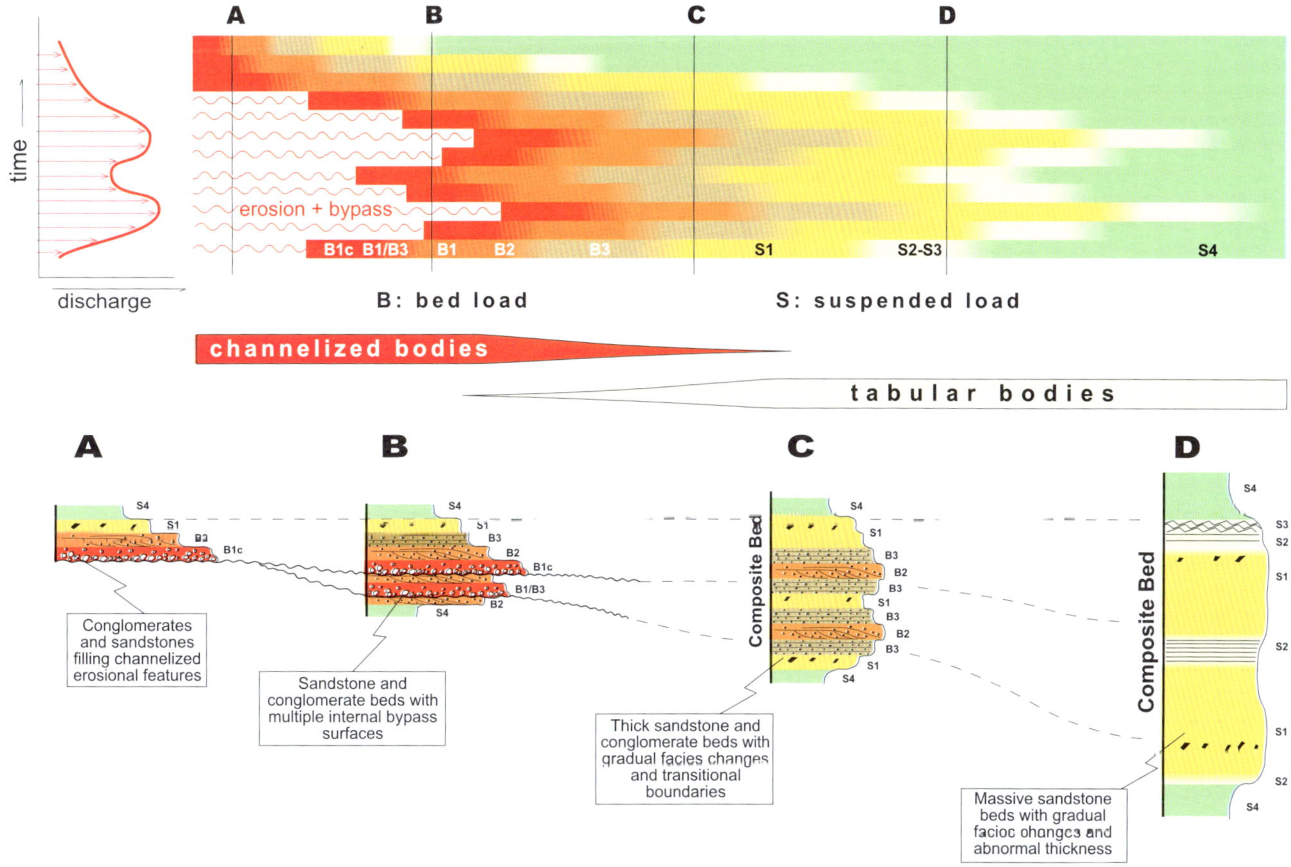

FIGURE 13. **The origin of a composite bed is related to variations in flow velocity and sediment concentration during a single long-lived hyperpycnal discharge. Note that the erosional surfaces do not always bound different events and grade basinward into nonerosional/transitional levels.**

sedimentary body is a composite bed (Zavala et al., 2007a). A composite bed is an internally complex depositional body accumulated during a single long-lived hyperpycnal discharge. Commonly, this body internally shows a facies arrangement that strongly departs from that expected for surgelike or episodic flows. Composite beds are typical and diagnostic features of long-lived hyperpycnal sedimentation because they indicate a deposit continuously fed by a long-lasting, fluctuating, and dilute sediment-gravity flow.

Composite beds typically display a vertical succession of different lithofacies showing cyclical recurrences and transitional to sharp passages between them. These cyclical and gradual changes between different facies are the result of near-continuous deposition from a quasi-steady turbulent flow. Composite beds can be very thick (up to 45 m [148 ft]) depending on the duration of the related hyperpycnal flow and the available accommodation space (Arcuri and Zavala, 2007, 2008). Other common attributes of composite beds include internal and laterally discontinuous erosional surfaces, scarce burrows, and a basal coarsening-upward interval. In marine environments, composite beds evolve laterally into packages of lofting rhythmites. Composite beds related to cyclical velocity variations are commonly characterized by recurrent facies changes that suggest different and recurrent equilibrium flow conditions (Zavala et al.,

FIGURE 12. **This figure depicts the whole conceptual evolution of a single sustained hyperpycnal discharge. In (1), the low density of the incoming flow results in a hypopycnal flow with an associated buoyant plume. In (2), the density of the incoming flow has reached that of seawater and consequently a homopycnal flow develops. In (3) and (4), the bulk density of the incoming flow has exceeded that of seawater. Consequently, fluvial waters sink, dragging by suction the buoyant plume previously developed, then the overlying waters. In (5) to (8), the density of the incoming flow progressively decreases. Consequently, the LP progressively moves landward and the deltaic system returns to the hypopycnal initial conditions. C = relative concentration of fluvial discharge; *t* = time.**

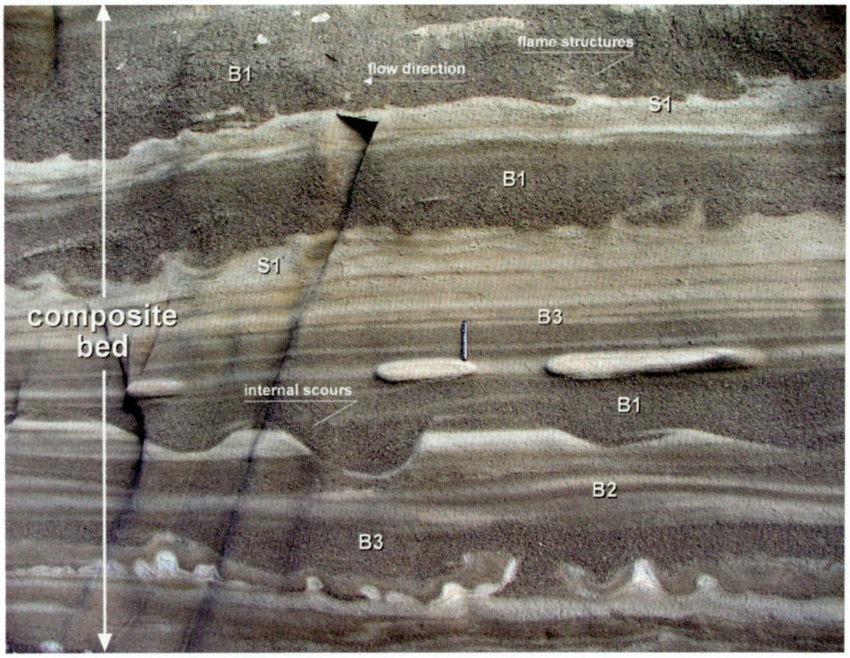

FIGURE 14. Some internal characteristics of composite beds. This composite bed lacks any bioturbation and is characterized by an impressive facies recurrence with abundant load casts. Flame structures indicate a flow direction from right to left. Deep-marine deposits, Miocene of the Austral Basin, Argentina.

2006a; Lamb et al., 2008; Lamb and Mohrig, 2009). As an example, the gradual vertical passage between climbing ripples and plane beds could be related to a velocity increase in the overpassing hyperpycnal flow (Zavala et al., 2006a). More commonly, composite beds related to sustained hyperpycnal flows show a very complicated internal arrangement, reflecting cyclic depositional changes between suspended-load- and bed-load-dominated facies associated with different velocity and fallout rates. The vertical anisotropy in facies and the relatively rapid accumulation result in the common occurrence of load cast and flame structures (Figure 14).

SUMMARY AND CONCLUSIONS

The hyperpycnal system is a relatively new category of depositional system that corresponds to the subaqueous extension of the fluvial system. A single hyperpycnal system can extend for hundreds of kilometers away from the river mouth and will develop a predictable path of genetically related facies during its travel basinward. Three main categories of genetic facies are recognized in hyperpycnal systems corresponding to B (bed-load-related), S (suspended-load-related), and L (lofting-related) facies.

The B facies are related to bed-load processes (mostly sliding and rolling) developed at the base of a long-lived turbulent (hyperpycnal) flow and tend to dominate in proximal to medial parts of the hyperpycnal system. Three main types of B facies correspond to B1 (clast to matrix-supported massive conglomerates), B2 (pebbly sandstones with large-scale cross-bedding related to migrating climbing dunes), and B3 (pebbly sandstones with diffuse horizontal lamination).

The S facies comprise finer grained deposits related to the gravitative collapse of materials transported in suspension within the turbulent (hyperpycnal) flow. Four categories of S facies are recognized: S1 (massive sandstones), S2 (parallel to low-angle laminated sandstones), S3 (fine-grained sandstones with climbing ripples), and S4 (massive mudstones). The S4 facies are more common in lacustrine environments because in marine environments, muddy sediment is lifted up (lofted) with the less dense interstitial water as the flow wanes.

The L facies is composed of a rhythmic intercalation of thin beds of massive sands, silts, micas, and plant debris having great lateral continuity. The L facies develops exclusively in marine and saline environments and is related to the fallout of fine-grained sands, silts, micas, and plant debris from lofting plumes. Lofting plumes originate from the buoyant inversion induced by the lift-up of the less dense fresh water contained in the hyperpycnal flow as it progressively loses part of the suspended load by deposition.

All these facies categories are genetically related and occupy a definite position within the hyperpycnal system. The B facies is diagnostic of proximal areas and progressively disappears as the flow enters the area of lobe deposition. The S facies is the consequence of the loss of flow capacity and is typical of the medium to distal parts of the system. The L facies results from the flow inversion, which is diagnostic of flow-margin areas (both down the depositional axis and laterally along the axis).

This proposed facies tract is a first attempt to try to achieve a better understanding of hyperpycnal systems. Nevertheless, our understanding of these deposits is still very limited, and considerable effort will be required during future years to try to improve upon facies models, develop new sedimentologic tools, and assess the real importance of hyperpycnal deposits in the accumulation of ancient strata, including those associated with hydrocarbon reservoirs.

ACKNOWLEDGMENTS

We thank all those people who generously contributed during the years to field discussions related to hyperpycnal sedimentation, particularly E. Mutti, E. Remacha, G. Davoli, D. DiBiase, H. Maretto, J.J. Ponce, H. Gamero, C. Contreras, M. Sanguinetti, J. Marcano, and J Carvajal. Field discussions with S. Pattison, C. Olariu, J. Bhattacharya, T. Mulder, S. Dasgupta, B. Romans, M. Perlmutter, and R. Higgs deeply contributed to improving this chapter. Roger Slatt carefully reviewed the manuscript, providing helpful comments and corrections. This research was partially funded by the Consejo Nacional de Investigaciones Científicas y Técnicas (CONICET) of Argentina.

REFERENCES CITED

Amy, L. A., B. Kneller, and W. D. McCaffrey, 2007, Facies architecture of the Grès de Peïra Cava, SE France: Landward stacking patterns in ponded turbiditic basins: Journal of the Geological Society (London), v. 164, p. 143–162, doi:10.1144/0016-76492005-019.

Antobreh, A. A., and S. Krastel, 2006, Morphology, seismic characteristics and development of Cap Timiris Canyon, offshore Mauritania: A newly discovered canyon preserved off a major arid climatic region: Marine and Petroleum Geology, v. 23, p. 37–59, doi:10.1016/j.marpetgeo.2005.06.003.

Arcuri, M., and C. Zavala, 2007, Very thick massive sandstone bodies: Origin and internal architecture: AAPG Annual Conference, session 3853, effects of active structural growth and confined basins on sandbody architecture I: AAPG Search and Discovery Article 90063, March 30 to April 4, 2007, Long Beach, California, 1 p., http://www.searchanddiscovery.net/abstracts/html/2007/annual/abstracts/lbArcuri.htm?q=%2Btext%3Aarcuri (accessed August 3, 2010).

Arcuri, M., and C. Zavala, 2008, Hyperpycnal shelfal lobes—Some examples of the Lotena and Lajas formations, Neuquen Basin, Argentina, *in* J. J. Ponce and E. B. Olivero, conveners, Sediment transfer from shelf to deepwater—Revisiting the delivery mechanisms: Conference Proceedings, AAPG Search and Discovery Article 50066, AAPG Hedberg Conference, March 3–7, 2008, Ushuaia-Patagonia, Argentina, 4 p., http://www.searchanddiscovery.net/documents/2008/08018arcuri/index.htm?q=%2Btext%3Aarcuri (accessed August 3, 2010)

Arnott, R. W. C., and B. M. Hand, 1989, Bedforms, primary structures and grain fabric in the presence of suspended sediment rain: Journal of Sedimentary Petrology, v. 69, p. 1062–1069.

Bates, C., 1953, Rational theory of delta formation: AAPG Bulletin, v. 37, p. 2119–2162.

Bouma, A. H., 1962, Sedimentology of some flysch deposits: A graphic approach to facies interpretation: Amsterdam, Netherlands, Elsevier, 168 p.

Bourget, J., S. Zaragosi, T. Mulder, J. L. Schneider, T. Garlan, A. Van Toer, V. Mas, and N. Ellouz-Zimmermann, 2010, Hyperpycnal-fed turbidite lobe architecture and recent sedimentary processes: A case study from the Al Batha turbidite system, Oman margin: Sedimentary Geology, v. 229, p. 144–159, doi:10.1016/j.sedgeo.2009.03.009.

Bruhn, C. H. L., and R. G. Walker, 1997, Internal architecture and sedimentary evolution of coarse-grained, turbidite channel-levee complexes, early Eocene Regencia Canyon, Espírito Santo Basin, Brazil: Sedimentology, v. 44, p. 17–46, doi:10.1111/j.1365-3091.1997.tb00422.x.

Camacho, H., C. J. Busby, and B. Kneller, 2002, A new depositional model for the classical turbidite locality at San Clemente State Beach, California: AAPG Bulletin, v. 86, p. 1543–1560.

Cheng-Shing, C., and Y. Ho-Shing, 2008, Evidence of hyperpycnal flows at the head of the meandering Kaoping Canyon off SW Taiwan: Geo-Marine Letters, v. 28, p. 161–169, doi:10.1007/s00367-007-0098-7.

Collins, M. B., 1986, Processes and controls involved in the transfer of fluviatile sediments to the deep ocean: Journal of the Geological Society (London), v. 143, p. 915–920, doi:10.1144/gsjgs.143.6.0915.

Deptuck, M. E., G. S. Steffens, M. Barton, and C. Pirmez, 2003, Architecture and evolution of upper fan channel-belts on the Niger Delta slope and in the Arabian Sea: Marine and Petroleum Geology, v. 20, p. 649–676, doi:10.1016/j.marpetgeo.2003.01.004.

De Rooij, F., and S. B. Dalziel, 2001, Time and space resolved measurements of deposition under turbidity currents, *in* B. McCaffrey, B. Kneller, and J. Peakall, eds., Particulate gravity currents: International Association of Sedimentologists Special Publication 31, p. 207–215.

Gamero, H., C. Zavala, and C. Contreras, 2005, A reinterpretation of the Misoa facies types: Implications of a new depositional model, Maracaibo Basin, Venezuela (abs.): AAPG Search and Discovery Article 90046,

2005 AAPG International Conference and Exhibition, Evolving Stratigraphic Techniques and Interpretation III, Paris, 1 p., http://www.searchanddiscovery.net/documents/abstracts/2005intl_paris/gamero.htm (accessed August 3, 2010).

Gamero, H., J. Reader, C. Izatt, C. Zavala, and C. Contreras, 2007, Herrera sandstones in the southern basin area, Trinidad: Evidence of hyperpycnites deposited away from ancient Oficina delta systems in eastern Venezuela: AAPG Search and Discovery Article 90063, Annual AAPG Convention, April 2–5, 2007, Long Beach, California, 1 p., http://www.searchanddiscovery.net/abstracts/html/2007/annual/abstracts/lbGamero.htm?q=%2Btext%3Agamero (accessed August 3, 2010).

Gamero, H., N. Lewis, R. Welsh, C. Zavala, and C. Contreras, 2008, Evidences of a shelfal hyperpycnal deposition in the Pliocene sandstones in the Oilbird field, SE coast, Trinidad: Impact on reservoir distribution and field redevelopment, *in* J. J. Ponce and E. B. Olivero, conveners, Sediment transfer from shelf to deepwater—Revisiting the delivery mechanisms: Conference Proceedings, AAPG Search and Discovery Article 90079, AAPG Hedberg Conference, March 3–7, 2008, Ushuaia-Patagonia, Argentina, 4 p., http://www.searchanddiscovery.net/abstracts/html/2008/hedberg_argentina/extended/gamero/gamero.htm?q=%2Btext%3Azavala (accessed August 3, 2010).

Gervais, A., T. Mulder, B. Savoye, and E. Gonthier, 2006, Sediment distribution and evolution of sedimentary processes in a small sandy turbidite system (Golo system, Mediterranean Sea): Implications for various geometries based on core framework: Geo-Marine Letters, v. 26, p. 373–395, doi:10.1007/s00367-006-0045-z.

Grimm, K. A., and K. B. Föllmi, 1994, Doomed pioneers: Allochthonous crustacean tracemakers in anaerobic basinal strata, Oligo-Miocene San Gregorio Formation, Baja California Sur, Mexico: Palaios, v. 9, p. 313–334, doi:10.2307/3515054.

Guzmán, O., and C. Campos, 2008, The Pampatar Formation (Margarita Island, Venezuela): A result of gravity flows in deep marine water, *in* J. J. Ponce and E. B. Olivero, conveners, Sediment transfer from shelf to deepwater—Revisiting the delivery mechanisms: Conference Proceedings, AAPG Search and Discovery Article 50070, AAPG Hedberg Conference, March 3–7, 2008, Ushuaia-Patagonia, Argentina, 7 p., http://www.searchanddiscovery.com/abstracts/html/2008/hedberg_argentina/extended/guzman/guzman.htm (accessed August 3, 2010).

Harms, J. C., J. B. Southard, D. R. Spearing, and R. G. Walker, 1975, Depositional environments as interpreted from primary sedimentary structures and stratification sequences: SEPM Short Course Notes no. 2, 161 p.

Harms, J. C., J. B. Southard, and R. G. Walker, 1982, Structures and sequences in clastic rocks: SEPM Short Course Notes no. 9, 249 p.

Hesse, R., and S. Khodabakhsh, 1998, Depositional facies of late Pleistocene Heinrich events in the Labrador Sea: Geology, v. 26, p. 103–106, doi:10.1130/0091-7613(1998)026<0103:DFOLPH>2.3.CO;2.

Hesse, R., and S. Khodabakhsh, 2006, Significance of fine-grained sediment lofting from melt-water generated turbidity currents for the timing of glaciomarine sediment transport into the deep sea: Sedimentary Geology, v. 186, p. 1–11, doi:10.1016/j.sedgeo.2005.10.006.

Hesse, R., H. Rashid, and S. Khodabakhsh, 2004, Fine-grained sediment lofting from meltwater-generated turbidity currents during Heinrich events. Geology, v. 32, p. 449–452, doi:10.1130/G20136.1.

Johnson, K. S., C. K. Paull, J. P. Barry, and F. P. Chavez, 2001, A decadal record of underflows from a coastal river into the deep sea: Geology, v. 29, p. 1019–1022, doi:10.1130/0091-7613(2001)029<1019:ADROUF>2.0.CO;2.

Jopling, A. V., and R. G. Walker, 1968, Morphology and origin of ripple-drift cross lamination, with examples of Pleistocene of Massachusetts: Journal of Sedimentary Petrology, v. 38, p. 971–984.

Joseph, P., and S. A. Lomas, 2004, Deep-water sedimentation in the Alpine Foreland Basin of SE France: New perspectives on the Grés d'Annot and related systems—An introduction, *in* P. Joseph and S. A. Lomas, eds., Deep-water sedimentation in the Alpine Basin of SE France: New perspectives on the Grés d'Annot and related systems: Geological Society (London) Special Publication 221, p. 1–16.

Kassem, A., and J. Imran, 2001, Simulation of turbid underflows generated by the plunging of a river: Geology, v. 29, p. 655–658, doi:10.1130/0091-7613(2001)029<0655:SOTUGB>2.0.CO;2.

Klaucke, I., B. Savoye, and P. Cochonat, 2000, Patterns and processes of sediment dispersal on the continental slope of Nice, SE France: Marine Geology, v. 162, p. 405–422, doi:10.1016/S0025-3227(99)00063-8.

Knapp, R. T., 1943, Density currents: Their mixing characteristics and their effect on the turbulence structure of the associated flow: Proceedings of the Second Hydraulics Conference: University of Iowa Studies in Engineering, Bulletin, v. 27, p. 289–306.

Kneller, B., and M. Branney, 1995, Sustained high-density turbidity currents and the deposition of thick massive sands: Sedimentology, v. 42, p. 607–616, doi:10.1111/j.1365-3091.1995.tb00395.x.

Lamb, M. P., and D. Mohrig, 2009, Do hyperpycnal-flow deposits record river-flood dynamics?: Geology, v. 37, p. 1067–1070, doi:10.1130/G30286A.1.

Lamb, M. P., P. M. Myrow, C. Lukens, K. Houck, and J. Strauss, 2008, Deposits from wave-influenced turbidity currents: Pennsylvanian Minturn Formation, Colorado, U.S.A.: Journal of Sedimentary Research, v. 78, p. 480–498, doi:10.2110/jsr.2008.052.

Lowe, D. R., 1979, Sediment gravity flows: Their classification and some problems of application to natural flows and deposits: SEPM Special Publication 27, p. 75–82.

Lowe, D. R., 1982, Sediment gravity flows: II. Depositional

models with special reference to the deposits of high-density turbidity currents: Journal of Sedimentary Petrology, v. 52, p. 279–297.

Mansurbeg, H., M. A. K. El-ghali, S. Morad, and P. Plink-Björklund, 2006, The impact of meteoric water on the diagenetic alterations in deep-water, marine siliciclastic turbidites: Journal of Geochemical Exploration, v. 89, p. 254–258, doi:10.1016/j.gexplo.2006.02.001.

Manville, V., and J. D. L. White, 2003, Incipient granular mass flows at the base of sediment-laden floods, and the roles of flow competence and flow capacity in the deposition of stratified bouldery sands: Sedimentary Geology, v. 155, p. 157–173, doi:10.1016/S0037-0738(02)00294-4.

Migeon, S., B. Savoye, N. Babonneau, and F. Spy Andersson, 2004, Processes of sediment-wave construction along the present Zaire deep-sea meandering channel: Role of meanders and flow stripping: Journal of Sedimentary Research, v. 74, p. 580–598, doi:10.1306/091603740580.

Migeon, S., T. Mulder, B. Savoye, and F. Sage, 2006, The Var turbidite system (Ligurian Sea, northwestern Mediterranean) morphology, sediment supply, construction of turbidite levee and sediment waves: Implications for hydrocarbon reservoirs: Geo-Marine Letters, v. 26, p. 361–371, doi:10.1007/s00367-006-0047-x.

Mulder, T., and J. Alexander, 2001, The physical character of subaqueous sedimentary density flows and their deposits: Sedimentology, v. 48, p. 269–299, doi:10.1046/j.1365-3091.2001.00360.x.

Mulder, T., and J. P. M. Syvitski, 1995, Turbidity current generated at river mouths during exceptional discharges to the world oceans: Journal of Geology, v. 103, p. 285–299, doi:10.1086/629747.

Mulder, T., S. Migeon, B. Savoye, and J. C. Faugéres, 2001a, Inversely graded turbidite sequences in the deep Mediterranean: A record of deposits from flood-generated turbidity currents?: Geo-Marine Letters, v. 21, p. 86–93, doi:10.1007/s003670100071.

Mulder, T., S. Migeon, B. Savoye, and J. M. Jouanneau, 2001b, Twentieth century floods recorded in the deep Mediterranean sediments: Geology, v. 29, p. 1011–1014, doi:10.1130/0091-7613(2001)029<1011:TCFRIT>2.0.CO;2.

Mulder, T., J. P. M. Syvitski, S. Migeon, J. C. Faugéres, and B. Savoye, 2003, Marine hyperpycnal flows: Initiation, behavior and related deposits: A review: Marine and Petroleum Geology, v. 20, p. 861–882, doi:10.1016/j.marpetgeo.2003.01.003.

Mutti, E., 1992, Turbidite sandstones: AGIP-Istituto di Geologia Università di Parma, San Donato Milanese, 275 p.

Mutti, E., and F. Ricci Lucchi, 1972, Le torbiditi dell'Appennino Settentrionale: Introduzione all'analisi di facies: Memorie della Società Geologica Italiana, v. 11, p. 161–199.

Mutti, E., G. Davoli, and R. Tinterri, 1994, Flood-related gravity-flow deposits in fluvial and fluvio-deltaic depositional systems and their sequence-stratigraphic implications, *in* H. W. Posamentier and E. Mutti, eds., Second high-resolution sequence stratigraphy conference, Tremp, Abstract Book: Italy, Istituto di Geologia, Universita di Parma, p. 137–143.

Mutti, E., G. Davoli, R. Tinterri, and C. Zavala, 1996, The importance of ancient fluvio-deltaic systems dominated by catastrophic flooding in tectonically active basins: Memorie di Scienze Geologiche, Universita di Padova, v. 48, p. 233–291.

Mutti, E., N. Mavilla, S. Angella, and L. L. Fava, 1999, An introduction to the analysis of ancient turbidite basins from an outcrop perspective: AAPG Continuing Education Course Notes no. 39, p. 1–98.

Mutti, E., R. Tinterri, G. Benevelli, D. Di Biase, and G. Cavanna, 2003, Deltaic, mixed and turbidite sedimentation of ancient foreland basins: Marine and Petroleum Geology, v. 20, p. 733–755, doi:10.1016/j.marpetgeo.2003.09.001.

Mutti, E., D. Bernoulli, F. Ricci Lucchi, and R. Tinterri, 2009, Turbidites and turbidity currents from Alpine "flysch" to the exploration of continental margins: Sedimentology, v. 56, p. 267–318, doi:10.1111/j.1365-3091.2008.01019.x.

Myrow, P. M., M. P. Lamb, C. Lukens, and K. Houck, 2006, Paleohydraulic interpretations of wave-modified hyperpycnal flow deposits: American Geophysical Union, Fall Meeting 2006, Abstract PP23B-1763, v. 87, no. 52, 1 p.

Myrow, P. M., M. P. Lamb, C. Lukens, K. Houck, and J. Strauss, 2008, Proximal to distal facies relationships in deposits of wave-influenced hyperpycnal flows, *in* J. J. Ponce and E. B. Olivero, conveners, Sediment transfer from shelf to deepwater—Revisiting the delivery mechanisms: Conference Proceedings, AAPG Hedberg Conference, March 3–7, 2008, Ushuaia-Patagonia, Argentina, 5 p.

Nakajima, T., 2006, Hyperpycnites deposited 700 km away from river mouths in the Central Japan Sea: Journal of Sedimentary Research, v. 76, p. 59–72.

Normark, W. R., and D. J. Piper, 1991, Initiation processes and flow evolution of turbidity currents: Implications for the depositional record: SEPM Special Publication 46, p. 207–230.

Olariu, C., and J. P. Bhattacharya, 2006, Terminal distributary channels and delta front architecture of river-dominated delta systems: Journal of Sedimentary Research, v. 76, p. 212–233, doi:10.2110/jsr.2006.026.

Pattison, S., 2005, Isolated highstand shelf sandstone body of turbiditic origin, lower Kenilworth Member, Cretaceous Western Interior, Book Cliffs, Utah: Sedimentary Geology, v. 177, p. 131–144, doi:10.1016/j.sedgeo.2005.02.005.

Pattison, S., 2008, Role of wave-modified underflows in the across-shelf transport of fine-grained sediments: Examples from the Book Cliffs, Utah, *in* J. J. Ponce and E. B. Olivero, conveners, Sediment transfer from shelf to deepwater—Revisiting the delivery mechanisms: Conference Proceedings, AAPG Search and Discovery

Article 90079, AAPG Hedberg Conference , March 3–7, 2008, Ushuaia-Patagonia, Argentina, 4 p., http://www.searchanddiscovery.com/abstracts/html/2008/hedberg_argentina/extended/pattison/pattison.htm (accessed August 3, 2010).

Peakall, J., M. Felix, B. McCaffrey, and B. Kneller, 2001, Particulate gravity currents: Perspectives, *in* B. McCaffrey, B. Kneller, and J. Peakall, eds., Particulate gravity currents: International Association of Sedimentologists, Special Publication 31, p. 1–8.

Petter, A. L., and R. J. Steel, 2005, Deepwater-slope channels and hyperpycnal flows from the Eocene of the central Spitsbergen Basin: Predicting basin-floor sands from a shelf edge/upper slope perspective: AAPG Search and Discovery Article 90039, AAPG Annual Meeting, June 19–22, 2005, Calgary, Alberta, 1 p., http://www.searchanddiscovery.net/abstracts/html/2005/annual/abstracts/petter.htm?q=%2Btext%3Apetter (accessed August 3, 2010).

Petter, A. L., and R. J. Steel, 2006, Hyperpycnal flow variability and slope organization on an Eocene shelf margin, Central Basin, Spitsbergen: AAPG Bulletin, v. 90, p. 1451–1472, doi:10.1306/04240605144.

Plink-Björklund, P., and R. J. Steel, 2004, Initiation of turbidite currents: Outcrop evidence for Eocene hyperpycnal flow turbidites: Sedimentary Geology, v. 165, p. 29–52, doi:10.1016/j.sedgeo.2003.10.013.

Ponce, J. J., E. B. Olivero, and D. R. Martinioni, 2008a, Syndepositional lateral migration of hyperpycnal lobes in submarine ramp systems—Early Miocene, Austral Basin, Argentina, *in* J. J. Ponce and E. B. Olivero, conveners, Sediment transfer from shelf to deepwater—Revisiting the delivery mechanisms: Conference Proceedings, AAPG Search and Discovery Article 50074, AAPG Hedberg Conference, March 3–7, 2008, Ushuaia-Patagonia, Argentina, 4 p., http://www.searchanddiscovery.net/documents/2008/08027ponce/index.htm?q=%2Btext%3A50074 (accessed August 3, 2010).

Ponce, J. J., E. B. Olivero, and D. R. Martinioni, 2008b, Deep-marine hyperpycnal channel-levee complexes in the Miocene of Tierra del Fuego, Artentina: Architectural elements and facies associations, *in* J. J. Ponce and E. B. Olivero, conveners, Sediment transfer from shelf to deepwater—Revisiting the delivery mechanisms: Conference Proceedings, AAPG Search and Discovery Article 50073, AAPG Hedberg Conference, March 3–7, 2008, Ushuaia-Patagonia, Argentina, 6 p., http://www.searchanddiscovery.net/documents/2008/08026ponce/index.htm?q=%2Btext%3A50073 (accessed August 3, 2010).

Posamentier, H., and V. Kolla, 2003, Processes and reservoir architecture of deep-water sinuous channels from the shelf edge to the basin floor, based on analyses of 3d seismic and sidescan imagery (abs.), *in* D. Hodgson, C. Edwards, and R. Smith, eds., Submarine slope systems: Processes, products and prediction. April 28–29, 2003, University of Liverpool, Conference Abstracts Volume, p. 67.

Prior, D. B., B. D. Bornhold, W. J. Wiseman, and D. R. Lowe, 1987, Turbidity current activity in a British Columbia fjord: Science, v. 237, p. 1330–1333, doi:10.1126/science.237.4820.1330.

Saccavino, L., L. Buatois, and C. Zavala, 2008, Integrating sedimentologic and ichnologic data in characterization of hyperpycnal flow deposits of Cretaceous fluvio-dominated deltas (Magallanes Basin, southern Argentina), *in* J. J. Ponce and E. B. Olivero, conveners, Sediment transfer from shelf to deepwater—Revisiting the delivery mechanisms: Conference Proceedings, AAPG Search and Discovery Article 90079, AAPG Hedberg Conference, March 3–7, 2008, Ushuaia-Patagonia, Argentina, 4 p., http://www.searchanddiscovery.net/abstracts/html/2008/hedberg_argentina/extended/saccavino/saccavino.htm?q=%2Btext%3Asaccavino (accessed August 3, 2010).

Saitoh, Y., 2004, Hyperpycnal-flow deposits from the Miocene continental delta complex in the Ryukyu island arc, southwest Japan (abs.): 5th International Conference on Asian Marine Geology, Abstract Volume, p. 176.

Saller, A., R. Lin, and J. Dunham, 2006, Leaves in turbidite sands: The main source of oil and gas in the deep-water Kutei Basin, Indonesia: AAPG Bulletin, v. 90, p. 1585–1608, doi:10.1306/04110605127.

Sanders, J. E., 1965, Primary sedimentary structures formed by turbidity currents and related sedimentation mechanisms, *in* G. V. Middleton, ed., Primary sedimentary structures and their hydrodynamic interpretation: SEPM Special Publication 12, p. 192–219.

Schumm, S. A., 1977, The fluvial system: New York, John Wiley and Sons, 338 p.

Schumm, S. A., 1981, Evolution and response of the fluvial system: Sedimentologic implications: SEPM Special Publication 31, p. 19–29.

Simons, D. B., E. V. Richardson, and C. F. Nordin, 1965, Sedimentary structures generated by flow on alluvial channels, *in* G. V. Middleton, ed., Primary sedimentary structures and their hydrodynamic interpretation: SEPM Special Publication 12, p. 34–52.

Southard, J. B., 1991, Experimental determination of bedform stability: Annual Review of Earth and Planetary Sciences, v. 19, p. 423–455, doi:10.1146/annurev.ea.19.050191.002231.

Soyinka, O., and R. M. Slatt, 2008, Identification and micro-stratigraphy of hyperpycnites and turbidites in Cretaceous Lewis Shale, Wyoming: Sedimentology, v. 55, p. 1117–1133, doi:10.1111/j.1365-3091.2007.00938.x.

Sparks, R. S. J., R. T. Bonnecaze, H. E. Huppert, J. R. Lister, M. A. Hallworth, J. Phillips, and H. Mader, 1993, Sediment-laden gravity currents with reversing buoyancy: Earth and Planetary Science Letters, v. 114, p. 243–257, doi:10.1016/0012-821X(93)90028-8.

Stow, D. A. V., and M. Mayall, 2000, Deep-water sedimentary systems: New models for the 21st century: Marine and Petroleum Geology, v. 17, p. 125–135, doi:10.1016/S0264-8172(99)00061 1.

Sumner, E. J., L. A. Amy, and P. J. Talling, 2008, Deposit structure and processes of sand deposition from decelerating sediment suspensions: Journal of Sedimentary Research, v. 78, p. 529–547, doi:10.2110/jsr.2008.062.

Walker, R., 1984, Facies models: Geoscience Canada, Reprint Series 1, 317 p.

Warrick, J. A., J. Xu, M. A. Noble, and H. J. Lee, 2008, Rapid formation of hyperpycnal sediment gravity currents offshore of a semi-arid California river: Continental Shelf Research, v. 28, p. 991–1009, doi:10.1016/j.csr.2007.11.002.

Zavala, C., 2000a, Nuevos avances en la sedimentología y estratigrafía secuencial de la Formación Mulichinco (Valanginiano) en la Cuenca Neuquina: Boletín de Informaciones Petroleras, v. 63, p. 40–54.

Zavala, C., 2000b, Stratigraphy and sedimentary history of the Plio-Pleistocene Sant'Arcangelo Basin, southern Apennines, Italy: Rivista Italiana di Paleontologia e Stratigrafia, v. 106, p. 399–416.

Zavala, C., 2005, Tracking sea bed topography in the Jurassic. The Lotena Group in the Sierra de la Vaca Muerta (Neuquén Basin, Argentina): Geologica Acta, v. 3, p. 133–145.

Zavala, C., 2008, Toward a genetic facies tract for the analysis of hyperpycnal deposits, *in* J. J. Ponce and E. B. Olivero, conveners, Sediment transfer from shelf to deepwater—Revisiting the delivery mechanisms: Conference Proceedings, AAPG Search and Discovery Article 50075, AAPG Hedberg Conference, March 3–7, 2008, Ushuaia-Patagonia, Argentina, 2 p., http://www.searchanddiscovery.net/documents/2008/jw0805zavala/index.html?q=%2Btext%3A50075 (accessed August 4, 2010).

Zavala, C., and R. Gonzalez, 2001, Estratigrafía del Grupo Cuyo (Jurásico inferior-medio) en la Sierra de la Vaca Muerta, Cuenca Neuquina: Boletín de Informaciones Petroleras, v. 65, p. 52–64.

Zavala, C., G. Azúa, H. Freije, and J. Ponce, 2000, Los sistemas fluvio-deltaicos del Grupo Curamalal (Paleozoico Inferior). Cuenca Paleozoica de Ventania, Provincia de Buenos Aires, Argentina: Revista de la Asociación Geológica Argentina, v. 55, p. 165–178.

Zavala, C., J. Ponce, D. Drittanti, M. Arcuri, H. Freije, and M. Asensio, 2006a, Ancient lacustrine hyperpycnites: A depositional model from a case study in the Rayoso Formation (Cretaceous) of west-central Argentina: Journal of Sedimentary Research, v. 76, p. 41–59, doi:10.2110/jsr.2006.12.

Zavala, C., M. Arcuri, and H. Gamero, 2006b, Towards a genetic model for the analysis of hyperpycnal systems (abs.): Geological Society of America Abstracts with Programs, v. 38, no. 7, p. 541.

Zavala, C., H. Gamero, and M. Arcuri, 2006c, Lofting rhythmites: A diagnostic feature for the recognition of hyperpycnal deposits (abs.): Geological Society of America Abstracts with Programs, v. 38, no. 7, p. 541.

Zavala, C., M. Arcuri, H. Gamero Díaz, and C. Contreras, 2007a, The composite bed: A new distinctive feature of hyperpycnal deposition (abs.): AAPG Annual Convention & Exhibition, v. 16, p. 157.

Zavala, C., J. Marcano, J. Carvajal, 2007b, Proximity and Laterality indexes: A new tool for the analysis of ancient hyperpycnal deposits in the subsurface, *in* Proceedings of the 4th Geological Society of Trinidad and Tobago Geological Conference, 3 p.

Zavala, C., J. Carvajal, J. Mercano, and M. Delgado, 2008a, Sedimentological indices: A new tool for regional studies of hyperpycnal systems, *in* J. J. Ponce and E. B. Olivero, conveners, Sediment transfer from shelf to deepwater—Revisiting the delivery mechanisms: Conference Proceedings, AAPG Search and Discovery Article 50076, AAPG Hedberg Conference, March 3–7, 2008, Ushuaia-Patagonia, Argentina, 4 p., http://www.searchanddiscovery.net/documents/2008/jw0806zavala/index.html?q=%2Btext%3A50076 (accessed August 3, 2010).

Zavala, C., L. Blanco Valiente, and Y. Vallez, 2008b, The origin of lofting rhythmites: Lessons from thin sections, *in* J. J. Ponce and E. B. Olivero, conveners, Sediment transfer from shelf to deepwater—Revisiting the delivery mechanisms: Conference Proceedings, AAPG Search and Discovery Article 50077, AAPG Hedberg Conference, March 3–7, 2008, Ushuaia-Patagonia, Argentina, 4 p., http://www.searchanddiscovery.net/documents/2008/jw0807zavala/index.html?q=%2Btext%3A50077 (accessed August 3, 2010).

Zavala, C., J. Marcano, J. Carvajal, and M. Delgado, 2011, Genetic indices in hyperpycnal systems: A case study in the late Oligocene–early Miocene Merecure Formation, Maturin Subbasin, Venezuela, *in* R. M. Slatt and C. Zavala, eds., Sediment transfer from shelf to deep water—Revisiting the delivery system: AAPG Studies in Geology 61, p. 53–73.

3

Zavala, C., J. Marcano, J. Carvajal, and M. Delgado, 2011, Genetic indices in hyperpycnal systems: A case study in the late Oligocene–early Miocene Merecure Formation, Maturin Subbasin, Venezuela, *in* R. M. Slatt and C. Zavala, eds., Sediment transfer from shelf to deep water—Revisiting the delivery system: AAPG Studies in Geology 61, p. 53–73.

Genetic Indices in Hyperpycnal Systems: A Case Study in the Late Oligocene–Early Miocene Merecure Formation, Maturin Subbasin, Venezuela

Carlos Zavala
GCS Argentina, Universidad Nacional del Sur, Bahía Blanca, Argentina

Jose Marcano, Jair Carvajal, and Manuel Delgado
Petróleos de Venezuela Sociedad Anónima (PDVSA), Laboratorio Geológico El Chaure, Puerto La Cruz, Venezuela

ABSTRACT

The mapping and analysis of ancient turbidite deposits using geographically dependent sedimentologic indices (or proximality) determined by genetic facies relationships and grain-size variations were mostly applied during the early years of turbidite research. Nevertheless, the understanding that fine-grained deposits (thin-bedded turbidites) can also occur in proximal (interchannel) areas, and the impossibility of differentiating them from real distal deposits, revealed many inconsistencies in this approach to the development of turbidite fan facies models. More recently, a genetic facies tract has been proposed for hyperpycnal systems, which distinguishes between thin-bedded deposits accumulated in interchannel areas from those deposited on distal areas.

The genetic facies tract proposed for hyperpycnal systems in marine settings comprises three main facies categories related to bed-load, suspended-load, and lofting processes, respectively. Bed-load-related facies are coarse grained and related to drag forces provided by the overpassing turbulent flow. Suspended-load-related facies are composed of fine-grained sandstones and originate from the gravitational collapse of the suspended load as the long-lived flow progressively wanes. Lofting-related facies are the result of the fallout of very fine-grained sands,

Copyright ©2011 by The American Association of Petroleum Geologists.
DOI:10.1306/13271350St613438

silts, plant debris, and micas from lofting plumes mainly in flow-margin areas. Proximal (Pt) and lateral (Lt) facies indices have been defined based on the relative abundance of B, S, and L facies within a stratigraphic interval. The Pt and Lt indices are dimensionless numbers that range from 100 to 0. The Pt index measures how proximal a given stratigraphic section is with respect to the entire hyperpycnal system. The Lt index provides an indication of the stratigraphic lateral distance of a given section with respect to the flow axis. These genetic indices were applied to hyperpycnal deposits of the late Oligocene–early Miocene Merecure Formation, Maturin Subbasin, Venezuela.

Results suggest that the Merecure Formation may be more extensive than previously considered with a main sediment source from cratonic areas located in the south and southeast. Index mapping also suggests a syntectonic accumulation during the Oligocene, which controlled the subaqueous topography and, thus, the distribution of sand bodies.

INTRODUCTION

The fact that turbidites change facies with an associated decreasing grain size from source to the basin was known early in turbidite research (Natland and Kuenen, 1951). The recognition and analysis of the grain size, depositional features, and facies types considered as typical of proximal to distal settings led to establishing several guidelines aimed at distinguishing between proximal and distal turbidites. During the 1960s, many attempts were made to quantify these facies and associated textural changes, which led to the proposal of the proximality index (P index) in the analysis of turbidite deposits (Walker, 1967). The P index analysis was applied with some early success, but the understanding that fine-grained turbidites could also occur in interchannel (levee) proximal areas led to confusing interpretations. The description of fine-grained proximal deposits resulted in the progressive abandonment of the P index analysis, which was replaced by facies analysis (Mutti and Ricci Lucchi, 1972). The controversy over the origin of fine-grained facies in turbidite deposits resulted in the concept of thin-bedded turbidites, which was far more descriptive and not directly dependent on its position in a proximal-to-distal profile.

In recent years, the increased understanding that many turbulent flows in the marine setting could originate by direct discharge from rivers in flood (or hyperpycnal flows; Bates, 1953; Normark and Piper, 1991; Mulder and Syvitski, 1995; Mutti et al., 1996, 1999) has opened a new research frontier in shelfal to deep-marine clastic sedimentation. Hyperpycnal flows and their deposits, hyperpycnites (Mulder et al., 2003), originate when rivers discharge a turbulent mixture of water and sediment having a bulk density that exceeds that of water in the receiving basin (Bates, 1953). In the case of a river flow entering a marine basin, the fluvial discharge should have a suspended-load concentration exceeding 35 to 45 kg/m^3 to overcome the density contrast between fresh water and seawater (Mulder and Syvitski, 1995). This land-generated and relatively dense flow will sink in coastal areas, generating a turbulent flow that moves basinward along the bottom of the basin, allowing the transfer of a large volume of fresh water and sediments from the river to the basin floor. Depending on different hydrologic and physiographic factors, hyperpycnal flows can develop either in the shelf or in deep-water settings, and their deposits can be considered as true shelfal (see also the concept of shelfal sandstone lobes in Mutti et al., 1996) to deep-water turbidites. Hyperpycnite beds related to sustained flows in marine settings can be recognized by some common and distinguishing characteristics such as reverse-normal grading (Mulder et al., 2003), gradual facies recurrence (Zavala et al., 2006a), abundant plant debris (Plink-Björklund and Steel, 2004), and associated silt-sand couples with abundant plant debris and mica known as lofting rhythmites (Zavala et al., 2006c, 2008b).

Recently, a new genetic facies tract was proposed for the analysis of hyperpycnal deposits in lacustrine to marine settings (Zavala et al., 2006b, 2011; Zavala, 2008). This facies tract classifies hyperpycnal-related facies according to the main processes active during the accumulation of hyperpycnites, from bed load (facies B), suspended load (facies S), and lofting (facies L). Bed-load facies accumulate in proximal areas because of shear forces provided by the overpassing turbulent flow, whereas L facies mainly develop at flow-margin (levee) areas. Suspended-load facies, however,

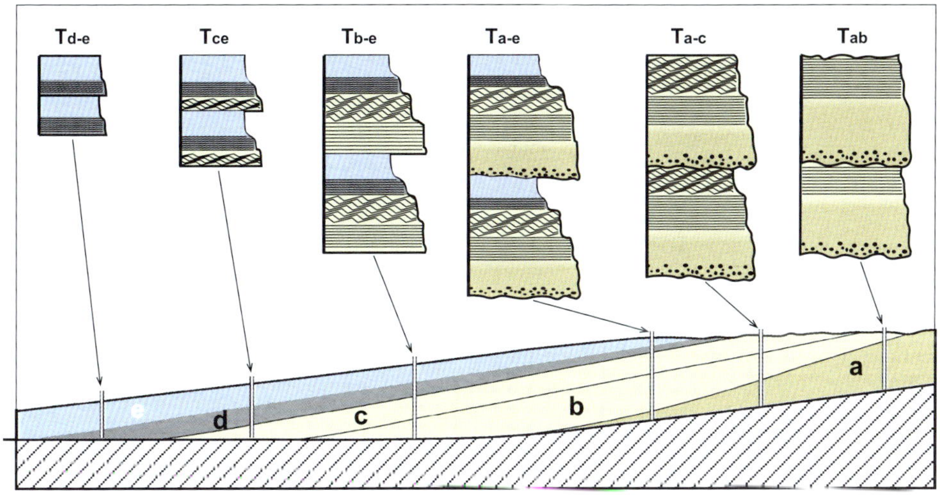

FIGURE 1. Variations in the typical Bouma facies sequence in a downcurrent direction (from right to left). Note that in proximal facies sequences, the upper intervals are missing (Tc–Te), whereas distal turbidite beds progressively lose the lower intervals (Ta–Tb). These facies changes in the basinward direction can be used to predict the occurrence of certain facies types in a given area. Modified after Bouma, 1962.

are related to the progressive collapse and accumulation of turbulent-suspended materials when the flow progressively wanes in medium to distal areas. Consequently, quantifying the relationship between these three categories of genetically related facies allows determination of the position of a given hyperpycnite section within the entire depositional system. This geographically oriented analysis is possible by the determination of two partially dependent genetic indices, termed proximal (Pt) and lateral (Lt) facies indices. The Lt index solves the interpretation problem of relatively low Pt of levee areas that was encountered in the earlier studies of turbidite deposits (Macdonald, 1986).

This chapter provides an example of this new methodology applied to a subsurface study conducted in the Oligocene–early Miocene Merecure Formation, located in the Maturin Subbasin, Venezuela. Because the Merecure Formation is one of the main oil-bearing units of eastern Venezuela, this methodology has substantially improved the facies analysis in the prediction of distribution and quality of clastic reservoirs.

INDICES (PROXIMALITY) IN TURBIDITE STUDIES

General Overview

When Natland and Kuenen (1951) first documented that the Ventura Basin (California) turbidite beds thinned in a downcurrent direction, several attempts were made to quantify the main characteristics of proximal-to-distal facies in turbidites. Bouma (1962) published perhaps the most famous facies sequence in the Peira Cava flysch (Oligocene) composed of five divisions, termed Ta to Te. The divisions form a fining-upward sequence with a basal, massive, and graded sandstone (Ta) that is overlain by parallel-laminated sandstones (Tb) and fine-grained rippled sandstones (Tc), and capped by silts and hemipelagic mud (Td and Te). Bouma (1962) also noted that in a proximal setting, the facies sequence tends to lack the upper divisions, whereas toward the distal setting, the basal intervals are progressively absent (Figure 1). This sequence was quantified by Walker (1967), who defined a P index based on the percentage of beds within a group of beds that began with a particular Bouma division. According to Walker (1967), the P index can range from 100 (most proximal) to 0% (most distal) and could be used to estimate the location of any given section relative to its source direction.

The concept of proximality in turbidite deposition was supported by the premise that a turbidity current tends to lose energy with distance along its travel path, accounting for the basinward decrease in erosion and the deposition of progressively finer grained sediments. Although the idea of depletive flows (in the sense of Kneller and Branney, 1995) in turbidity currents is valid, flows in proximal areas are commonly channelized and exhibit a strong lateral anisotropy in flow velocity that results in the accumulation of fine-grained deposits in flow-margin areas (levee deposits) having similar characteristics to those accumulated in distal areas (Haner, 1971; Mutti and Ricci Lucchi, 1972). These inconsistencies were discussed in several subsequent articles (Lovell, 1969, 1970; McCabe and Waugh, 1973) that demonstrated the P index is a poor guide to proximality.

The increased understanding of internal characteristics of turbidite fan models revealed that fine-grained deposits can develop in different fan settings, and consequently, many authors suggested that

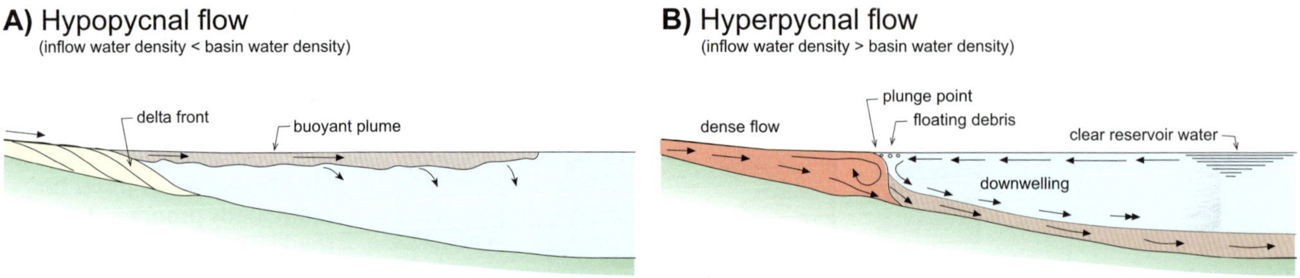

FIGURE 2. Comparison between (A) hypopycnal and (B) hyperpycnal flows (original concept by Bates, 1953). Note that in the case of the hyperpycnal flow, the fluvial discharge sinks beneath the less dense water body, continuing its travel basinward as a quasi-steady underflow. Panel B was redrawn from the pioneer work of Knapp (1943). After Zavala et al. (2008a).

the terms "proximal" and "distal" should be avoided (Nelson et al., 1978) for facies classifications, and all fine-grained sediments should be considered as thin-bedded turbidites without any implication for proximal or distal setting (Walker, 1978; Nilsen, 1980). Consequently, modern use of the terms proximal or distal turbidites refers only to the distance from the source area and should not be used to describe facies, the characteristics of which can sometimes be controversial (Macdonald, 1986).

Genetic Facies Tracts and Hyperpycnal Sedimentation: Hyperpycnal Systems

Recent advances in the understanding of a relatively new category of depositional system, termed "hyperpycnal system" (Zavala et al., 2006a), offer new perspectives to understanding the distribution of shelfal to deep-water sandstone strata. A hyperpycnal system is the subaqueous extension of the fluvial system (the fluvial system in the sense of Schumm, 1977, 1981) and originates when a river in flood discharges a relatively dense turbulent mixture of fresh water and sediment with a bulk density that exceeds that of the receiving water body (Mulder and Alexander, 2001). When these fluvial-derived dense flows reach the coast, they plunge to the sea floor and move basinward as a sustained and quasi-steady turbulent flow (Figure 2B). If the hyperpycnal flow can be maintained for a long period of time, these flows have the capacity to travel hundreds of kilometers basinward (Nakajima, 2006; Cheng-Shing and Ho-Shing, 2008; Bourget et al., 2010) even in relatively low gradient settings and to deposit very thick successions, especially in topography-controlled depocenters. Hyperpycnal system deposits commonly show some characteristic patterns erroneously considered as typical of fluvial deposition, including bed load, channelization, and meandering (Cheng-Shing and Ho-Shing, 2008). Contrary to conventional models for turbidity current sedimentation (Mutti et al., 1999, 2009), coarse-grained materials are not transported at the head in long-lived hyperpycnal flows but are transported as bed load (Figure 3) because of shear forces provided by the overpassing turbulent flow

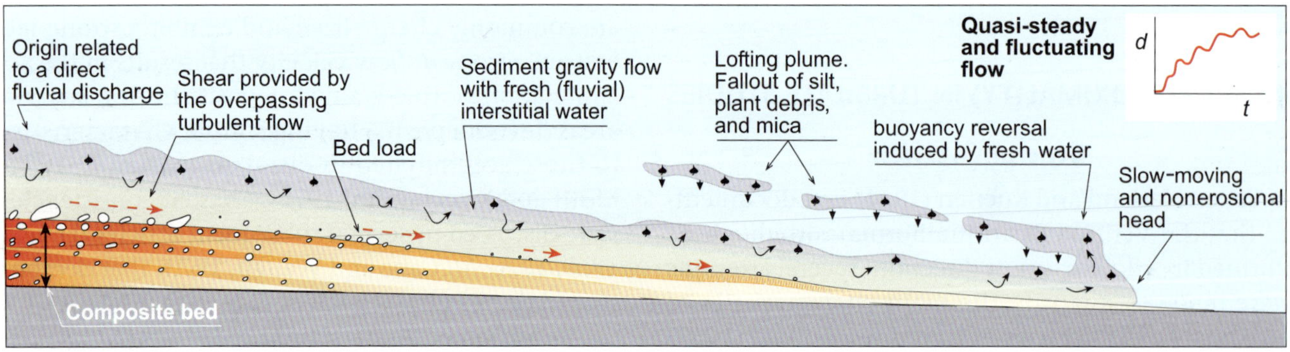

FIGURE 3. Main characteristics of long-lived hyperpycnal flows and their typical deposits. The complexity of the river (outflow) hydrograph (drawn on the right upper corner) results in the accumulation of composite beds (Zavala et al., 2007a). d = discharge; t = time.

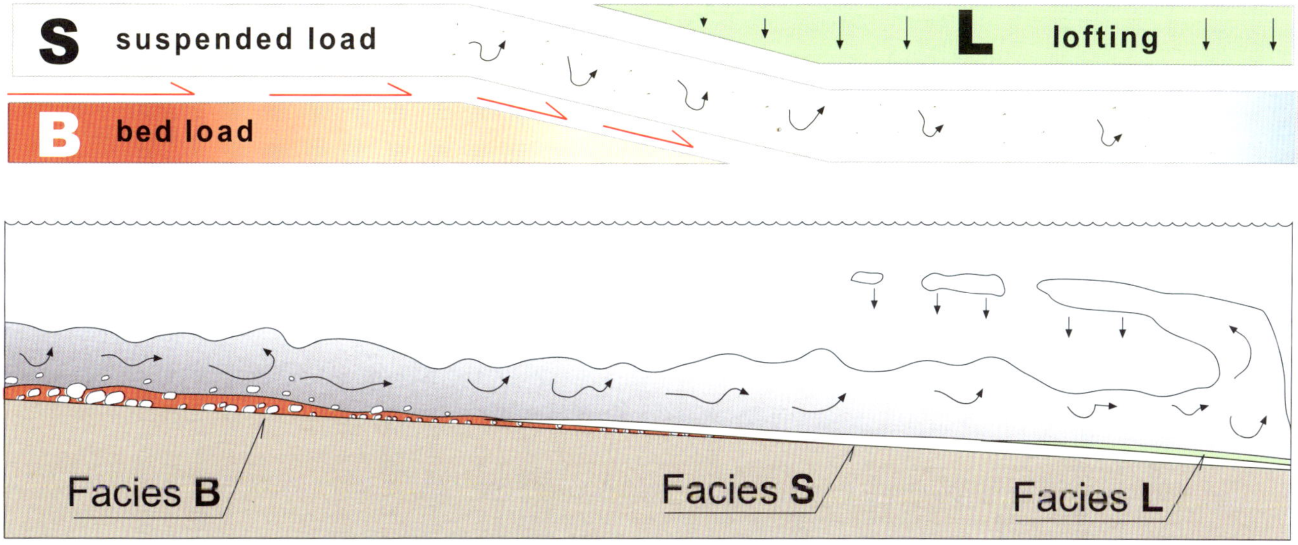

FIGURE 4. Conceptual diagram for facies analysis of hyperpycnal deposits with associated bed load in marine settings. Facies B includes all those facies related to bed-load processes developed at the base of an overpassing long-lived turbulent flow. Facies S corresponds to the gravitational settling of sand-size materials transported in suspension by the turbulent flow (suspended-load facies). Facies L is related to the fallout of fine-grained materials lifted into the interstitial fresh water contained in the flow once its sand-size suspended load is deposited. After Zavala et al., 2006b, 2011; Zavala, 2008.

(Plink-Björklund and Steel, 2004; Zavala et al., 2006b; Zavala, 2008).

Facies Analysis of Hyperpycnal Systems: A Genetic Approach

The basic scheme of facies classification in this chapter has been applied and tested during the last 7 yr in several hyperpycnal stratigraphic units (Zavala et al., 2006b, 2011; Zavala, 2008). This facies classification is based on the distinction of three main categories of genetic facies, related to bed-load (facies B), suspended-load (facies S), and lofting (facies L) processes (Figure 4).

Facies B (bed load) includes the coarsest sediments of the facies tract that are transported by drag and shear forces provided by the overpassing turbulent hyperpycnal flow. Consequently, B facies are dominant in proximal positions. Three main facies types are recognized in this category (Figure 5), termed B1 (massive conglomerates), B2 (pebbly sandstone with asymptotic low-angle cross-stratification), and B3 (pebbly sandstone with diffuse planar lamination). Facies S is fine grained and is deposited because of the gravitational collapse of turbulent suspended load in the main body of the hyperpycnal flow. Four facies types are recognized within this category (Figure 5), termed S1 (massive sandstone), S2 (laminated sandstone), S3 (sandstone with climbing ripples), and S4 (massive siltstone and mudstone). Facies L (lofting) accumulates from a lofting plume, which is a typical feature of hyperpycnal inflows in marine or saline environments. In this situation, the hyperpycnal discharge contains a fluid (fresh water) that is less dense than ambient seawater. Consequently, when the flow progressively loses parts of its suspended load by deposition, the current will lift from the substrate through buoyancy reversal (Sparks et al., 1993; Kneller and Buckee, 2000; Hesse et al., 2004), forming lofting plumes charged with fine-grained sediments, plant debris, and micas. These fine-grained sediments are gradually deposited on the sea bottom, forming repeated silt-sand couplets of great lateral extent known as lofting rhythmites (Zavala et al., 2006c, 2008b). Individual sand-silt levels in lofting rhythmites are commonly up to 2 mm (0.08 in.) thick and show good lateral continuity. Silt-sandstone couplets and their intercalations integrate laminated and decimeter-thick packages. These laminated packages lack or show rare and poorly diverse ichnofaunas. Lofting rhythmites appear isolated between mudstone successions or located between or toward the top of massive-to-laminated tabular sandstone beds. The recognition of lofting facies in marine environments is therefore extremely important because it allows the diagnosis of a direct fluvial connection and a hyperpycnal origin

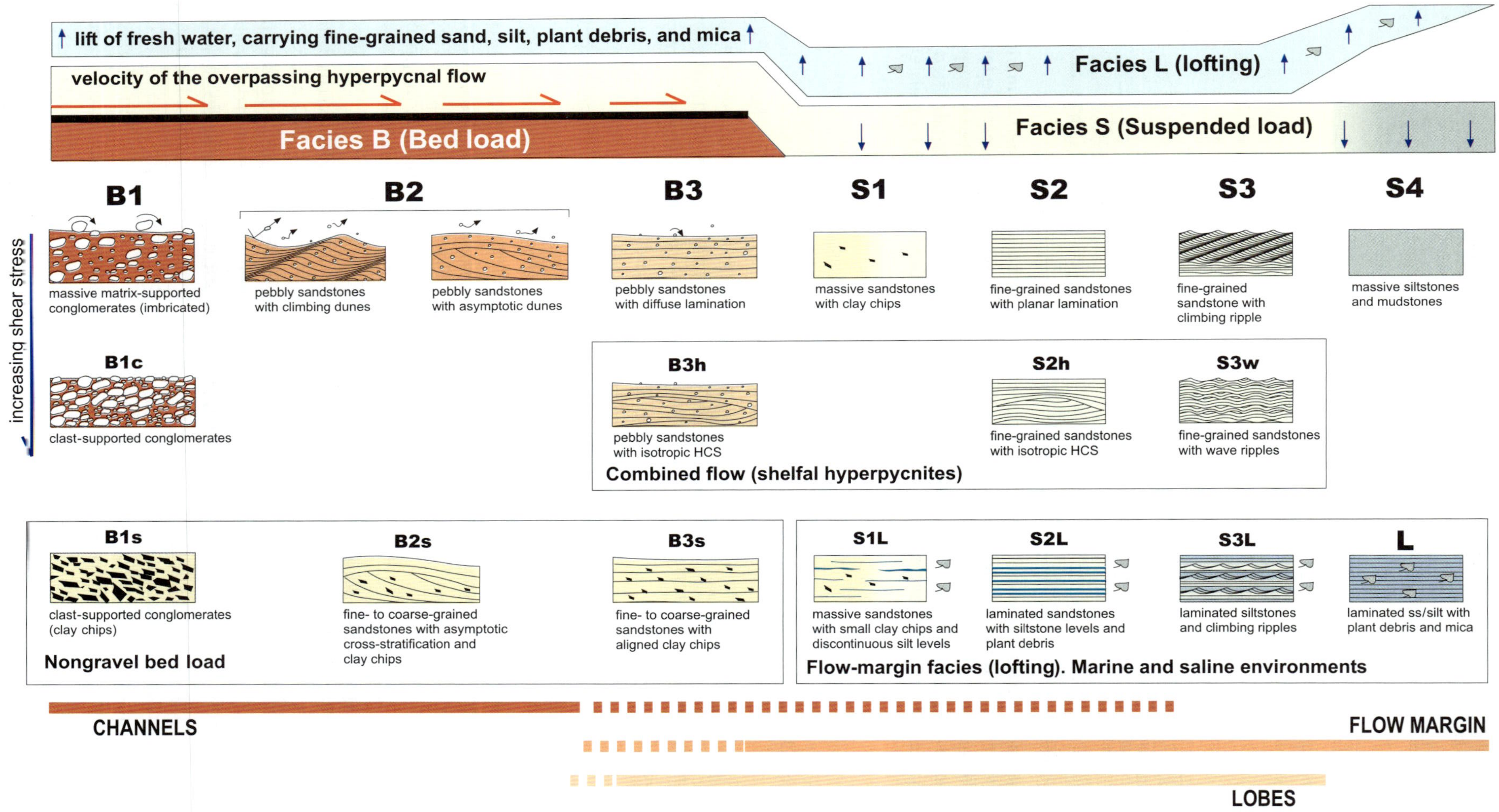

FIGURE 5. Conceptual scheme for the genetic interpretation of clastic facies in hyperpycnal systems. Type B facies relates to bed-load processes at the base of an overpassing long-lived turbulent flow. Type S facies originate from the gradual accumulation of sand-size suspended materials carried in the turbulent flow. Facies L is composed of very fine-grained sandstones interbedded with laminated silts with abundant plant debris and mica deposited by fallout from a lofting plume. Facies L is exclusive of marine/saline environments. After Zavala et al., 2006b, 2011; Zavala, 2008. HCS = hummocky cross-stratification; ss/silt = sandstone/silt

for associated deposits. Lofting rhythmites should not be confused with tidal rhythmites. Some distinguishing characteristics of lofting rhythmites include (1) lack of true mud couplets—the separating levels between sand-silt intervals are composed of plant debris and micas instead of mud; (2) absence of clear internal organization in neap-spring cycles; (3) close association and transitional (lateral) gradation with massive (facies S1L), laminated (facies S2L), and rippled (facies S1L) sandstones; (4) lack or very scarce bioturbation dominated by *Palaeophycus*; (5) well-developed normal grain-size grading in individual laminae (Zavala et al., 2008b).

Because the hyperpycnal flows show a drastic velocity decrease from the flow axis toward the laterals (Zavala et al., 2006a), facies L will be mainly accumulated in flow-margin (levee) areas.

Hyperpycnites commonly compose very complex beds showing internal erosional surfaces and gradual facies recurrences related to deposition from long-lived and highly dynamic fluctuating flows (Zavala et al., 2006a; Lamb et al., 2008; Jackson and Johnson, 2009; Lamb and Mohrig, 2009). This complex behavior results in the accumulation of composite beds (Figure 3) (Zavala et al., 2007a) having an internal facies arrangement that strongly departs from conventional facies models deposited from surgelike turbidity flows (Bouma, 1962; Mutti et al., 2003).

Facies B characterizes transfer zones, and its occurrence allows prediction of sandstone deposits (facies S) basinward. Facies L are mostly developed in marginal flow areas.

Genetic Indices: Proximal and Lateral Facies Indices

The genetic-oriented analysis applied to the study of hyperpycnal systems allows facies mapping and recognition of bypass, depositional, and marginal areas in the subsurface. Two main indices were considered in this study, termed proximal (Pt) and lateral (Lt) facies indices (Zavala et al., 2007b, 2008a). These indices should be calculated individually for each analyzed section.

The Proximal Facies Index

The Pt index is a dimensionless number that measures how proximal a given hyperpycnal sediment body is with respect to the entire hyperpycnal system. The Pt index is based on the relative dominance of bed-load-related facies (B facies) in proximal positions and the basinward increase of suspended-load facies (S facies) as the long-lived hyperpycnal flow progressively wanes with the subsequent collapse of suspended materials. The Pt index is calculated as follows:

$$\mathrm{Pt} = 100\mathrm{B}/(\mathrm{B} + \mathrm{S}) \qquad (1)$$

where Pt is the proximal facies index, B is the total thickness of bed-load facies, and S is the total thickness of suspended-load facies in a given stratigraphic section or core. Note that this analysis only applies for hyperpycnal facies.

The Pt index is a dimensionless number that ranges between 0 and 100. The greater the Pt index, the more proximal the deposit is with respect to its source area. Proximal indices between 100 and 50 characterize proximal parts of the system. Proximal indices between 50 and 0 suggest intermediate positions in the system. A proximality of 0 marks the end of the channel-lobe transition zone (CLTZ) (Mutti and Normark, 1987; Wynn et al., 2002) and the beginning of the distal part of the system (lobe) with tabular sandstone beds almost entirely composed of S facies. In addition, the decay rate of the Pt index between different locations can be used as a proxy to estimate the dimensions of the hyperpycnal system under consideration.

The Lateral Facies Index

The Lt index is a dimensionless number that ranges from 0 to 100, which is a measure of the relative abundance of L facies with respect to B and S facies in a given sedimentary section. In other words, the Lt index will give an indication of the relative lateral location of the analyzed well or outcrop from the main flow axis, which will be dominated by B or S facies. The lateral facies index can be obtained as follows:

$$\mathrm{Lt} = 100\mathrm{L}/(\mathrm{L} + \mathrm{B} + \mathrm{S}) \qquad (2)$$

where Lt is the lateral facies index, L is the total thickness of lofting facies, B is the total thickness of bed-load facies, and S is the total thickness of suspended-load hyperpycnal facies in an analyzed core or outcrop.

A hyperpycnal flow is a sediment gravity flow, and the accumulation of coarse-grained facies will be very sensitive to any subaqueous topography. Facies B and S tend to develop infilling the lowermost positions of the submarine landscape. On the contrary, L facies mostly characterizes relatively elevated (levee) areas located laterally with respect to the main axis of the hyperpycnal flows. On relatively flat depositional areas, individual hyperpycnal deposits commonly generate a depositional relief, resulting in compensation cycles (Mutti and Sonnino, 1981; Postma et al., 1993; Bourget et al., 2010), composed of simple vertical aggradation and lateral migration of the deposits. The normal stacking of hyperpycnal deposits will result in an alternation of B, S, and minor L facies types. On these conditions, Lt indices tend to be low, typically less than 15.

If the basin floor is affected by growing structures (synsedimentary deformation related to tectonics, salt, or mud diapirs), hyperpycnal sedimentation will be highly controlled by the resulting basin-floor topography. In relatively elevated areas, L facies will be abnormally common, resulting in Lt indices exceeding 35. High Lt indices in a given section could be useful criteria to recognize growing structures and to predict the occurrence of laterally equivalent thick sandstone deposits.

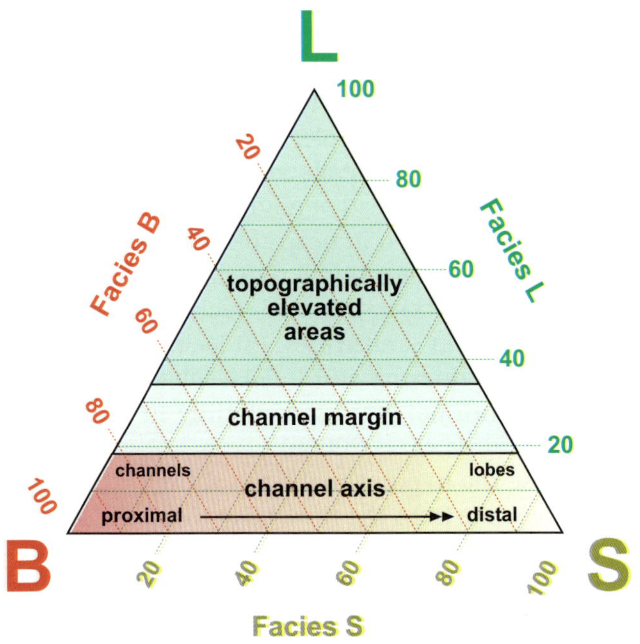

FIGURE 6. The use of ternary diagrams illustrates the relationships between bed-load, suspended-load, and lofting facies and shows the position of studied strata with respect to the system as a whole. In a single well, ternary analysis can be useful to determine sedimentary trends (progradation/retrogradation) between different depositional sequences.

Ternary Indices

In addition to the Pt and Lt facies indices, ternary indices and diagrams are useful to depict the different proportions between the three main facies categories used in this genetic analysis (B, S, and L facies) (Figure 6).

The B, S, and L indices are calculated comparing the total thickness of each category with respect to the total thickness of hyperpycnal facies (B + S + L = 100). The ternary diagrams provide several more detailed fields that are useful to analyze the position of the strata with respect to the entire hyperpycnal system.

THE MERECURE FORMATION

Overview

With a cumulative production of 50 billion bbl of oil, the upper Oligocene–lower Miocene Merecure Formation (Funkhouser et al., 1948) is one of the main oil-bearing formations in the eastern Venezuela Basin. The origin and evolution of this basal succession was highly controversial (González de Juana et al., 1980; Porras et al., 2002; Zavala et al., 2007b). These deposits are characterized by thick bedded, massive, or cross-stratified fine- to coarse-grained sandstones, with scarce and poorly diagnostic trace and body fossils, abundant plant debris, and some lignite levels. Consequently, these sandstones were previously interpreted as (1) brackish to open marine strata deposited within a fluvial-deltaic setting (González de Juana et al., 1980; Campos et al., 1985), (2) estuarine to shallow-marine deposits (Porras et al., 2002), and more recently (3) shallow (shelfal) hyperpycnal systems (Zavala et al., 2007b). The hyperpycnal interpretation of the Merecure Formation is derived from careful conventional facies analysis of two cores located in the Travi and Santa Barbara fields. Main observations that support this interpretation are as follows: (1) the common occurrence of very thick (>5 m [16 ft]) massive and laminated sandstones showing gradual vertical facies changes and facies recurrence, which suggest an accumulation from long-lived fluctuating turbulent flows (Zavala et al., 2006a; Lamb et al., 2008; Jackson and Johnson, 2009; Lamb and Mohrig, 2009); (2) the association of these coarse-grained facies with lofting rhythmites with abundant plant debris and mica, which are considered diagnostic features of marine hyperpycnal deposition (Zavala et al., 2006c, 2008b); and (3) the associated occurrence of upper bathyal foraminifers (Delgado, 2007, personal communication), which support a relatively deep fully marine environment. Abundant plant debris locally showing entire leaves is also considered to be evidence of a direct fluvial connection (Plink-Björklund and Steel, 2004; Petter and Steel, 2006; Zavala et al., 2006a).

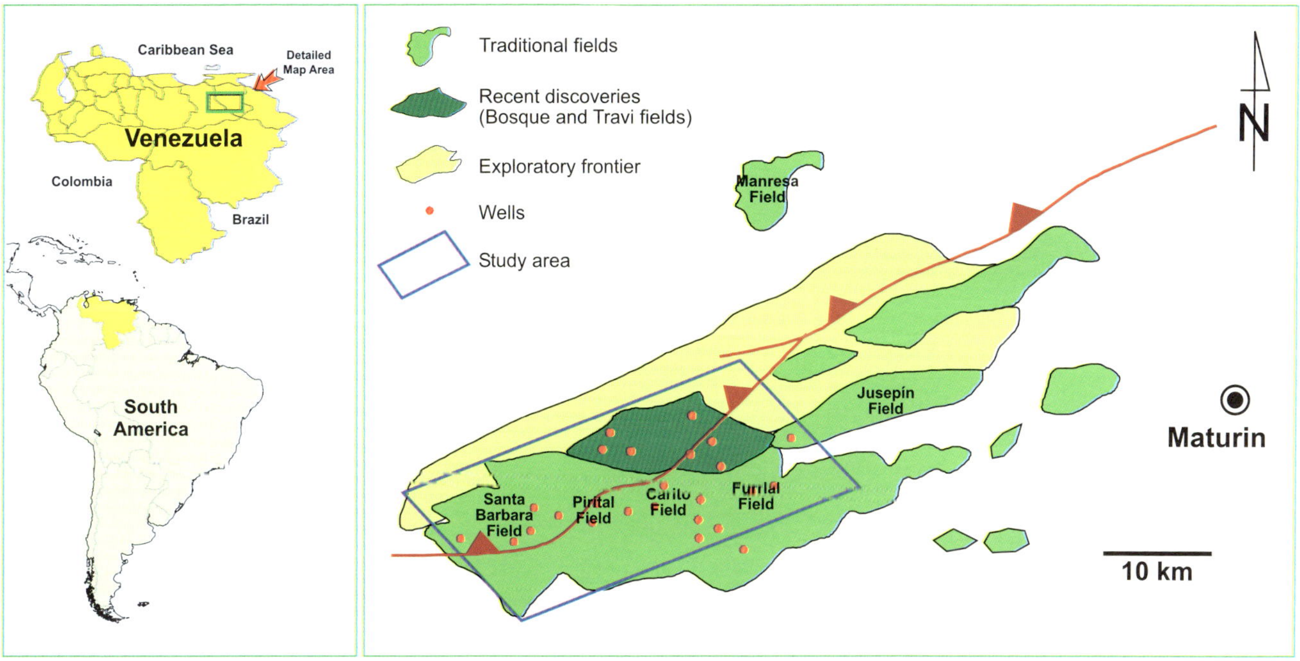

FIGURE 7. Location map of the Maturin area showing the study area and main oil fields.

This chapter provides an example of sedimentologic indices analysis applied to the Oligocene-early Miocene Merecure Formation in the northern part of the Monagas state, Venezuela. A total thickness of 3876.75 m (12,718.99 ft) of core was described and integrated with electrical logs. Facies description and classification was done according to the genetic facies scheme proposed for hyperpycnal deposits (Zavala et al., 2006b, 2011; Zavala 2008).

Geological Framework

The study area covers approximately 1000 km^2 (386 mi^2) and is located within the Maturin Subbasin, which forms the eastern part of the eastern Venezuela Basin (Figure 7). The sedimentary fill of this basin consists of about 9000 m (29,500 ft) of marine and continental deposits accumulated in a passive margin setting (Cretaceous–Eocene) followed in post-Eocene by a compressive foreland basin (Erlich and Barrett, 1994) related to the oblique collision of the Caribbean plate against the South American plate. This oblique collision zone migrated progressively eastward during the late Oligocene–earliest Miocene, dividing the foreland basin into three areas: (1) a southern platform area, (2) a central foredeep, and (3) a northern region overthrust area (Parnaud et al., 1995).

As a consequence of major foreland basin subsidence, a very thick succession of clastic sediments accumulated during the Oligocene and Miocene following a roughly east-west elongate depocenter. These clastic deposits consist of a basal coarse-grained succession (Merecure Formation) followed in sharp contact with deep-marine fine-grained accumulations (Carapita Formation). The Merecure Formation is composed of up to 500 m (1640 ft) of alternations of fine- to coarse-grained sandstones and shales. These sandstones of the Merecure Formation grade northward into a fine-grained succession (Areo Formation), interpreted as outer-shelf deposits (Stainforth, 1971; Parnaud et al., 1995). These clastic deposits were mostly sourced from the cratonic Guyana shield located to the south (Schlumberger, 1997). The Merecure Formation is partially equivalent to the Naricual Formation (Socas, 1991), which corresponds to coarse-grained deposits supplied from the north that accumulated close to an early uplifted and deformed mountain front or "serrania" (Figure 8) (Parnaud et al., 1995).

Detailed analysis of well logs and regional correlation of more than 50 wells allowed interpretation of a sequence-stratigraphic framework for the upper Oligocene–lower Miocene deposits in the area. A total of three main depositional sequences (M1, M2, and M3) were recognized and correlated. Each sequence is composed of an overall thinning and fining-upward lithogenetic unit (Figure 9). Each major sequence is in turn composed of up to three lower hierarchical elementary sequences. Main depositional sequences

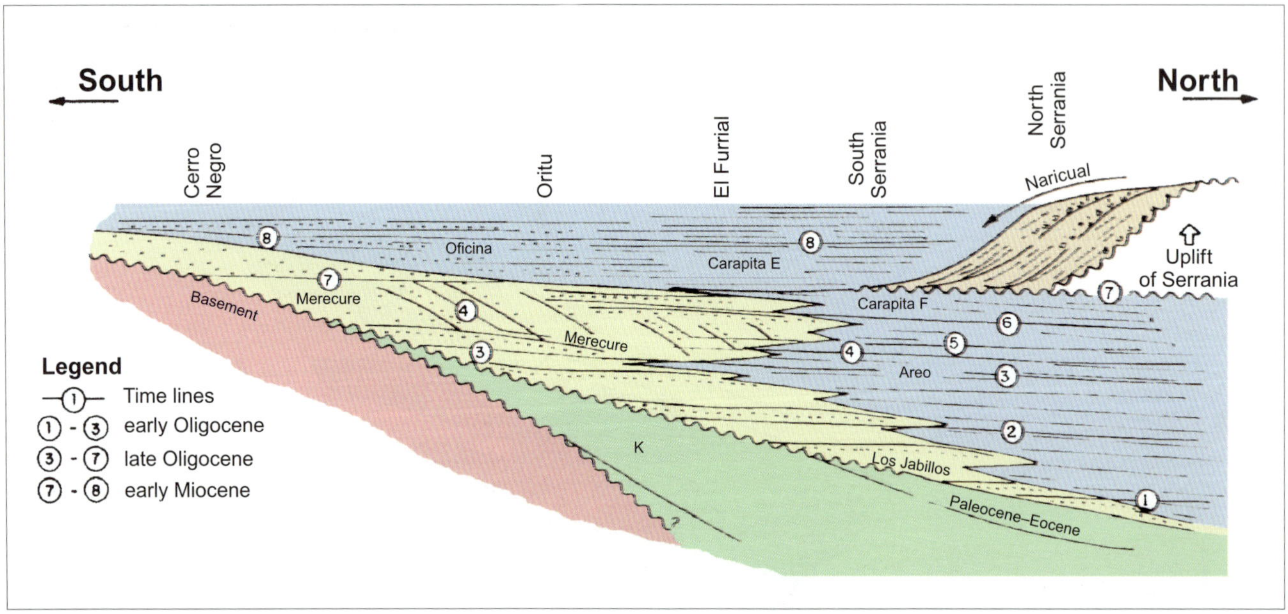

FIGURE 8. Conceptual regional cross section showing the deposition of Oligocene and early Miocene sediments. After Parnaud et al. (1995). K = Cretaceous.

start with thick deposits of medium- to coarse-grained sandstones, which evolve upward into intercalated siltstones and fine-grained sandstones, ending with massive and bioturbated mudstones. The definition of the hierarchical framework of bounding surfaces was applied in this chapter to build the basic sequence-stratigraphic schema derived from a log analysis of the internal cyclicity complemented by core analysis. In the Merecure Formation, main sequence boundaries seem not to be related to relative changes in sea level because no evidence of subaerial exposure or shallow-water indicators was found above sequence boundaries. Major sequence boundaries commonly develop over Glossifungites ichnofacies and are in turn marked by a drastic increase in the grain size and thickness of sandstone bodies. These characteristics suggest a highly variable and cyclically ordered sediment yield to the basin via hyperpycnal flows, possibly controlled by climatic and tectonic cycles (Mutti et al., 2009).

The Merecure Formation: Facies Analysis

A regional subsurface study of core and well log data from 30 wells over an area of 1000 km^2 (386 mi^2) reviewed the origin, stratigraphy, and sediment distribution of the Merecure Formation. As previously discussed, the origin of the Merecure Formation has been related to brackish to open marine conditions in fluvial-deltaic (González de Juana et al., 1980; Campos et al., 1985) to estuarine shallow-marine environments (Porras et al., 2002), although no detailed facies analysis or diagnostic features were provided to support these interpretations. In the study area, core data lack evidences of coastal to shallow-marine structures such as tidal bundles, root marks, desiccation cracks, or extensive-wave reworked strata. Most sand bodies (>80%) are composed of massive or crudely stratified beds, evidencing an origin mainly related to sustained long-lived turbulent (sediment gravity) flows (Kneller and Branney, 1995; Zavala et al., 2007b, 2008a; Sumner et al., 2008). These deposits are associated with outer neritic to upper bathyal marine fauna (Campos et al., 1985) and abundant plant remains.

A total of 3876.75 m (12,718.99 ft) of core were described and analyzed in detail, from which 13 sedimentary facies were defined (Figure 10). Main facies include massive to cross-stratified pebbly sandstones that grade basinward into massive to laminated fine-grained sandstones. With the only exception of facies P, which is related to shelfal accumulation under nonflood conditions, all other recognized facies are associated with hyperpycnal flow discharges. Facies classification was done using the genetic and predictive facies analysis of hyperpycnal systems (Figure 3) (Zavala et al., 2006b, 2011; Zavala, 2008). The 12 hyperpycnal facies types can in turn be grouped into B, S, and L processes and deposits. The characteristics of the identified facies are summarized in Figure 10, together with their average porosity.

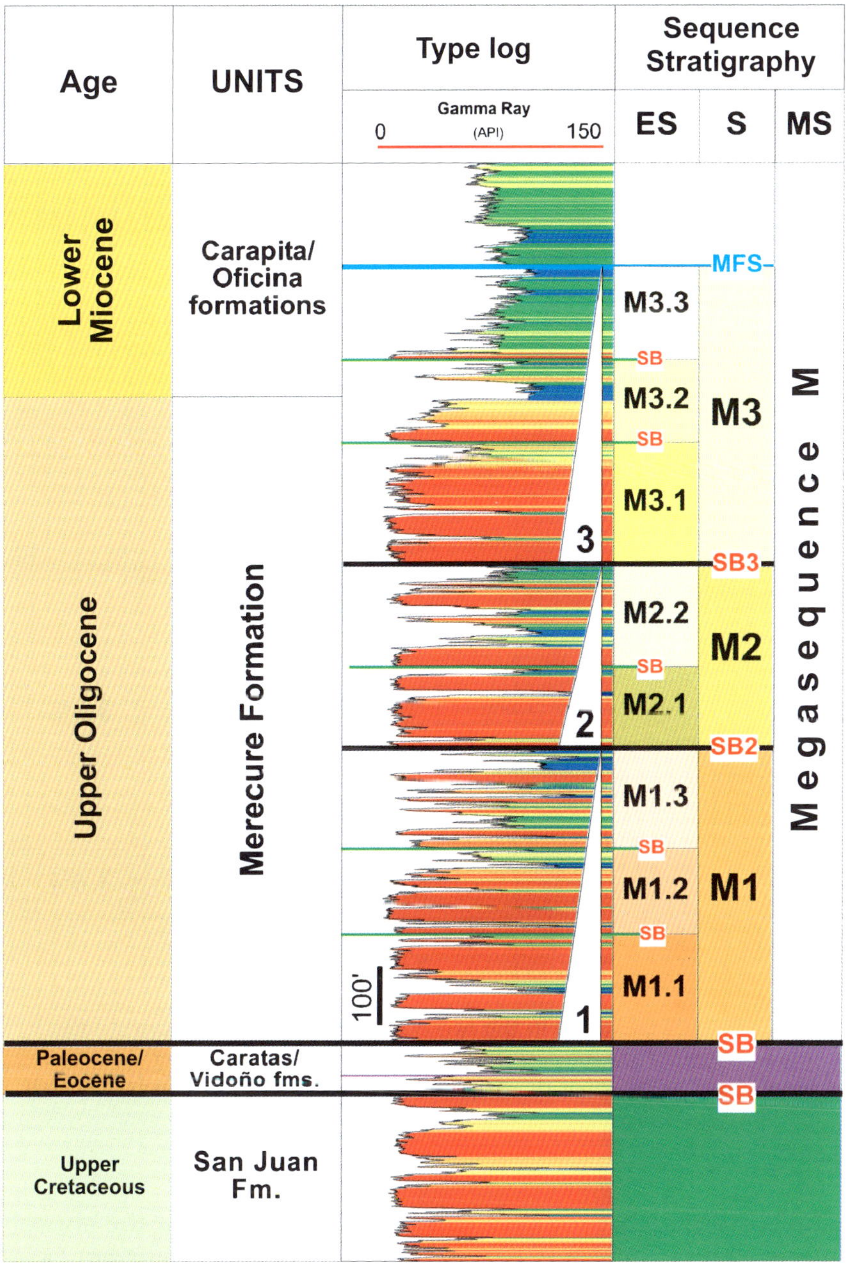

FIGURE 9. Type log of the Merecure Formation with proposed sequence stratigraphy. The 100-ft (30-ft) scale bar was indicative. ES = elementary sequence; MFS = maximum flooding surface; MS = megasequence; S = sequence; SB = sequence boundary; Fm. = Formation.

The B-type facies category includes coarse-grained sandstones and fine-grained conglomerates with abundant sandstone matrix. Main facies types of this category are massive (facies B1; Figure 11A, B), low-angle cross-bedded (facies B2; Figure 11C, D), and roughly laminated (facies B3; Figure 11E, F) coarse-grained deposits, commonly composing thick beds showing gradual and recurrent facies changes.

The S-type facies category is composed of fine-grained sandstones. Main types of S facies include massive (facies S1; Figure 12A, B), laminated (facies S2), and rippled (facies S3) deposits. Massive sandstones are very common, typically displaying isolated *Ophiomorpha* and *Thalassinoides* burrows related to doomed pioneers (Grimm and Föllmi, 1994). Laminated sandstones commonly show small-scale hummocky cross-stratification (HCS) (facies S2h; Figure 12C), which suggest combined flow conditions in shelfal areas.

The L-type facies category (Figure 12E–G) is the result of the fallout of very fine-grained sands, silts, plant debris, and micas from lofting plumes mainly in flow-margin areas. The transition from the flow axis to flow margin is characterized by massive sandstone beds interbedded with beds containing abundant plant debris (facies S1L; Figure 12D).

	Facies	Lithology	Sedimentary Structures	Origin	Thickness (m)	Porosity (0–20)
Bed Load	B1	Matrix-supported fine-grained conglomerates	Massive, clay chips	Bed load related to the overpassing of long-lived turbulent flows	0.1 – 0.9	8.0% (1; 8)
Bed Load	B2	Fine-grained conglomerates and conglomeratic sandstones. Clay chips	Low-angle cross-stratification	Bed load. Migration of asymptotic dunes	0.2 – 7	10.4% (5; 15)
Bed Load	B2s	Medium- to coarse-grained sandstones. Abundant clay chips	Low-angle cross-stratification	Bed load. Migration of asymptotic dunes	0.2 – 1.5	
Bed Load	B3	Fine-grained conglomerates and conglomeratic sandstones. Clay chips	Diffuse horizontal lamination, Aligned clasts	Bed load. Drag and rolling	0.12 – 14.75	13.2% (35; 25)
Bed Load	B3h	Fine-grained conglomerates and conglomeratic sandstones. Clay chips	Low-angle lamination (hummocky-like)	Bed load. Drag and rolling. Combined flow	0.2 – 3	
Traction Plus Fallout	S1	Medium- to fine-grained sandstones	Massive, small clay chips, dish	Progressive aggradation from quasi-steady turbulent flows	0.1 – 9.5	6.8% (20; 18)
Traction Plus Fallout	S1L	Medium- to fine-grained sandstones. Discontinuous silt levels	Massive, clay chips, dish, load structures, plastic clay chips	Progressive aggradation from quasi-steady turbulent flows (levee transition)	0.1 – 3.6	
Traction Plus Fallout	S2	Medium- to fine-grained sandstones	Planar lamination	Progressive aggradation from quasi-steady turbulent flows	0.1 – 3	3.3% (3; 8)
Traction Plus Fallout	S2h	Medium- to fine-grained sandstones	Low-angle lamination (hummocky-like)	Progressive aggradation from quasi-steady flows (combined flows)	0.1 – 1.5	
Traction Plus Fallout	S2L	Fine-grained sandstones. Levels of silts and plant debris	Planar lamination	Progressive aggradation from quasi-steady turbulent flows (levee transition)	0.1 – 0.3	
Traction Plus Fallout	S3	Fine-grained sandstones	Climbing ripples	Traction plus fallout in quasi-steady turbulent flows	0.1 – 0.3	
Lofting	L	Fine-grained sandstones and siltstones, plant debris, and micas	Massive to laminated	Fallout from lofting plumes	0.1 – 6	3.4% (10; 15)
	P	Fine-grained sandstones and mudstones	Massive to laminated. Highly bioturbated	Fallout (offshore/prodelta - shoreface)	0.1 – 0.3	

FIGURE 10. Main characteristics of the sedimentary facies identified in the Merecure Formation from core studies. The column on the right shows the average porosity measured for each facies. The number in the lower right corner indicates the maximum value of porosity for all samples. The number of analyses conducted for each facies is indicated in the lower left.

Individual sandstone bodies can be up to 10 m (33 ft) thick and are arranged into different hierarchical orders. Facies analysis suggests an origin related to the accumulation from sustained and long-lived turbulent flows derived from subaerial discharges (hyperpycnal flows) in a shelfal marine setting.

The analysis of the relative abundance of B, S, and L facies allowed the definition of Pt and Lt facies indices for facies prediction (Zavala et al., 2007b, 2008a).

Analysis of Genetic Indices in the Merecure Formation

The description and classification of core facies was performed according to the genetic facies tract illustrated in Figure 5 using Excel spreadsheets especially designed for this project. Excel spreadsheets were very useful because they allowed the fast calculation of the total thickness of the different facies categories under study. The thickness of B, S, and L facies, and the corresponding determination of the Pt and Lt facies indices, were done within the proposed sequence-stratigraphic framework. The use of a sequence-stratigraphic framework allowed the step-by-step analysis of deposition of the entire succession.

The genetic facies tract description was used in construction of genetic facies logs, which were integrated with lithologic electric logs (gamma ray [GR], neutron log [NPHI], and density log [RHOB]; Figure 13). This step was crucial to determine if the core thicknesses were representative to characterize each sequence, whether or not it was cored.

In general, the genetic facies log shows a direct correspondence with the GR log. The lowest GR values

FIGURE 11. The B-type facies in the Merecure Formation. (A, B) Massive fine-grained conglomerates and coarse-grained sandstones with imbricated clasts (facies B1). (C, D) Coarse-grained sandstones and fine-grained conglomerates with low-angle and asymptotic cross-bedding, interpreted as foreset (f) lamination of migrating dunes (facies B2). (E, F) Coarse-grained deposits with levels of aligned clasts (facies B3) related to bed-load processes at the base of an overpassing turbulent flow. Black and white boxes in the scale bar are 5 mm each.

characterize type B facies. Consequently, in cases where the analyzed core interval was not considered representative or in intervals lacking core, the Pt and Lt indices were estimated from the GR log shape of the nearest wells having core data in the same interval.

Table 1 shows the final Pt and Lt indices calculated for each depositional sequence in the 30 analyzed wells. White boxes display data calculated using long cores, which are considered most representative of the sequence. The yellow boxes contain data from cores having a recorded thickness that is not considered representative and consequently has been adjusted by comparison with adjacent cored wells. The blue boxes contain estimates by log analysis from wells without core data.

The sedimentologic indices (Pt and Lt) were then plotted as maps for each sequence (Figure 14). Because Tertiary structural deformation has displaced the wells into different fault blocks, paleogeographic maps were constructed using Oligocene coordinates calculated from palinspastic reconstructions (Table 1). The analysis of genetic indices shows that Lt facies indices range between 3 and 42. An Lt facies index greater than 35 suggests the local occurrence of tectonically uplifted areas (Figure 6). Proximal indices range from 73 to 30 and suggest a proximal-to-medial position along the depositional profile (Figure 6). The analysis of genetic indices revealed that the Merecure Formation could be more extensive than previously considered, with a main sediment source from cratonic areas located toward the south and southeast (Figure 14). Mapping of the lateral facies and proximal indices suggests the existence of at least three zones of clastic input, which persist with some variations for the three identified sequences.

FIGURE 12. The S- and L-type facies in the Merecure Formation. (A, B) Massive fine-grained sandstones (facies S1) deposited from sustained turbulent flows. Note in panel B the *Ophiomorpha* burrows (Oph), which are interpreted as doomed pioneers. (C) Fine-grained sandstones with small-scale hummocky cross-stratification (facies S2h). (D) Massive fine-grained sandstones (facies S1) interbedded with carbonaceous levels (c) with micas (facies S1L) deposited in the flow transition zone. (E, F, G) Thin intercalations of fine-grained sandstones and siltstones with abundant plant debris (pd) and micas (facies L) deposited at flow margins. Sandstone levels show load structures (arrow) indicating soft sediment deformation. Bioturbation is scarce and characterized by rare *Palaeophycus.* Black and white boxes in the scale bar are 5 mm each.

These zones of clastic supply are located toward the southwest, south, and southeast.

Index mapping also suggests a syntectonic accumulation during the Oligocene, which controlled the subaqueous topography and the distribution of sandstone bodies. Figure 15 shows a possible depositional scenario for the accumulation of the hyperpycnal deposits of the Merecure Formation.

Petrographic data and conventional analysis of rock samples suggest a close relation between reservoir properties and facies. The B facies types show the best overall petrophysical properties, with porosities up to 25% (Figure 10). The S1 facies have good porosities related to their original open packing, but permeability is reduced by compaction and infiltration of fine-grained materials into pore spaces. In decreasing order of petrophysical quality, the best facies are the B3, followed by facies B2, B1, S1, L, and S2–S3. As B-type facies have the best petrophysical properties, the facies maps of Figure 14 can be used to predict reservoir quality in undrilled areas, with an important role in diminishing risks of new ventures.

CONCLUSIONS

The application of geographically controlled sedimentologic indices to quantify the relationships between hyperpycnal-flow-related facies has proven to be a very useful technique in the analysis of Merecure Formation reservoir strata.

When analyzed in the context of a genetic facies tract, hyperpycnal-flow-related facies have characteristics that allow the quantification of geographically controlled depositional processes, which can be determined from field and core studies. The determination

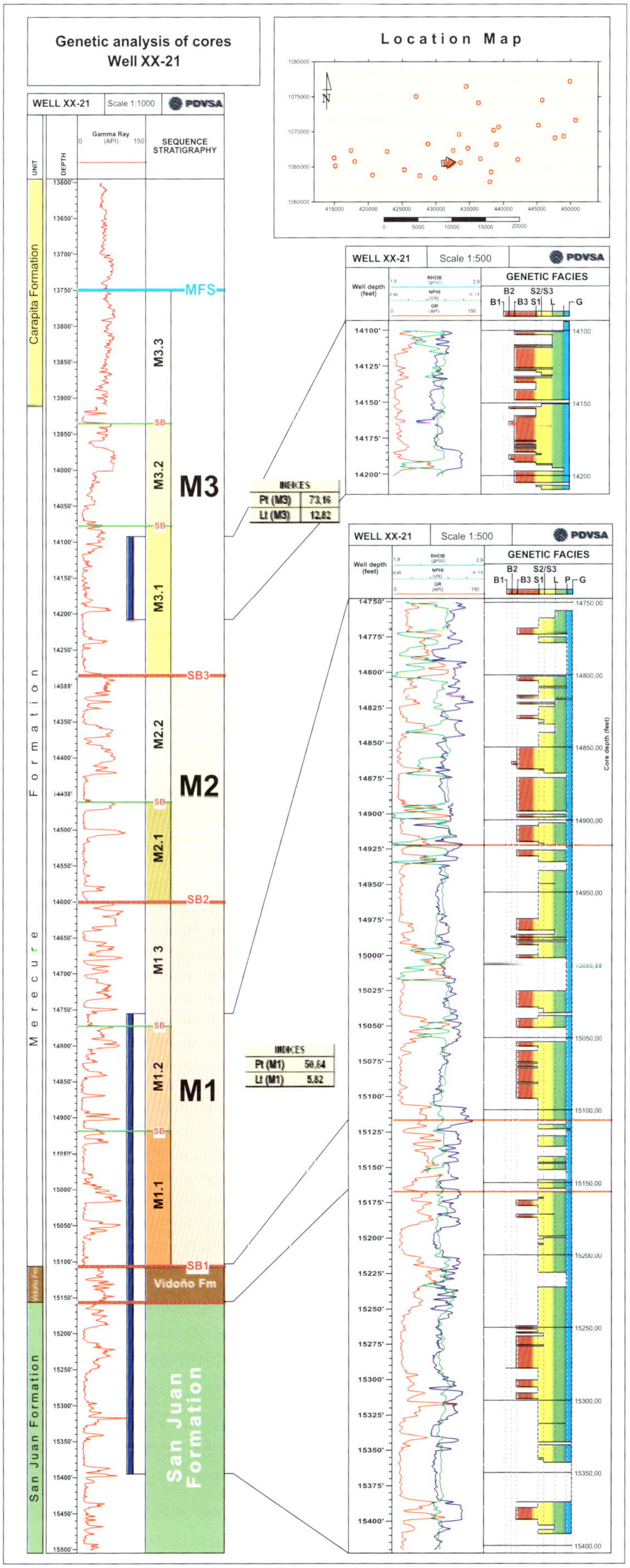

FIGURE 13. Example of a sedimentologic sheet used for the genetic indices analysis. The log on the left shows the cored intervals (purple bars), which are displayed on the genetic facies log of the right. SB = sequence boundary; MFS = maximum flooding surface; Fm = Formation; Gr = gamma ray; NPHI = neutron log; RHOB = density log.

Table 1. Calculated (white), corrected (yellow) and estimated (blue) proximal (Pt) and lateral facies (Lt) indices of the M1, M2 and M3 sequences in the analyzed wells. The two columns on the right display the coordinates during the Oligocene calculated from palinspastic reconstructions.

WELL	Pt (M1)	Lt (M1)	Pt (M2)	Lt (M2)	Pt (M3)	Lt (M3)	R COORD X	R COORD Y
1	46.67	5.36	60.61	18.01	71.72	7.82	413492.38	1078465.55
2	46.67	6.00	61.00	18.01	72.00	7.82	413603.66	1078141.77
3	54.00	14.00	45.00	16.00	72.00	8.00	416417.68	1077624.09
4	53.85	9.00	45.00	16.00	65.00	8.00	419190.85	1074409.15
5	55.00	15.00	51.05	26.28	60.00	9.00	415822.87	1080796.65
6	26.88	43.01	47.67	10.85	37.99	13.08	420742.38	1085084.76
7	no well log		61.00	18.00	59.89	5.68	426878.66	1084021.65
8	58.54	25.90	56.32	11.00	55.00	8.00	423085.37	1082084.76
9	36.03	3.58	55.00	10.00	51.00	10.00	425316.46	1080330.49
10	37.66	6.00	60.00	10.00	48.00	10.00	427487.20	1079065.85
11	45.00	12.00	40.00	24.00	29.52	15.00	425020.12	1090661.59
12	50.00	10.00	56.00	11.00	73.39	9.00	434077.44	1092925.61
13	50.00	10.00	56.00	10.00	73.39	9.55	431722.56	1095463.11
14	55.00	10.00	40.00	24.00	39.39	16.80	431499.00	1086249.00
15	68.37	11.04	52.07	14.38	38.52	5.67	436962.20	1089980.79
16	60.98	6.32	52.00	14.00	30.00	20.00	436923.17	1088394.51
17	63.72	17.92	52.00	12.00	35.00	8.00	437112.80	1086606.71
18	45.00	15.00	43.00	24.00	40.00	12.00	430703.96	1081758.84
19	62.71	8.21	55.00	20.00	65.00	12.00	434398.48	1083513.41
20	62.52	4.08	40.62	24.48	63.47	12.58	432556.10	1084565.55
21	50.84	5.82	50.00	11.00	73.16	12.82	432181.10	1081875.91
22	41.33	4.36	56.00	10.00	65.00	9.30	435728.66	1080481.71
23	60.00	10.00	75.60	2.79	65.00	9.30	435892.99	1082344.51
24	55.00	15.00	74.60	9.15	38.85	6.08	439954.27	1084228.66
25	62.00	7.00	64.96	0.00	82.00	3.88	442880.18	1089617.99
26	62.90	7.20	79.00	2.16	90.00	3.78	445299.09	1087482.01
27	62.19	11.27	81.98	5.29	83.67	3.59	446577.13	1087885.06
28	60.18	8.10	48.18	6.81	55.50	1.66	448351.22	1091022.26
29	60.00	8.00	58.36	7.00	80.00	2.00	444580.49	1094153.35
30	30.00	7.00	58.00	7.00	0.00	3.00	448890.55	1101730.18

and mapping of Pt and Lt indices allow prediction of facies distribution, source area, topographic controls on sedimentation, and when integrated with petrophysical analysis, reservoir quality.

The genetic indices procedure described in this chapter was exclusively developed for the analysis of hyperpycnal systems. Consequently, the interpretation of a hyperpycnal system as the main depositional system must first be documented by careful and detailed conventional facies analysis. In addition, the analysis of genetic indices must be done within a sequence-stratigraphic framework to identify strata from the same stratigraphic (coeval) interval. It is necessary to point out that the genetic facies analysis and classification do not replace conventional core description and facies analysis, which are crucial to determine the environment of deposition. Also, it is important that cores analyzed are of representative thicknesses of the sequence studied.

Facies analysis conducted on 3876.75 m (12,718.99 ft) of Oligocene to lower Miocene Merecure Formation cores from 30 wells located in the Maturin Subbasin, Venezuela, provided the database to review its origin, facies distribution, and internal architecture. Facies

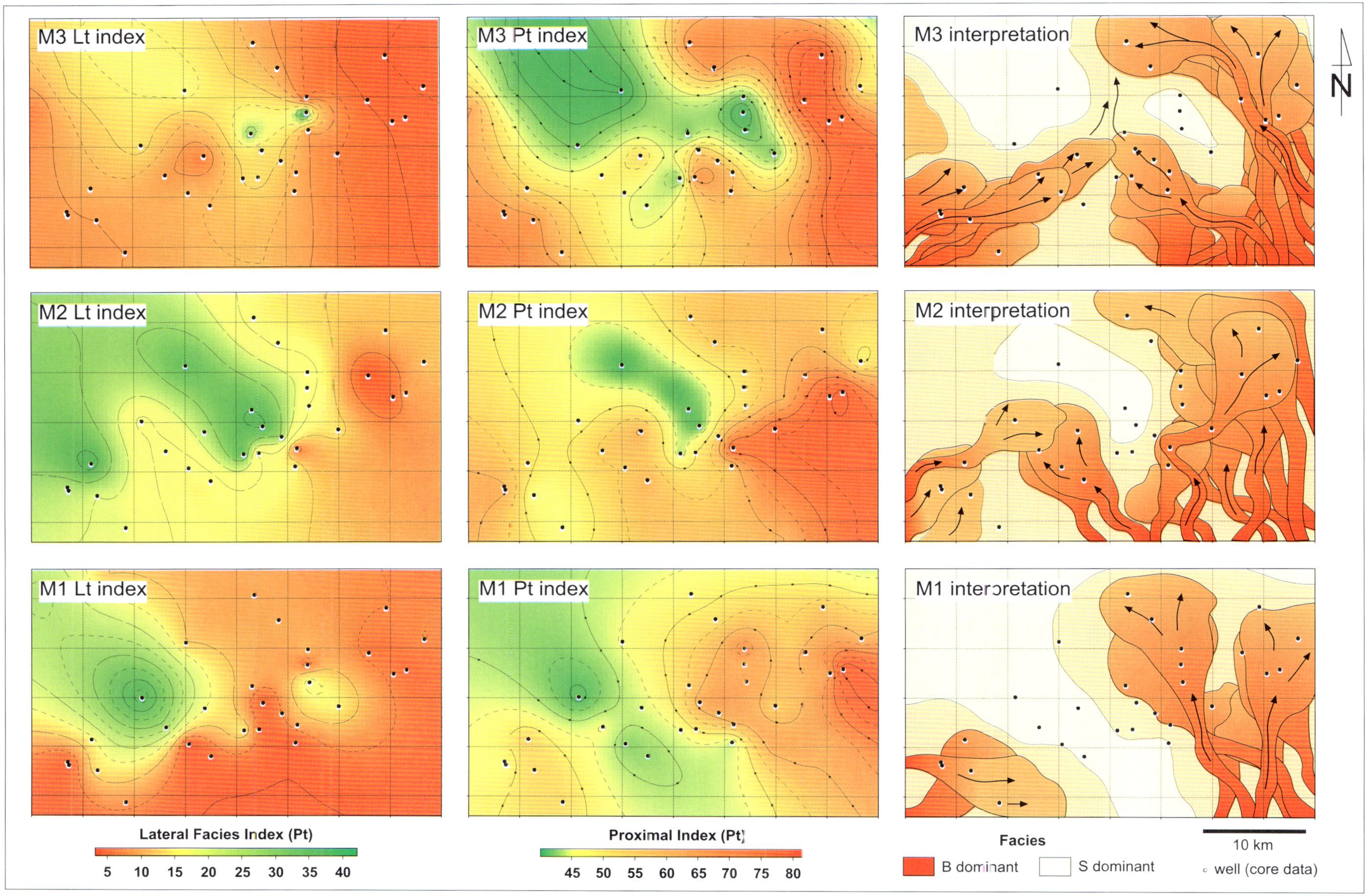

FIGURE 14. Lateral facies (Lt) and proximal indices maps (left) and their related interpretation (right) showing the distribution of B-type facies. Because B-type facies display the best petrophysical properties, these maps can be used as a proxy to predict the distribution of high-quality reservoirs.

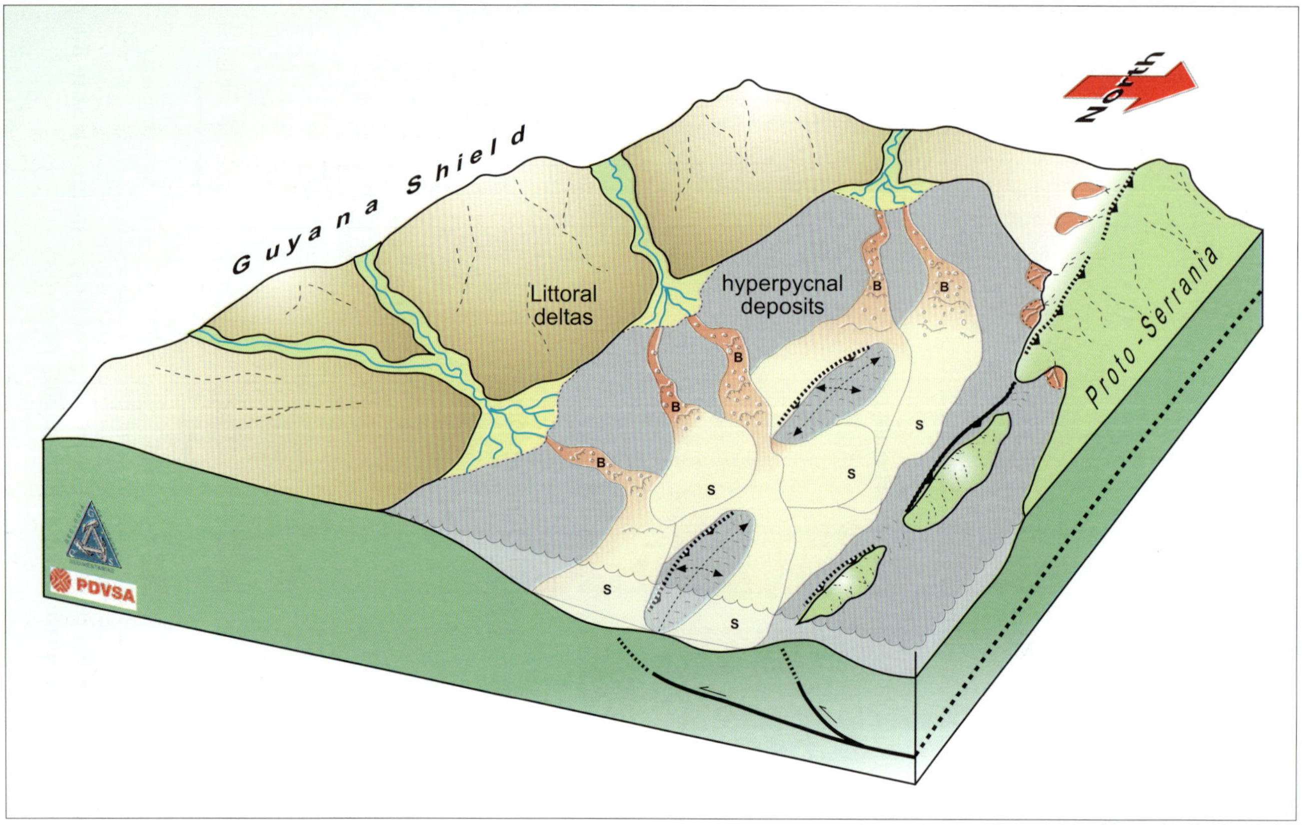

FIGURE 15. Proposed depositional model for the accumulation of sandstone bodies of the Merecure Formation by hyperpycnal flows triggered from littoral deltas located toward the south and southwest. The accumulation of these sandstones was controlled by a subaqueous topography resulting from early growing tectonic structures associated with the "proto-serrania."

analysis suggests that the strata were accumulated in a shelfal marine environment by long-lived turbulent flows linked to direct fluvial discharges (hyperpycnal systems). Twelve hyperpycnal-related sedimentary facies were recognized, which can be grouped into three main facies categories related to B, S, and L processes. The Pt and Lt indices for all three identified sequences (M1, M2, and M3) suggest the existence of three zones of clastic input located toward the southwest, south, and southeast, which persist with some variations for the three identified sequences. The Pt indices range from 73 to 30 and suggest for the entire system a proximal-to-medial position along the depositional profile. The Lt indices can be up to 42 in some specific areas, indicating synsedimentary growth structures that deeply controlled the location of main sandstone bodies. The analysis of the genetic indices suggests that the system may be larger than previously considered. Because in the system under study B facies types display the best petrophysics, the Pt map could be used to estimate the reservoir quality in undrilled areas.

ACKNOWLEDGMENTS

We thank PDVSA for permitting publication of the methodology used for the analysis of the Merecure Formation in the subsurface. Roger Slatt and an anonymous reviewer made important suggestions that substantially improved the contents of this chapter.

REFERENCES CITED

Bates, C., 1953, Rational theory of delta formation: AAPG Bulletin, v. 37, p. 2119–2162.

Bouma, A. H., 1962, Sedimentology of some flysch deposits: A graphic approach to facies interpretation: Amsterdam, Netherlands, Elsevier, 168 p.

Bourget, J., S. Zaragosi, T. Mulder, J. L. Schneider, T. Garlan, A. Van Toer, V. Mas, and N. Ellouz-Zimmermann, 2010, Hyperpycnal-fed turbidite lobe architecture and recent sedimentary processes: A case study from the Al Batha turbidite system, Oman margin: Sedimentary Geology, v. 229, p. 144–159, doi:10.1016/j.sedgeo.2009.03.009.

Campos, V., R. Lander, and S. Cabrera, 1985, Evolución

estructural en el noreste de Anzoátegui y su relación con el norte de Monagas: Memorias, VI Congreso Geológico, Venezolano, Caracas, Sociedad Venezolana de Geología, v. 4, p. 2397–2414.

Cheng-Shing, C., and Y. Ho-Shing, 2008, Evidence of hyperpycnal flows at the head of the meandering Kaoping Canyon off SW Taiwan: Geo-Marine Letters, v. 28, p. 161–169, doi:10.1007/s00367-007-0098-7.

Erlich, R. N., and S. F. Barrett, 1994, Petroleum geology of the eastern Venezuela foreland basin: AAPG Memoir 55, p. 341–362.

Funkhouser, H. J., L. C. Sass, and H. D. Hedberg, 1948, Santa Ana, San Joaquín, Guárico and Santa Rosa oil fields (Anaco fields), central Anzoátegui, Venezuela: AAPG Bulletin, v. 32, p. 1851–1908.

González de Juana, C., J. Iturralde de Arozena, and X. Picard, 1980, Geología de Venezuela y de sus Cuencas Petrolíferas: Caracas, Venezuela, Ediciones Foninves, 1021 p.

Grimm, K. A., and K. B. Föllmi, 1994, Doomed pioneers: Allochthonous crustacean tracemakers in anaerobic basinal strata, Oligo-Miocene San Gregorio Formation, Baja California Sur, Mexico: Palaios, v. 9, p. 313–334, doi:10.2307/3515054.

Haner, B. E., 1971, Morphology and sediments of Redondo submarine fan, southern California: Geological Society of America Bulletin, v. 82, p. 2413–2432, doi:10.1130/0016-7606(1971)82[2413:MASORS]2.0.CO;2.

Hesse, R., H. Rashid, and S. Khodabakhsh, 2004, Fine-grained sediment lofting from meltwater-generated turbidity currents during Heinrich events: Geology, v. 32, p. 449–452, doi:10.1130/G20136.1.

Jackson, C. A. L., and H. D. Johnson, 2009, Sustained turbidity currents and their interaction with debrite-related topography; Labuan Island, offshore NW Borneo, Malaysia: Sedimentary Geology, v. 219, p. 77–96, doi:10.1016/j.sedgeo.2009.04.008.

Knapp, R. T., 1943, Density currents: Their mixing characteristics and their effect on the turbulence structure of the associated flow: Proceedings of the Second Hydraulics Conference, Bulletin 27, University of Iowa Studies in Engineering, p. 289–306.

Kneller, B., and M. Branney, 1995, Sustained high-density turbidity currents and the deposition of thick massive sands: Sedimentology, v. 42, p. 607–616, doi:10.1111/j.1365-3091.1995.tb00395.x.

Kneller, B., and C. Buckee, 2000, The structure and fluid mechanics of turbidity currents: A review of some recent studies and their geological implications: Sedimentology, v. 47, p. 62–94, doi:10.1046/j.1365-3091.2000.047s1062.x.

Lamb, M. P., and D. Mohrig, 2009, Do hyperpycnal-flow deposits record river-flood dynamics?: Geology, v. 37, p. 1067–1070, doi:10.1130/G30286A.1.

Lamb, M. P., P. M. Myrow, C. Lukens, K. Houck, and J. Strauss, 2008, Deposits from wave-influenced turbidity currents: Pennsylvanian Minturn Formation, Colorado: Journal of Sedimentary Research, v. 78, p. 480–498, doi:10.2110/jsr.2008.052.

Lovell, J. P. B., 1969, Tyee Formation: A study of proximality in turbidites: Journal of Sedimentary Petrology, v. 39, p. 935–953.

Lovell, J. P. B., 1970, The palaeogeographical significance of lateral variations in the ratio of sandstone to shale and other features of the Aberystwyth Grits: Geological Magazine, v. 107, p. 147–158, doi:10.1017/S0016756800055503.

Macdonald, D. I. M., 1986, Proximal to distal sedimentological variation in a linear turbidite trough: Implications for the fan model: Sedimentology, v. 33, p. 243–259, doi:10.1111/j.1365-3091.1986.tb00534.x.

McCabe, P. J., and B. Waugh, 1973, Wenlock and Ludlow sedimentation in the Austwick and Horton-in-Ribblesdale Inlier, NW Yorkshire: Proceedings of the Yorkshire Geological Society, v. 39, p. 445–470.

Mulder, T., and J. Alexander, 2001, The physical character of subaqueous sedimentary density flows and their deposits: Sedimentology, v. 48, p. 269–299, doi:10.1046/j.1365-3091.2001.00360.x.

Mulder, T., and J. P. M. Syvitski, 1995, Turbidity current generated at river mouths during exceptional discharges to the world oceans: Journal of Geology, v. 103, p. 285–299, doi:10.1086/629747.

Mulder, T., J. P. M. Syvitski, S. Migeon, J. C. Faugéres, and B. Savoye, 2003, Marine hyperpycnal flows: Initiation, behavior and related deposits, A review: Marine and Petroleum Geology, v. 20, p. 861–882, doi:10.1016/j.marpetgeo.2003.01.003.

Mutti, E., and W. R. Normark, 1987, Comparing examples of modern and ancient turbidite systems: Problems and concepts, *in* J. K. Leggett and G. G. Zuffa, eds., Marine clastic sedimentology: Concepts and case studies: London, Graham and Trotman, p. 1–38.

Mutti, E., and F. Ricci Lucchi, 1972, Le torbiditi dell'Appennino Settentrionale: Introduzione all'analisi di facies: Memorie della Società Geologica Italiana, v. 11, p. 161–199.

Mutti, E., and M. Sonnino, 1981, Compensation cycles: A diagnostic feature of turbidite sandstone lobes (abs.): Bologna, Italy, 2d European Regional Meeting of International Association of Sedimentologists, p. 120–123.

Mutti, E., G. Davoli, R. Tinterri, and C. Zavala, 1996, The importance of ancient fluvio-deltaic systems dominated by catastrophic flooding in tectonically active basins: Memorie di Scienze Geologiche, Universita di Padova, Padova, Italy, v. 48, p. 233–291.

Mutti, E., N. Mavilla, S. Angella, and L. Fava, 1999, An introduction to the analysis of ancient turbidite basins from an outcrop perspective: AAPG Continuing Education Course Note 39, p. 1–98.

Mutti, E., R. Tinterri, G. Benevelli, D. Di Biase, and G. Cavanna, 2003, Deltaic, mixed and turbidite sedimentation of ancient foreland basins: Marine and Petroleum Geology, v. 20, p. 733–755, doi:10.1016/j.marpetgeo.2003.09.001.

Mutti, E., D. Bernoulli, F. Ricci Lucchi, and R. Tinterri, 2009,

Turbidites and turbidity currents from Alpine "flysch" to the exploration of continental margins: Sedimentology, v. 56, p. 267–318, doi:10.1111/j.1365-3091.2008.01019.x.

Nakajima, T., 2006, Hyperpycnites deposited 700 km away from river mouths in the Central Japan Sea: Journal of Sedimentary Research, v. 76, p. 59–72.

Natland, M. L., and P. H. Kuenen, 1951, Sedimentary history of the Ventura Basin, California, and the action of turbidity currents, *in* J. L. Haugh, ed., Turbidity currents and the transportation of course sediments into deep water: SEPM Special Publication 2, p. 76–107.

Nelson, C. H., W. R. Normark, A. H. Bouma, and P. R. Carlson, 1978, Thin-bedded turbidites in modern submarine canyons and fans, *in* D. J. Stanley and G. Kelling, eds., Sedimentation in submarine canyons, fans and trenches: Stroudsburg, Pennsylvania, Dowden, Hutchinson and Ross, p. 177–189.

Nilsen, T. H., 1980, Modern and ancient submarine fans: Discussion of papers by R. G. Walker and W. R. Normark: AAPG Bulletin, v. 64, p. 1094–1101.

Normark, W. R., and D. J. W. Piper, 1991, Initiation processes and flow evolution of turbidity currents: Implications for the depositional record, *in* R. H. Osborne, ed., From shoreline to abyss: Contribution in marine geology in honor of Francis Parker Shepard: SEPM Special Publication 46, p. 207–230.

Parnaud, F., Y. Gou, J.-C. Pascual, I. Truskowski, O. Gallango, H. Passalacqua, and F. Roure, 1995, Petroleum geology of the central part of the eastern Venezuela Basin, *in* A. J. Tankard, R. S. Suárez, and H. J. Welsink, eds., Petroleum basins of South America: AAPG Memoir 62, p. 741–756.

Petter, A. L., and R. J. Steel, 2005, Deepwater-slope channels and hyperpycnal flows from the Eocene of the Central Spitsbergen Basin: Predicting basin-floor sands from a shelf edge/upper slope perspective, AAPG Annual Meeting, June 19–22, 2005, Calgary, Alberta.

Plink-Björklund, P., and R. J. Steel, 2004, Initiation of turbidite currents: Outcrop evidence for Eocene hyperpycnal flow turbidites: Sedimentary Geology, v. 165, p. 29–52, doi:10.1016/j.sedgeo.2003.10.013.

Porras, J. S., E. L. Vallejo, D. Marchal, and C. Selva, 2002, Extensional folding in the eastern Venezuela basin: Examples from fields of Oritupano-Leona block: AAPG Annual Meeting Program, March 10–13, 2002, Search and Discovery Article 50003 (2003), 7 p.

Postma, G., F. J. Hilgen, and W. J. Zachariasse, 1993, Precession punctuated growth of a late Miocene submarine-fan lobe on Gavdos (Greece): Terra Nova, v. 5, p. 438–444, doi:10.1111/j.1365-3121.1993.tb00281.x.

Schlumberger, 1997, WEC, Venezuela, Evaluación de Pozos: Caracas, Venezuela, Schlumberger, 397 p.

Schumm, S. A., 1977, The fluvial system: New York, John Wiley and Sons, 338 p.

Schumm, S. A., 1981, Evolution and response of the fluvial system: Sedimentologic implications: SEPM Special Publication 31, p. 19–29.

Socas, M. M., 1991, Estudio sedimentológico de la Formación Naricual, estado Anzoátegui: Tesis de Grado, Universidad Central de Venezuela, Caracas, 302 p.

Sparks, R. S. J., R. T. Bonnecaze, H. E. Huppert, J. R. Lister, M. A. Hallworth, J. Phillips, and H. Mader, 1993, Sediment-laden gravity currents with reversing buoyancy: Earth and Planetary Science Letters, v. 114, p. 243–257, doi:10.1016/0012-821X(93)90028-8.

Stainforth, R. M., 1971, La Formación Carapita de Venezuela oriental: IV Congreso Geológico Venezolano, Memoria, Boletín Geológico, Caracas, Publicación Especial 5, v. 1, p. 433–463.

Sumner, E. J., L. A. Amy, and P. J. Talling, 2008, Deposit structure and processes of sand deposition from decelerating sediment suspensions: Journal of Sedimentary Research, v. 78, p. 529–547, doi:10.2110/jsr.2008.062.

Walker, R. G., 1967, Turbidite sedimentary structures and their relationship to proximal and distal depositional environments: Journal of Sedimentary Petrology, v. 37, p. 25–43.

Walker, R. G., 1978, Deep water sandstone facies and ancient submarine fans: Models for exploration for stratigraphic traps: AAPG Bulletin, v. 62, p. 932–966.

Wynn, R. B., N. H. Kenyon, D. G. Masson, D. A. V. Stow, and P. P. E. Weaver, 2002, Characterization and recognition of deep-water channel-lobe transition zones: AAPG Bulletin, v. 86, p. 1441–1462.

Zavala, C., 2008, Toward a genetic facies tract for the analysis of hyperpycnal deposits, *in* J. J. Ponce and E. B. Olivero, conveners, Sediment transfer from shelf to deepwater—Revisiting the delivery mechanisms: Conference Proceedings, AAPG Search and Discovery Article 50075, AAPG Hedberg Conference, March 3–7, 2008, Ushuaia-Patagonia, Argentina, 2 p., http://www.searchanddiscovery.com/documents/2008/jw0805zavala/ (accessed July 15, 2011).

Zavala, C., J. Ponce, M. Arcuri, D. Drittanti, H. Freije, and M. Asensio, 2006a, Ancient lacustrine hyperpycnites: A depositional model from a case study in the Rayoso Formation (Cretaceous) of west-central Argentina: Journal of Sedimentary Research, v. 76, p. 40–58.

Zavala, C., M. Arcuri, and H. Gamero, 2006b, Toward a genetic model for the analysis of hyperpycnal systems (abs.): Geological Society of America Abstracts with Programs, v. 38, no. 7, p. 541.

Zavala, C., H. Gamero, and M. Arcuri, 2006c, Lofting rhythmites: A diagnostic feature for the recognition of hyperpycnal deposits (abs.): Geological Society of America Abstracts with Programs, v. 38, no. 7, p. 541.

Zavala, C., M. Arcuri, H. Gamero, and C. Contreras, 2007a, The composite bed: A new distinctive feature of hyperpycnal deposition (abs.): AAPG Annual Convention and Exhibition, v. 16, p. 157.

Zavala, C., J. Marcano, and J. Carvajal, 2007b, Proximity and laterality indices: A new tool for the analysis of ancient hyperpycnal deposits in the subsurface: Proceedings of the 4th Geological Society of Trinidad and Tobago Geological Conference, 3 p.

Zavala, C., J. Carvajal, J. Marcano, and M. Delgado, 2008a, Sedimentological indices: A new tool for regional studies of hyperpycnal systems, *in* J. J. Ponce and E. B. Olivero, conveners, Sediment transfer from shelf to deepwater—Revisiting the delivery mechanisms: Conference Proceedings, AAPG Search and Discovery Article 50076, AAPG Hedberg Conference, March 3–7, 2008, Ushuaia-Patagonia, Argentina, 4 p., http://www.searchanddiscovery.com/documents/2008/jw0806zavala/ (accessed July 15, 2011).

Zavala, C., L. Blanco Valiente, and Y. Vallez, 2008b, The origin of lofting rhythmites: Lessons from thin sections, *in* J. J. Ponce and E. B. Olivero, conveners, Sediment transfer from shelf to deepwater—Revisiting the delivery mechanisms: Conference Proceedings, AAPG Search and Discovery Article 50077, AAPG Hedberg Conference, March 3–7, 2008, Ushuaia-Patagonia, Argentina, 4 p., http://www.searchanddiscovery.com/documents/2008/jw0807zavala/ (accessed July 15, 2011).

Zavala, C., M. Arcuri, H. Gamero Diaz, C. Contreras, and M. Di Meglio, 2011, A genetic facies tract for the analysis of sustained hyperpycnal flow deposits, *in* R. M. Slatt and C. Zavala, eds., Sediment transfer from shelf to deep water—Revisiting the delivery system: AAPG Studies in Geology 61, p. 31–51.

4

Ponce, J. J., and N. B. Carmona, 2011, Miocene deep-marine hyperpycnal channel levee complexes, Tierra del Fuego, Argentina: Facies associations and architectural elements, *in* R. M. Slatt and C. Zavala, eds., Sediment transfer from shelf to deep water—Revisiting the delivery system: AAPG Studies in Geology 61, p. 75–93.

Miocene Deep-marine Hyperpycnal Channel Levee Complexes, Tierra Del Fuego, Argentina: Facies Associations and Architectural Elements

Juan José Ponce and Noelia B. Carmona

Instituto Argentino de Oceanografía, Consejo Nacional de Investigaciones Científicas y Técnicas (IADO-CONICET), Bahía Blanca, Argentina

ABSTRACT

In the Miocene beds of Cabo Viamonte, deep-marine hyperpycnal channel levee complexes occur at the toe of the depositional slope in sigmoidal clinoforms. The hyperpycnal channel fill is composed of thick packages of transitional and recurrent (vertical and lateral) sedimentary structures without rheologic boundaries, with variable textural ranges, recurrence of inverse-to-normal grading, multiple internal reactivation surfaces, and abundant organic content. This complex internal arrangement allows recognition of three facies associations (FAs) in the hyperpycnal channel fill. The FA1 comprises the coarsest clastic fractions, composed of intraformational boulders, matrix- and/or clast-supported conglomerate, and gravelly sandstone, which results from the combination of inertial flows and turbidity currents generated during the circulation of hyperpycnal flows. The FA2 comprises conglomerate, gravelly sandstone, and coarse- to fine-grained sandstone that results from rapid aggradation from a hyperpycnal turbidity current capable of transporting gravel as bed load and particles up to the size of coarse sand in turbulent suspension. The FA3 comprises heterolithic deposits that result from progressive accumulation by traction and suspension processes of low-density hyperpycnal turbidity currents. The coarse-grained lower channel fill is interpreted as a single sedimentation event. This is suggested by multiple internal erosion surfaces without lateral continuity, the lack of lithologic contrasts between sediment below and above the reactivation

Copyright ©2011 by The American Association of Petroleum Geologists.

DOI:10.1306/13271351St613439

surfaces, and the absence of biogenic structures. On the contrary, the fine-grained upper channel fill, deposited mainly during the waning stages of the hyperpycnal flow, may involve more than one sedimentation event. This interpretation is supported by the presence of isolated beds with low-diversity trace fossils within the associated mudstones.

INTRODUCTION

The deposits generated by gravity flows, associated with intrabasinal resedimentation processes and/or hyperpycnal discharges in tectonically active basins, occur within a great variety of environments and deposits that include submarine canyons, slope channels, slump scars, channels and levees, sediment waves, lateral spillovers, and frontal spillovers or lobes (Posamentier, 2003; Posamentier and Walker, 2006). Submarine canyons and slope channels are the principal conduits that transfer sediment basinward (Posamentier and Walker, 2006).

Submarine canyons show a sharp V-shaped morphology with depths that generally do not exceed 1 km (0.62 mi) and a width that can reach 10 km (6.2 mi). In submarine canyons, the flows are confined within the conduit and, therefore, levees are not produced (Posamentier and Walker, 2006). The canyon fill is commonly dominated by fine-grained sediments accumulated after abandonment of the canyon, which occurs nowadays in the Mississippi Canyon (Goodwin and Prior, 1989). Nevertheless, submarine canyon systems with unusual coarse-grained clastic accumulations have been reported, such as the Carapeba-Pargo system offshore Brazil (Bruhn and Walker, 1995).

Slope channels show a U-shaped morphology and highly variable dimensions (tens of meters deep and hundreds of meters wide). In contrast to submarine canyons, large levees (more than 1000 m [3280 ft] wide) occur along these channels because the flow is not well confined (Posamentier and Walker, 2006). The slope channels can be physically connected to fluvial discharge systems (Posamentier, 2003; Plink-Björklund and Steel, 2004), suggesting a genetic relationship between hyperpycnal flows and the construction of these systems. The slope channels exhibit a predominantly aggradational fill of basal debris flow deposits overlain by amalgamated turbidites (Mayall and Stewart, 2000). The levees form thick heterolithic successions mainly of Bouma (1962) Tc–d-like intervals.

In both submarine canyons and slope channels, the morphology of the slope is the main parameter that controls the geometry and arrangement of the sedimentary facies (Fonnesu, 2003; Jennette et al., 2003; Prather, 2003). In the Miocene beds of Cabo Viamonte (Figure 1B), deep-marine channel levee complexes (CLCs) were recognized at the toe of a depositional slope in sigmoidal clinoforms (Ponce et al., 2008a, b, c). In contrast to earlier models of submarine fans (Normark, 1970; Mutti and Ricci Lucchi, 1972; Walker, 1978), detailed outcrop studies and preliminary observation of available seismic lines in Tierra del Fuego indicate that the CLCs are not associated with canyons. These channels are characterized by large-scale lenticular geometries. Their fill is composed of transitional and recurrent (vertical and lateral) sedimentary structures without rheologic boundaries, variable textural ranges, recurrence of inverse-to-normal grading, multiple internal reactivation surfaces, and abundant organic content. These features suggest that the channel and levee deposits of the Cabo Viamonte beds accumulated from sustained and pulsating gravity flows generated by extraordinary fluvial discharges entering the sea as hyperpycnal flows (Ponce et al., 2008a, b). In terms of geometries, facies arrangements, and sedimentary processes, the complexity observed in these deposits is still poorly understood. Therefore, the main purposes of this chapter are (1) to characterize the facies associations (FAs) and the architectural elements of the hyperpycnal CLCs located at the toe of the depositional slope, and (2) to discuss the sedimentary mechanisms involved in the generation of their complex fill.

GEOLOGIC FRAMEWORK

The upper Oligocene–middle Miocene deposits of the foreland Austral Basin of Tierra del Fuego comprise five depositional sequences (SI–SV; Figure 1A, B) bounded by unconformities or hiatuses of different magnitudes (Ponce et al., 2005, 2008a). These sequences form part of the Cabo Domingo Group defined by Malumián (1999) and represent the youngest foredeep fill developed in front of the adjacent fold and thrust belt (Ghiglione et al., 2002; Ponce et al.,

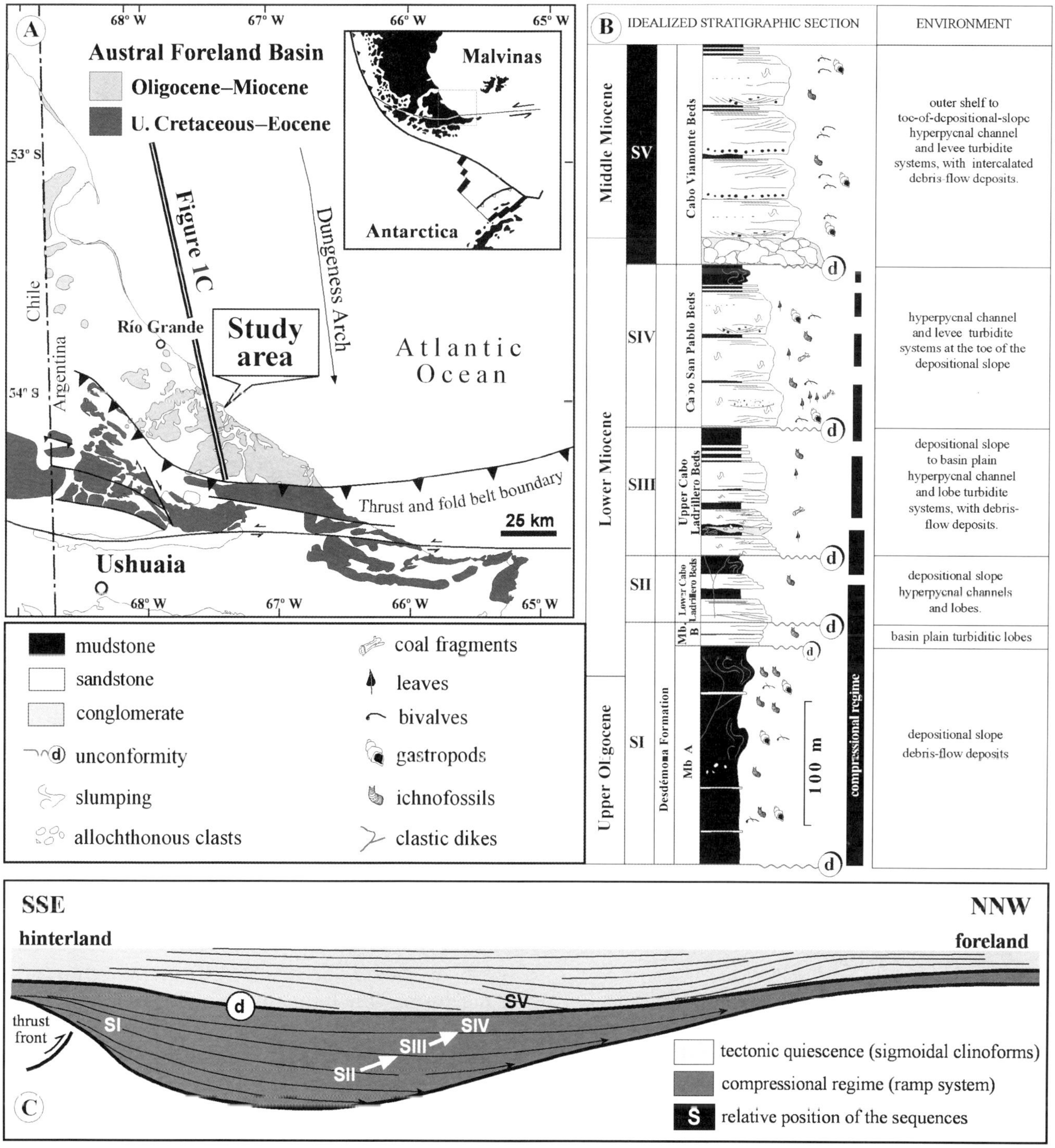

FIGURE 1. (A) Simplified geologic map of Tierra del Fuego showing the study area. (B) Vertical section of the five recognized sequences SI–SV and summary of their depositional environment (interval of interest SV in black). (C) Schematic transect to represent the clinoform geometries and the relative position of the sequences within the basin. U. = Upper; Mb. = Member.

2005, 2008a; Malumián and Olivero, 2006; Olivero and Malumián, 2008). They consist of a subhorizontal clastic succession of more than 600 m (1968 ft) of thickness. The Desdémona Formation (SI) includes two members: (1) mudstone dominated, deposited by cohesive flows; and (2) glauconite-rich sandstone dominated, deposited by episodic turbidity currents. The lower Cabo Ladrillero beds (SII), upper Cabo Ladrillero beds (SIII), Cabo San Pablo beds (SIV), and Cabo Viamonte beds (SV) are composed of complex arrangements of boulder to fine conglomerate, coarse- to fine-grained sandstone, and mudstone that were deposited mainly by hyperpycnal flows (Figure 1B) (Ponce et al., 2008a).

Two principal tectonic-sedimentary stages are recognized in the evolution of the infill of these sequences (Ponce et al., 2008a). The first stage involves deposition of the SI (late Oligocene–early Miocene) during the last phase of the compressional tectonic regime and the development of a ramp system close to the orogen (Figure 1C). This sequence was unconformably deposited on the frontal limb of a growing anticline (thrust front; Figure 1C). This first tectonic-sedimentary stage continues with the accumulation of SII to SIV, representing positions toward the foreland within a ramp system. The three sequences were deposited away from the orogenic area, thus leaving a more subtle imprint on the early Miocene compressional regime (Figure 1C).

The second tectonic-sedimentary stage, involving the Cabo Viamonte beds (SV), occurred during the uppermost lower–middle Miocene and represents sedimentation during tectonic quiescence with the development of sigmoidal clinoforms in the basin (Figure 1C). According to the physical expression seen in outcrops and interpreted for the subsurface, the basal unconformity of SV is interpreted as a major unconformity recorded in the upper Oligocene–middle Miocene succession along the Atlantic coast. It may correspond to a dramatic sea level drop recognized at 21 Ma in the global eustatic curve (Haq et al., 1987). The input of fresh basaltic clasts observed in SV also records a compositional break between these and the underlying strata (Malumián and Olivero, 2006; Ponce et al., 2008a).

HYPERPYCNAL CHANNEL LEVEE FACIES ASSOCIATIONS

The complex internal arrangement, with multiple erosion surfaces, lateral and vertical transition of sedimentary structures, textural variation, and recurrent inverse-to-normal graded bedding, can be grouped into three principal FAs in the hyperpycnal CLCs of the Cabo Viamonte beds.

These complex hyperpycnal FAs can be analyzed using the flow-boundary zone concept for hyperpycnal flow-sediment interaction (Figure 2). Branney and Kokelaar (1992, 2002) developed the flow-boundary zone concept to explain the transition between sedimentary structures, as well as the textural variations in pyroclastic density flow deposits. The flow-boundary zone includes the uppermost portion of the forming deposit, the boundary, and the lowermost portion of the density flow (Figure 2A). This zone varies progressively as the deposit aggrades or is removed by erosion under sustained flow conditions. Variations in the nature of this boundary zone depend on the flow velocity, particle concentration, and duration of the event. In this sense, Branney and Kokelaar (2002) identified four main types of flow-boundary zones that can be gradational between each other. At one end (Figure 2B) is a density flow in which particle concentration in the upper part of the forming deposit is much higher than at the basal levels of the flow, and therefore, the velocity and concentration profiles show a distinctive step (double inflection). Under these conditions, a rheologic contrast between the upper and basal part of the deposit occurs, allowing clasts of different sizes to move freely and to produce tractive sedimentary structures. The other end (Figure 2C) shows a density flow with a high concentration of loose particles at its top and a high concentration of particles at the base. This situation prevents development of a rheologic contrast between those particles immediately above and below the flow-boundary zone. The development of tractive sedimentary structures is inhibited under such conditions, resulting in a continuous aggradation of massive beds (Figure 2C). This last type of flow-boundary zone was applied by Kneller and Branney (1995) to explain the presence of thick packages of massive sandstone. The flow-boundary zone is considered to vanish during an increasingly erosive pulse (Figure 2D), when the density flow in continuous acceleration exceeds the range of velocities needed to produce erosion. Hence, the previously accumulated deposit can be partially or totally cannibalized, depending on the amount of time during which the flow remains erosive (t_1–t_2 in Figure 2D). The flow-boundary zone becomes reestablished with the progressive deceleration of the flow (t_2–t_3 in Figure 2D).

Facies Association 1

The FA_1 comprises the coarsest clastic fractions, composed of intraformational boulders of laminated mudstone up to 4 m (13 ft) long in their major axes; matrix- and/or clast-supported conglomerate, with clasts up to 0.80 m (2.62 ft) in diameter; and gravelly sandstone, displaying tabular and lenticular geometries. Individual beds can exceed 6 m (19 ft) in thickness, show important textural variations, and exhibit gradations between sedimentary structures (Figure 3).

The boulder conglomerate in FA1 comprises isolated lenses with a morphology resembling that of

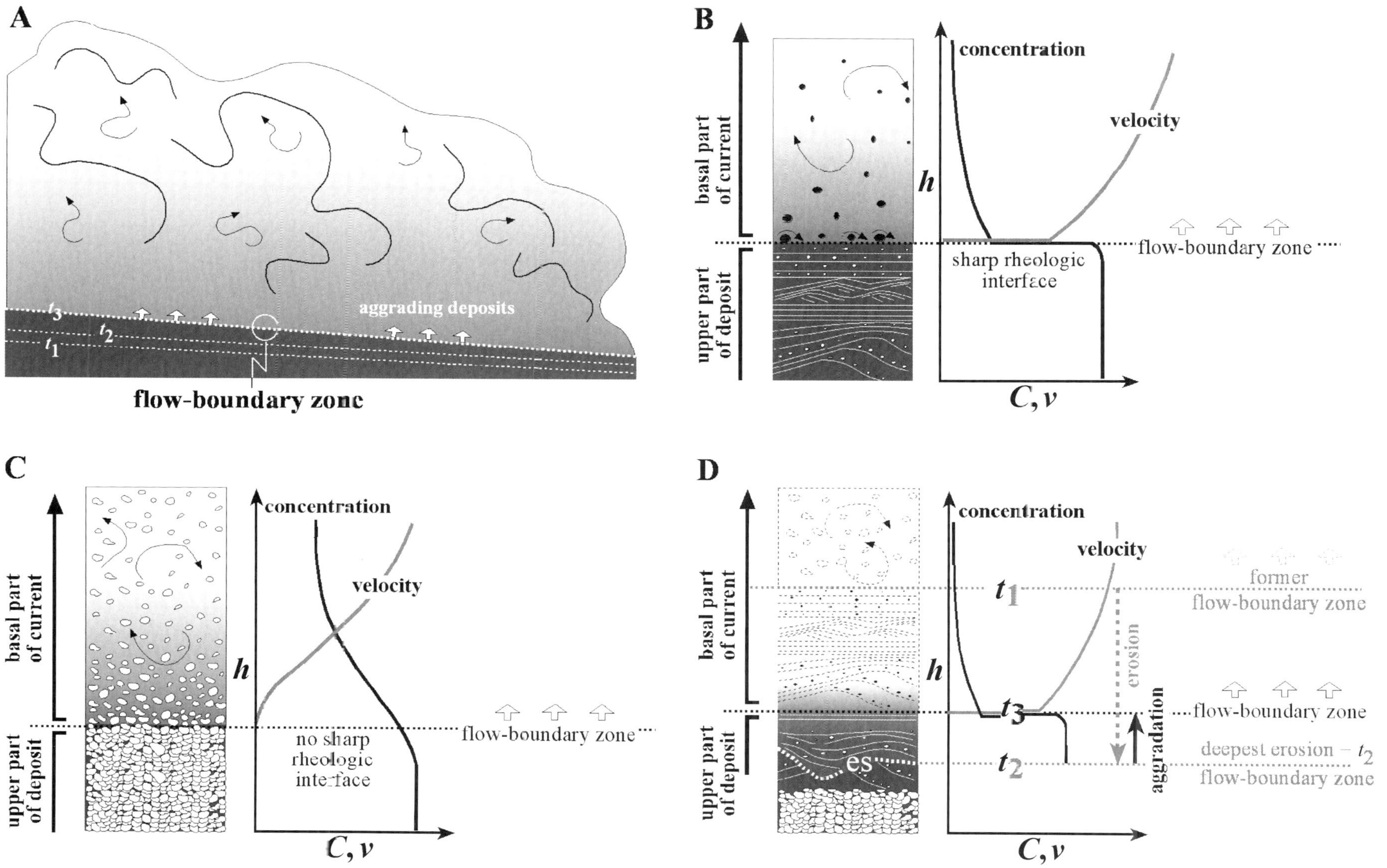

FIGURE 2. Sedimentation produced by a quasi-steady flow (modified from Branney and Kokelaar, 2002). (A) Progressive aggradation at the flow-boundary zone ($t_1-t_2-t_3$). (B) Aggradation of deposits with tractive sedimentary structures at a flow-boundary zone characterized by a high particle concentration in the upper part of the deposit in formation, and a low particle concentration at the lower part of the density flow. (C) Aggradation of massive sedimentary deposits at a flow-boundary zone with a high particle concentration in the upper part of the deposit in formation and in the lower part of the density flow. (D) Erosion surface (es) generated as a result of a hyperpycnal flow velocity increase. Under these conditions, the flow produces a partial or total cannibalization of the previously formed deposits (t_1-t_2). When the hyperpycnal flow experiences a progressive deceleration, the flow-boundary zone is reestablished (t_2-t_3). t = time; C = concentration; v = velocity; h = upper part of deposit + basal part of current.

fluvial bars (Figure 3A). In some cases, the boulders are arranged in diffuse surfaces or are reoriented by gravitational collapse (Figure 3B). Bed-load structures, constituted by climbing dunes (Figure 3C), are frequent in this FA. Armored intraclasts are found associated with diffuse surfaces generated in gravelly sandstone deposits (Figure 3D) or on top of laminae that constitute the sediment waves and the large-scale climbing dunes (Figure 3A, C). The clast-supported conglomerate forms isolated clusters on erosive surfaces. These conglomerates show clasts with their major axes arranged perpendicular to the main flow direction, and some of them are imbricated. The conglomerates frequently grade transitionally to massive or diffuse laminated gravelly sandstone. No bioturbation is observed in FA1.

Interpretation

The FA1 results from the combination of inertial flows and turbidity currents, generated during the circulation of hyperpycnal flows along steep slopes. The instability produced by the erosion of an overpassing hyperpycnal flow in a depositional slope is the principal triggering mechanism by which successive debris-fall avalanches are formed, subsequently adding mass to the hyperpycnal flow. The inertial forces that operate in these flows can transport—by sliding, saltation, and rolling—large boulders and armored intraclasts that accumulate through a granular impact in zones of slope change (cf. grain fall in Nemec, 1990), generating bodies with geometries similar to fluvial bars (t_1 in Figure 4). These deposits have low preservation potential because of the permanent reworking by the overpassing hyperpycnal flow. The accumulation by progressive aggradation with a flow-boundary zone dominated by traction, results in gravelly sandstones and armored intraclasts—transported as bed load—arranged on diffuse surfaces that copy the bar-shaped feature (Figure 4). In several cases, the morphology of the inertial deposits left behind constitutes a natural obstacle that favors the development and migration of large bed-load structures (e.g., sediment waves and climbing dunes). The thick deposits with transitional and recurrent passages between gravel with imbricated clasts, massive gravelly sandstone, and gravelly sandstone with diffuse lamination are the result of temporal and spatial flow velocity (concentration) variations. These variations in flow velocity occur mainly in those areas located behind and immediately in front of the avalanche deposits, where the flow experiences a transformation by expansion (t_2–t_3 in Figure 4).

Facies Association 2

The FA2 comprises conglomerate, gravelly sandstone, and coarse- and fine-grained sandstone arranged in tabular and lenticular bodies. Individual beds show thicknesses between 3 (9 ft) and 6 m (19 ft), marked textural variations, and a transition between sedimentary structures.

The conglomerate is massive or shows imbricated clasts, with a transition to massive gravelly sandstone deposits (Figure 5A), which can show intervals with soft-sediment deformation structures. The gravelly sandstone intervals, up to 1 m (3.28 ft) thick, dominate in FA2. Climbing dunes have been recognized in deposits constituted by gravelly sandstone and coarse-grained sandstone, where gravel clasts are associated with the bedding surfaces (Figure 5B, C). Cut-and-fill structures are more frequent in this FA than in FA1. These structures show a symmetric geometry in cross section (Figure 5C, D), and an asymmetric profile, with a more gentle slope, along the flow direction (Figure 6). The cut-and-fill structures are related (and show a transitional passage) to climbing dunes of different scales. Likewise, the climbing dunes can be transitional to thin deposits of massive, coarse sandstone, or to gravelly or coarse sandstone with diffuse parallel lamination (Figure 6). No bioturbation was observed in FA2.

Interpretation

The FA2 results from the rapid aggradation produced by a hyperpycnal turbidity current capable of transporting gravel as bed load and particles up to the size of coarse sand in turbulent suspension along steep slopes (Figure 4). This FA shows a greater selection in grain size than FA1, and it is dominated by tractive sedimentary structures. The development of thick layers of gravelly sandstones with recurrent and transitional passages between diffuse parallel lamination, climbing dunes, and cut-and-fill structures, and the recurrence of textural variations reflect the important fluctuations in velocity (concentration) experienced by the flow during its discharge. These fluctuations in velocity affect the nature of the flow-boundary zone and generate alternation between aggradational sedimentation and erosion pulses. Although aggrading, a rheologic contrast occurs at the flow-boundary zone (Figure 2B), producing upper-flow regime structures (e.g., climbing dunes) that transport gravel as bed load (t_2 in Figure 4). If the hyperpycnal flow velocity increases, exceeding the erosion threshold of gravelly sands, the climbing dune structures start to be truncated (erosion surfaces [es] in Figures 2D, 4). These

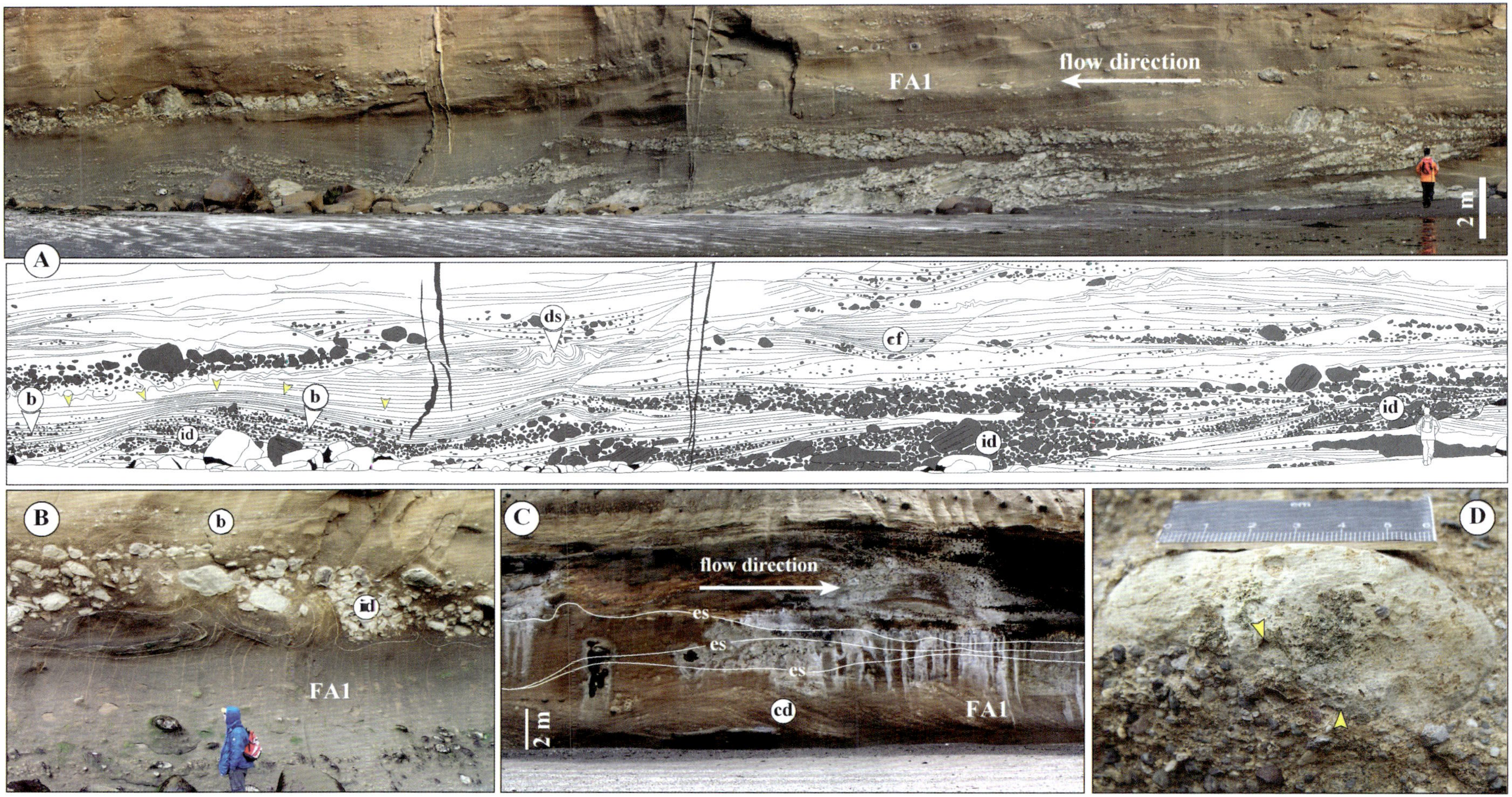

FIGURE 3. Main characteristics of the facies association generated by inertial flows and high-density hyperpycnal turbidity currents (FA1). (A) Panoramic view showing bar-shaped inertial deposits (id), reworked at the top by the overpassing hyperpycnal flow that generated sediment-wave bed-load migration (yellow arrows); cut-and-fill structures (cf); deformation structures (ds); and surfaces with armored mudstone intraclasts transported as bed load (b). Clastic dykes of different compositions and grain sizes cut across these sedimentary structures. (B) Inertial deposit with boulders that were reoriented by gravitational collapse. (C) Climbing dunes sedimentary structures (cd) truncated at the top by multiple erosion surfaces (es). (D) Detail of an armored mudstone intraclast (missing clasts cavities indicated by yellow arrows).

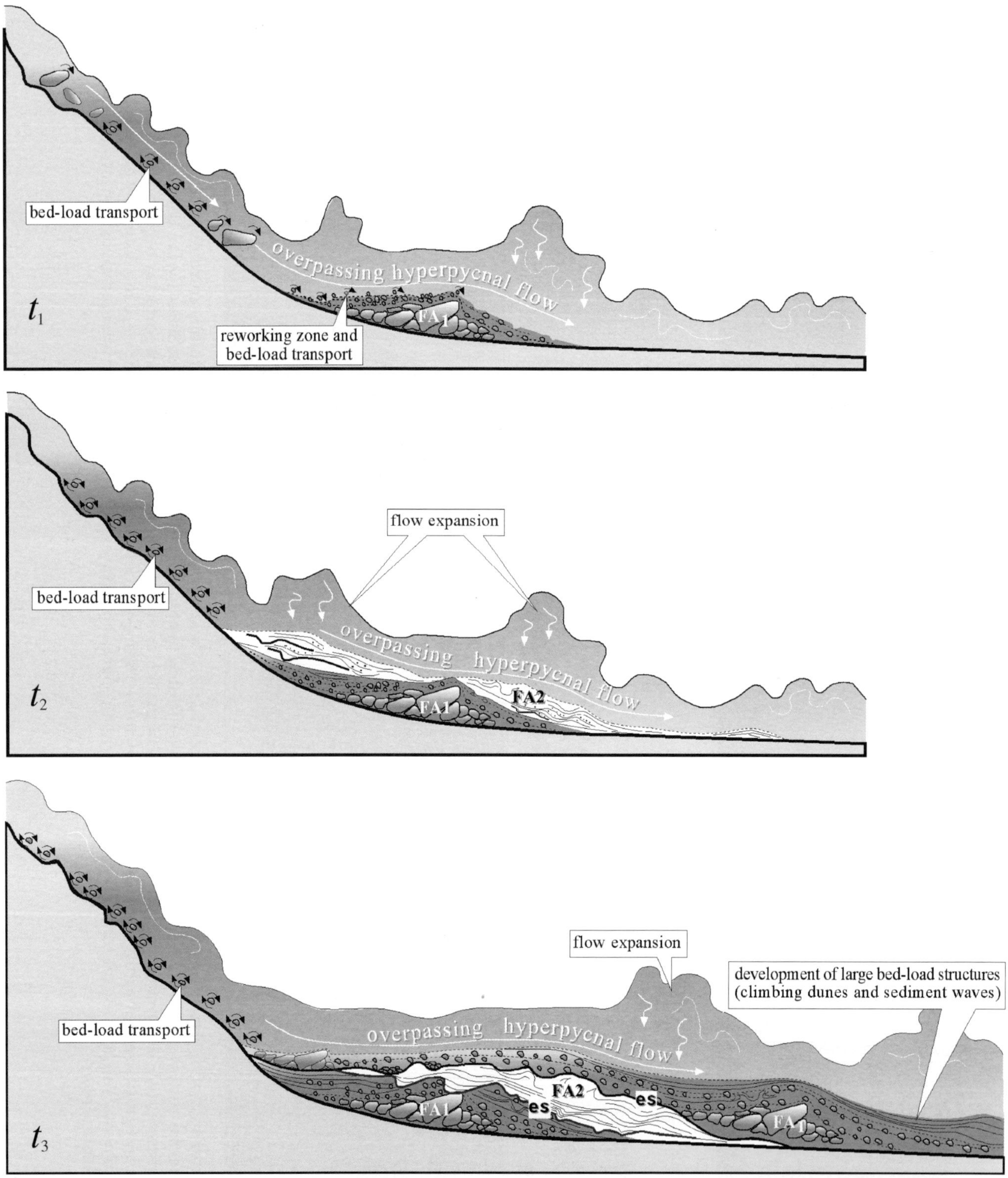

FIGURE 4. Longitudinal schematic sections showing the evolution of the hyperpycnal channel fill, representing the waxing stage of one event of sedimentation. At t_1, avalanche deposits accumulate and are subsequently reworked by the overpassing hyperpycnal flow (FA1). The continuing hyperpycnal flow produces deposits with a combination of complex aggradational and erosional features (FA2). This sedimentation at t_2 is controlled by the morphology of accumulations developed at t_1 (e.g., FA1) that, acting as obstacles, favor flow expansions. Cannibalization of previously generated deposits, with development of erosion surfaces (es), occurs by a progressive acceleration of the hyperpycnal flow. When the hyperpycnal flow decelerates, the flow-boundary zone is reestablished at t_3, generating the aggradational sedimentation with alternating or recurring FA1 and/or FA2. FA = facies association; t = time.

structures form when the flow exceeds the critical velocity to partially or totally erode and retransport the grain-size fractions accumulated during t_2 (Figure 4). On the contrary, if the density flow gradually diminishes in velocity, a flow-boundary zone, with development of tractive structures (mainly climbing dunes), is reestablished (t_2–t_3 in Figure 2D; t_3 in Figure 4). The recurrent and transitional passage between gravelly massive sandstone and coarse-grained sandstone with imbricated gravel clasts is interpreted as an intergradation between a flow-boundary zone with a high concentration of sediment in suspension (gravelly

FIGURE 5. Main sedimentary features of the facies association produced by hyperpycnal turbidity currents (FA2). (A) Massive gravel conglomerate (Cgm) with transitional passage to massive gravelly sandstone (Sgm), which shows toward the top imbricated clasts (black lines) that indicate bed-load transport during the hyperpycnal discharge. (B) Sandstone climbing dunes (Scd) with aligned gravel veneers between laminae. (C, D) Details of the lower channel-fill arrangements showing the transitional passage and recurrence between sedimentary structures. The internal erosion surfaces have a marked lateral variation and vertical recurrence (dashed lines). cf = cut-and-fill structures; Sl = laminated sandstone; Sm = massive sandstone.

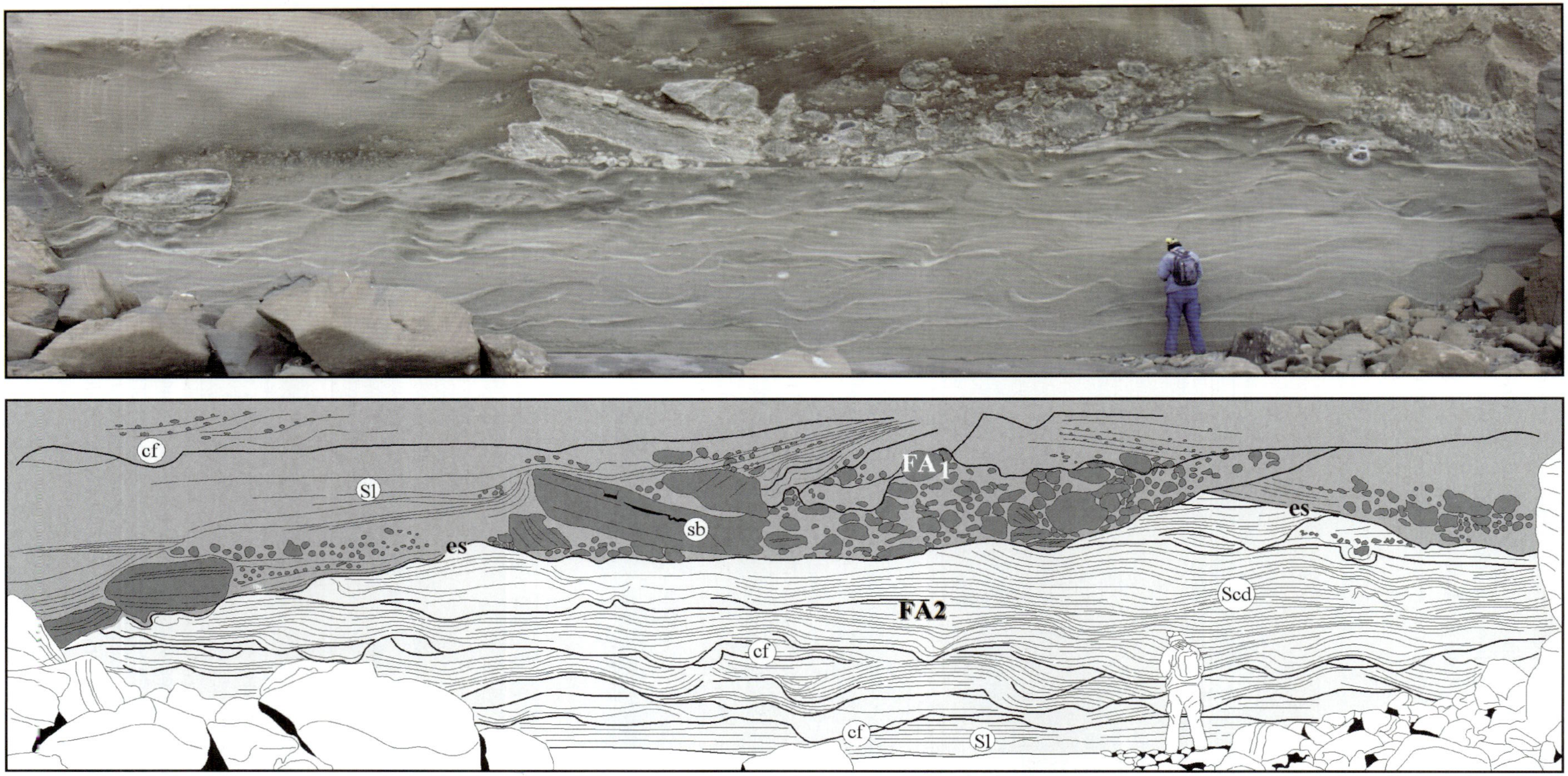

FIGURE 6. Interpretation of hyperpycnal lower channel-fill deposits. The lower part of the section is constituted by recurring parallel laminated sandstone (Sl) and sandstone with climbing dunes (Scd), showing multiple cut-and-fill structures (cf) that characterize FA2. The upper part of the section shows a reactivation surface (es) covered with inertial deposits (slide blocks [sb]) typical of FA1 that grade to similarly laminated sandstone. FA = facies association.

massive sandstone) to a flow-boundary zone with a low particle concentration (gravelly sandstone with tractive sedimentary structures).

Facies Association 3

The FA3 comprises heterolithic deposits arranged in tabular bodies. Individually, this FA is 0.30 (0.98 ft) to 1 m (3.28 ft) thick and has sharp lower contacts and transitional passages between parallel-laminated sandstone, climbing ripples, current ripples, lenticular bedding, and laminated and massive mudstone. Isolated levels with slump structures can also occur.

The FA3 forms successions with a rhythmic interbedding of thin (centimeter-thick) levels of fine to very fine sandstone and mudstone. Internally, this FA commonly shows Bouma *Ta*–e-like arrangements (Figure 7A) or transitional and recurrent passages between sandstone with parallel lamination, climbing ripples, current ripples, lenticular bedding, and laminated or massive mudstone, with important concentrations of phytodetritus and very well-preserved *Nothofagus* leaves (Figure 7B–D). Isolated syneresis cracks are present. The heterolithics that constitute the tabular bodies can be sandstone or mudstone dominated. The sandstone-dominated heterolithics comprise fine massive sandstone or sandstone with parallel lamination, climbing ripples, and wavy bedding, and interlayering of laminated or massive mudstone (Figure 7E). The mudstone-dominated heterolithics comprise laminated and massive mudstone, with soft-sediment deformation structures that, subordinately, incorporate fine- to very fine-grained sandstone with current ripples and lenticular bedding (Figure 7F). Beds of FA3 can be partially or completely obliterated by bioturbation, with deposit-feeding structures such as *Scolicia*, *Nereites*, *Phycosiphon*, and concentrically laminated burrows (Figure 7A–C, G). Pre-event trace fossils, such as *Paleodictyon*, have been recognized casted at the base of current-rippled sandstone (Figure 7H). Grazing and locomotion traces also appear associated with these heterolithic levels.

Interpretation

The FA3 results from progressive accumulation by traction and suspension processes produced in association with low-density hyperpycnal turbidity currents. The recurrent presence of thin contorted layering of massive mudstone associated with abundant phytodetritus, blanketing laminated and cross-laminated sandstone, suggests that flocculation occurs during the deposition of part of the mud-rich FA3 sediments. The rapid accumulation of mud-rich sediment produced by flocculation in these deposits may have been favored by the nucleation of organic particles (see Kranck, 1973; Mikkelsen and Pejrup, 1998), which are extremely abundant in the hyperpycnal flows. Moreover, Mikkelsen and Pejrup (1998) observed that the floccules with high content of organic material are much more resistant than those generated purely by saline contrasts because the organic material can act as glue for the fine particles, giving them more cohesion. The occurrence of bed-load structures, like climbing ripples and current ripples, involving mud and phytodetritus in their foresets suggests that floccules have formed aggregates with an equivalent grain-size to fine sand and, therefore, could be transported as bed load. Schieber et al. (2007) demonstrated that the transport of flocculated mud produces bed-load structures (e.g., floccule ripples) in the 10 cm/s to 26 cm/s velocity range. Therefore, the flocculation process in this FA helps to explain the generation of intervals with variable thicknesses composed of alternations of fine sandstone with climbing ripples, wavy and lenticular bedding, and massive mudstone with soft-sediment deformation structures, and high organic content, accumulated during the same depositional event.

Successions with Bouma *Ta*–e-like arrangements result from the progressive aggradation developed during deceleration of the hyperpycnal flow, with a gradational passage between two kinds of flow-boundary zones. One has a high particle concentration in the upper part of the deposit and the lower part of the flow, with massive fine-grained sediment accumulation (Bouma *Ta*-like interval). The other is characterized by a low particle concentration in the lower part of the flow and a high particle concentration in the upper part of the deposit, with tractive sedimentary structures (Bouma *Tb*–c-like interval). The presence of thin layers of laminated mudstone (Bouma *Td*–e-like interval) with *Nothofagus* leaves, with their major axes parallel to the bedding plane, indicates deposition by fallout from suspension.

HYPERPYCNAL CHANNEL LEVEE COMPLEX ARCHITECTURAL ELEMENTS

Based on the morphological, geometric, and lithologic features of the Cabo Viamonte beds, two main architectural elements are recognized: (1) multistory hyperpycnal channel and (2) levee deposits. The multistory hyperpycnal channels constitute typical architectural elements located at the toe of depositional slopes, associated with major sigmoidal clinoforms

A
Sm
clb
M
clb
M
Sr
Sl
Sr
Sl
Sm
FA3
t
B
Sl
M
M
Sr
Sl
M
S
FA3
Sr
2 cm
Sl
M
t
C
S
FA3
M
Sl
Sr
Sl
Sm
t
D
ASOCIACION
ESCALA FOTOGRAFICA
10 cm
FA3
E
Sl
Scr
FA3
Sl
Sm
Scr
t
F
FA3
t
G
9
10
5 mm
Ne
Ne
H
P
50
centavos
1994

(Figure 1C). Maximum dimensions of individual channels are commonly about 500 m (1640 ft) wide and 40 m (131 ft) deep. The channel fill is mainly aggradational, with its lower and upper parts characterized by distinct lithologies (Figure 8A). The lower channel fill consists mainly of sandy conglomerate and pebbly sandstone, with subordinate large (up to 1 m [3.28 ft] in diameter) intraformational boulders, blocks glided from the channel walls, and slump deposits (FA1 and FA2; Figure 8B–D). The upper channel fill consists of graded tabular beds, which markedly onlap the erosive channel margins (FA3; Figure 8). Subordinately, some of the hyperpycnal upper channel fill (Figure 8E) contains slump deposits.

The levee deposits are wedge-shaped and consist mainly of rhythmically interbedded heterolithic mudstone and fine sandstone (FA3; Figure 8D, E), with isolated slump deposits. Thickness is greatest in proximity to the channel margins and thins away distally. The minimum lateral extension observed in levee deposits is 250 m (820 ft) perpendicular to the channel axis. These levee deposits can be stacked with the tabular bedding of the upper channel fill.

HYPERPYCNAL CHANNEL FILL: SINGLE VERSUS MULTIPLE EVENTS

The hyperpycnal CLCs exposed in Cabo Viamonte show a multistory architecture. A point to consider is if the infill of individual channels in hyperpycnal CLCs has been produced by only one or several sedimentation events.

The thick lower channel-fill deposits of the hyperpycnal channels of the Cabo Viamonte beds (FA1 and FA2) have multiple internal erosion surfaces without lateral continuity. These surfaces have their maximum erosive expression nearby the channel axis and vanish toward the marginal zones of the channel, where they pass transitionally to bed-load transport deposits with armored mudstone intraclasts, imbricated clasts, parallel lamination, and climbing dunes. These lateral variations suggest that the lower channel fill accumulated during the same sedimentation event. This interpretation is supported by a weak to an almost absent lithologic contrast between the sediment accumulated below and above the reactivation surfaces. These surfaces resulted from fluctuations in the velocity of a single hyperpycnal flow and not from the amalgamation of beds produced by multiple events. Moreover, the absence of biogenic structures in FA2 indicates that there were not real pauses during deposition of this FA, precluding colonization of the infauna.

Conversely, the onlap tabular beds of the upper channel fill (FA3) are mainly related to the waning stage of the flow. They could represent the accumulation from more than one sedimentation event. This interpretation is supported by the presence of isolated levels with low-diverse trace fossil content (concentrically laminated burrows) associated with the mudstone of the fine heterolithics.

HYPERPYCNAL CHANNEL LEVEE COMPLEX INFILL STAGES

Two main infill stages are recognized in the development of the studied hyperpycnal CLCs (Figure 9). The first stage occurs during the waxing period of discharge of rivers in flood. It comprises the incision of the channels and deposition of the lower channel fill and levee.

The lower channel fill includes the coarsest deposits (FA1 and FA2) that, depending on sediment concentration and flow velocity, could be massive or could show complex current sedimentary structures. The inertial deposits generally constitute the

FIGURE 7. **Principal characteristics of the facies association produced by low-density hyperpycnal turbidity currents (FA3). (A) Bouma *Ta*–e-like arrangements constituted by massive (Sm), parallel-laminated (Sl), and rippled (Sr) sandstone with important phytodetritus concentrations, and laminated mudstone to massive mudstone (M) with concentrically laminated burrows (clb) at the topmost level. (B, C) Heterolithic deposits showing transitional and recurrent passages between sandstone with parallel lamination, current ripples, and laminated or massive mudstone. Thin contorted layers of laminated and massive mudstone with abundant phytodetritus are interpreted to be produced by flocculation processes. The observed trace fossils are *Scolicia* (*S*) and concentrically laminated burrows. (D) View of the bedding surface with excellent preservation of *Nothofagus* leaves in heterolithic levels. (E) Sandstone-dominated levee deposits, accumulated closer to the channel axis and constituted by a transitional passage between massive, parallel laminated, and climbing-ripple sandstone (Scr). (F) Heterolithic (distal) levee deposits, constituted by rhythmic sandstone and mudstone alternation; parallel-laminated, rippled, and wavy to lenticular bedded sandstone, interlayered with laminated mudstone. (G) Plane view of *Nereites* (*Ne*). (H) *Paleodictyon* (P) casted at the base of a sandstone bed. t = top of bed.**

FIGURE 8. (A) Main architectural elements of hyperpycnal channel levee complexes (CLCs). (B) Bidimensional view of hyperpycnal CLC. Numbers 1–8 indicate successive channel fills. Arrow shows flow direction of 2 to 7, which is almost orthogonal in 1 and 8. (C) Hyperpycnal CLC showing lower (gravelly and poorly sorted sandstone) and upper (heterolithic mudstone and fine sandstone) channel-fill deposits onlapping the basal erosion surface (white arrows). (D) Hyperpycnal CLC cross section. (E) Partial view of a hyperpycnal CLC constituted only by upper channel fill, with slump structures (s) and onlap relationships.

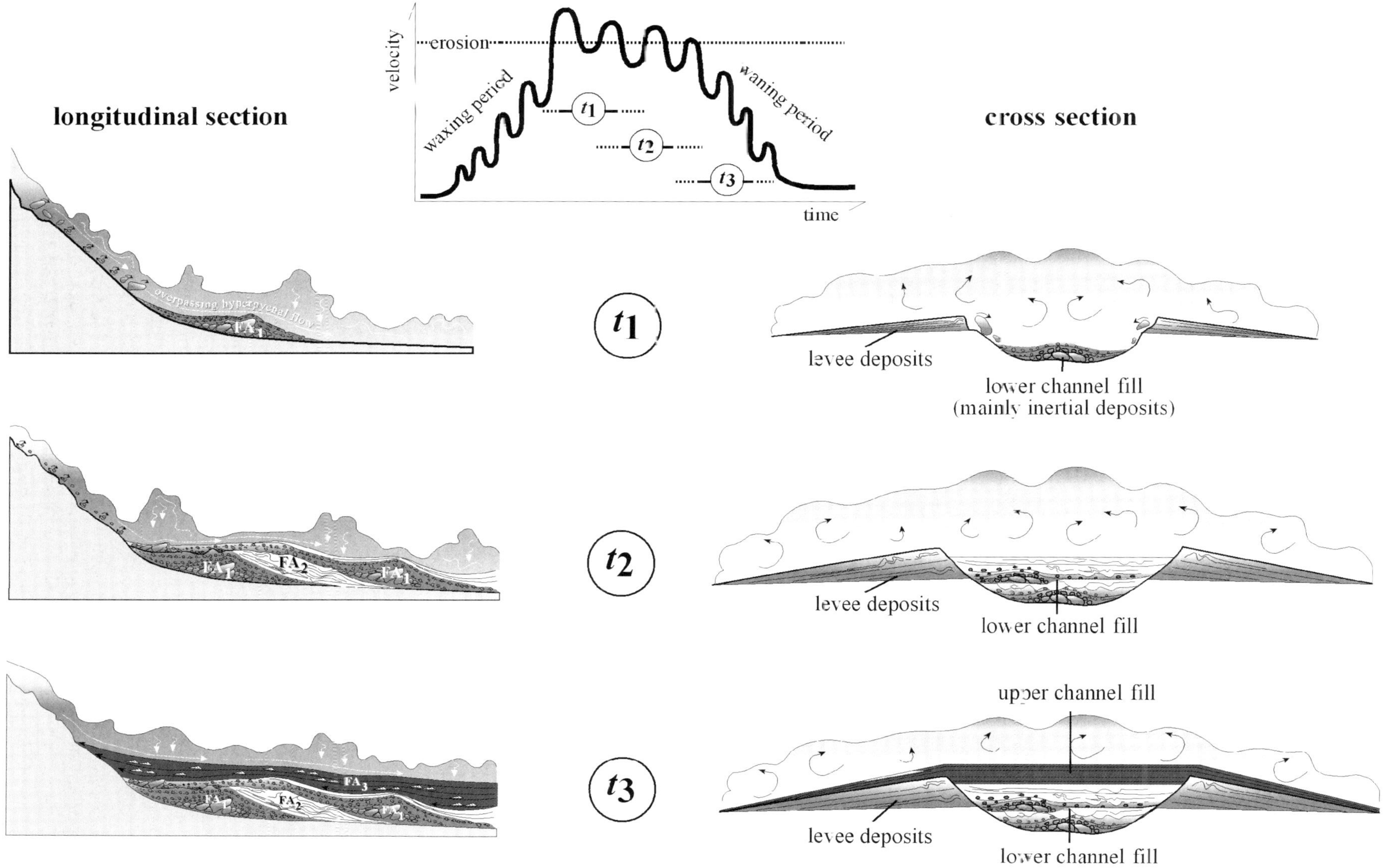

FIGURE 9. Schematic sections to illustrate the sedimentary evolution of the individual channel fills during the waxing and waning periods of a hyperpycnal discharge (velocity vs. time curve; modified from Zavala et al., 2006). Inertial deposits (FA1), with morphologies similar to those of fluvial bars, accumulate at the lowermost part of the hyperpycnal channel during t_1, and simultaneous aggradation occurs at the levees (FA3). These channel deposits are covered later (t_2) by pebbly coarse- to fine-grained sandstone (FA2) and/or by new avalanche deposits (FA1); meanwhile, sedimentation at the levees continues (FA3). Fine-grained tabular lobe deposits (FA3) accumulate within the channel and at levee positions during the waning period (t_3) of the hyperpycnal flow. FA = facies association.

first accumulations in the hyperpycnal CLC (t_1 in Figure 9). These deposits form localized obstacles that could subsequently control the sedimentary features of the successive deposits. If the flow has high sediment concentrations, after climbing the stoss side of the obstacle, it experiences a flow expansion at the lee side of the bar-shaped feature, producing the accumulation of thick packages of massive sandstone. If the flow has dilute sediment concentrations, the smaller intraclasts left behind in the tail of the flow move as bed load. Although moving over the obstacle, these intraclasts are forced to copy the undulating shape of the bar, resulting in large dune structures (Figure 3A). The continuous velocity fluctuations during the waxing stage of the hyperpycnal flow, together with a diminishing sediment concentration, favor the origin and transitional passage, without rheologic boundaries, of tractive sedimentary structures (e.g., parallel lamination and climbing dunes). In the case of the climbing dunes, their relative high angle of climb (>30°) suggests prevalence of aggradation over traction during the generation of these structures (Figure 5B, C). Rapid velocity fluctuations in the flow, particularly those initially characterized by a sudden increase in velocity, produce erosion with development of cut-and-fill structures, which are preserved because of rapid sediment filling of the scours during a subsequent decrease in velocity of the flow (t_2 in Figure 9). These cut-and-fill structures are observed repeatedly in lateral and vertical transition to climbing dunes; the latter are commonly developed immediately downcurrent of asymmetric scours (Figure 6). The main accumulation of coarse material is produced during this stage under sustained flow conditions. Deposition at the levee, constituted by FA3, also occurs during the waxing period of the hyperpycnal flow (t_1 and t_2 in Figure 9), and thus, it is simultaneous with the deposition of the lower channel fill (FA1 and FA2). The heterolithic successions within the levee show sandstone-dominated packages in positions closer to the channel axis, and mudstone-dominated packages farther away from the channel axis. In the sandstone-dominated packages, tractive structures prevail, and small-scale slumping is generated by gravitational instability during the enlargement of the levee. During deposition of the mudstone-dominated packages, layers with flocculated mudstone and important concentrations of phytodetritus occur.

The second stage includes the upper channel fill (FA3), particularly those deposits comprising the finer grained sediment (Figure 8; t_3 in Figure 9). The fine-grained clastic deposits reflect accumulation during the waning period of hyperpycnal flows. During this stage, the hyperpycnal system experiences a progressive retreat with velocity fluctuations. The upper channel fill comprises thick tabular successions of graded beds with complex internal arrangements or Bouma *Ta*–e-like divisions. The tabular beds accumulated in the upper channel fill also show isolated intervals with development of contorted massive mudstone. Occasionally, the infill of the hyperpycnal channels comprises only upper channel-fill deposits. In these cases, the main incision surface, produced during the waxing stage of the hyperpycnal flow without coarse-grained deposition, is colonized by large *Diplocraterion-Arenicolites* structures (Figures 8E, 10). These biogenic structures were produced in a firm-ground substrate, as indicated by the presence of delicate bioglyphs in their walls (Figure 10C).

CONCLUSIONS

The main factors that controlled the development of the complex sedimentary fill of the hyperpycnal CLCs in the Cabo Viamonte beds were the intrinsic pulsating and sustained character of the hyperpycnal flows and the changes in clinoform morphology, particularly those related to variations of the depositional slope angle.

The velocity (concentration) fluctuations of the hyperpycnal flow controlled the nature of the flow-boundary zone and favored the accumulation of thick sedimentary packages, without distinct rheologic boundaries, which show recurrence and transitional passages between sedimentary structures. A decrease in the slope angle, particularly at the toe of the depositional slope, favors a flow expansion and accumulation of inertial deposits that resemble a fluvial bar morphology. The instability of the depositional slope, caused by the circulation of the hyperpycnal flow, acts as the triggering mechanism of these inertial flow deposits (FA1). This interpretation contrasts with sedimentary models that present the development of inertial deposits as produced by a cohesive flow dilution.

Two main architectural elements that developed at the toe of the depositional slope are recognized in the hyperpycnal CLC of the Cabo Viamonte beds. Multistory hyperpycnal channels, 500 m (1640 ft) wide and 40 m (131 ft) deep, are characterized by a lower and an upper channel fill, and wedge-shaped

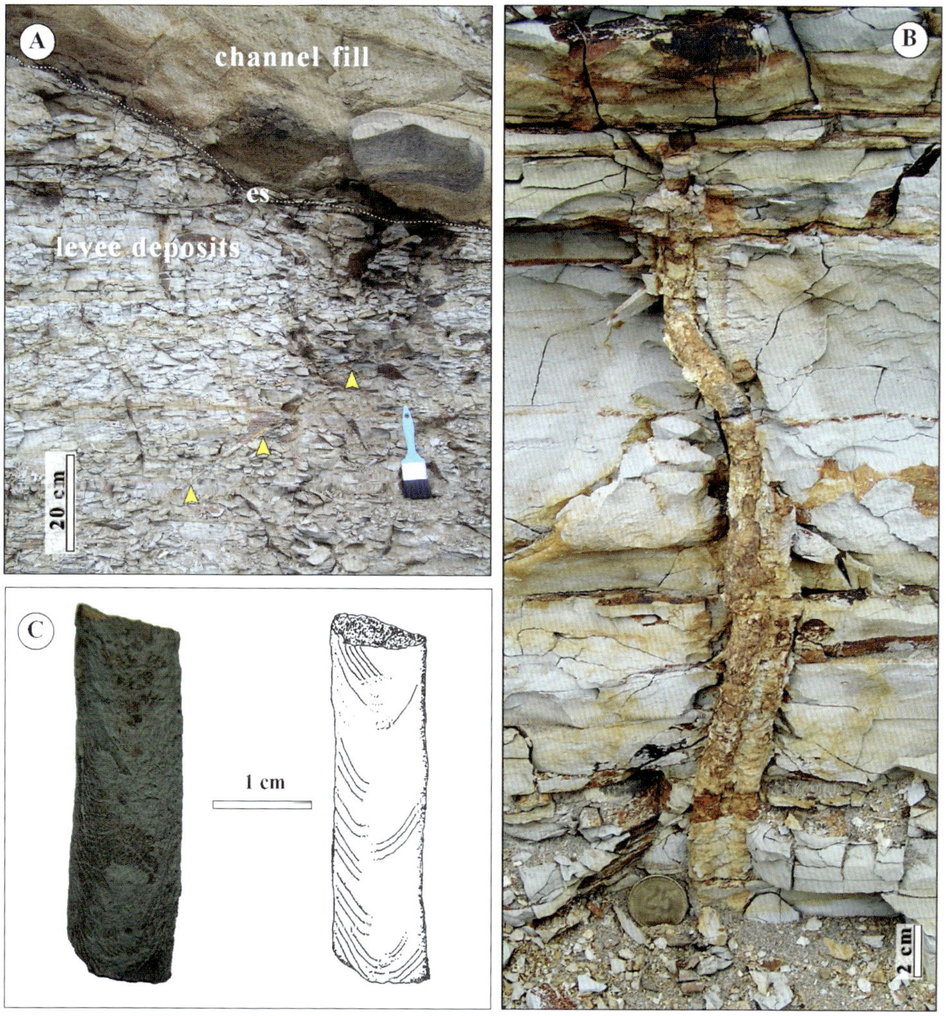

FIGURE 10. (A) Channelized erosion surface (es) colonized by *Arenicolites-Diplocraterion* tracemakers (yellow arrows). (B) Large specimen of *Arenicolites.* (C) Close-up view and schematic drawing of a partial limb of one specimen of *Arenicolites,* with an excellent preservation of bioglyphs, reflecting colonization of a firm substrate.

levee deposits of up to 250 m (820 ft) of lateral extension away from the channel axis.

The coarse-grained lower channel-fill FAs (FA1 and FA2) originated during the waxing period of discharge of the hyperpycnal flow, and include a complex arrangement of sedimentary structures. The preservation of these structures requires a particular set of key conditions: the sustained character of the flow, high sedimentation rates, and marked variations of sediment concentration, which are controlled by velocity fluctuations of the hyperpycnal flow.

The fine-grained upper channel-fill FA (FA3), composed of tabular graded beds, accumulated during waning stages of the hyperpycnal flow. In this context, the progressive deceleration of the flow generated a general backstepping of the system and produced only tabular graded bed deposits.

The levee deposits (FA3) accumulated during waxing stages of the hyperpycnal flow. The flocculation processes were very important for the development of these heterolithic facies, producing rapid sedimentation of aggregates of fine particles. The generation of floccules by nucleation is favored by the high concentration of organic material transported within the hyperpycnal flow. The occurrence of bed-load structures (current ripples), with abundant massive mudstone and phytodetritus in their foresets, suggests that floccules form aggregates with sizes equivalent to fine sand and therefore can be transported as bed load.

The multiple internal erosion surfaces without lateral continuity, the absence of lithologic contrasts between the sediment accumulated below and above the reactivation surfaces, and the lack of biogenic structures indicate that the lower hyperpycnal channel fill accumulated during a single sedimentation event. On the contrary, the presence of isolated levels with trace fossils in the upper channel fill (FA3) could suggest that these deposits accumulated from more than one sedimentation event.

ACKNOWLEDGMENTS

We thank Eduardo B. Olivero for his support and for the discussions on the subject. J. J. P. appreciates the kind offer of Carlos Zavala to lead the AAPG Hedberg Research Conference "Sediment Transfer from Shelf to Deepwater—Revisiting the Delivery Mechanisms" field trip with Olivero in Ushuaia, where the authors had the valuable chance to discuss the sedimentologic aspects on-site with conference participants. The study is part of the doctoral thesis of J. J. P. (Olivero and Zavala, supervisors). Preliminary versions of the work were presented at the AAPG Hedberg Conference (see Ponce and Olivero, 2008; Ponce et al., 2008b). Dany Martinioni is thanked for the assistance to the authors during the field work, the aid during the early writing of the chapter, and his critical contributions on its final version. Daniel Aureliano, Alvar Sobral, and José Contreras helped during field work. Roger Slatt is thanked for his useful review of this chapter. The owners of the estancias Inés and Viamonte granted access through their properties to the outcrops. Financial support for this study was provided by CONICET (PIP-5100 to E. B. Olivero), Agencia Nacional de Promoción Científica y Tecnológica (ANPCyT) (PICT-840 to N. B. Carmona), and Total Austral.

REFERENCES CITED

Bouma, A. H., 1962, Sedimentology of some flysch deposits: A graphic approach to facies interpretation: Amsterdam, Netherlands, Elsevier, 168 p.

Branney, M. J., and B. P. Kokelaar, 1992, A reappraisal of ignimbrite emplacement: Progressive aggradation and changes from particulate to non-particulate flow during emplacement of high-grade ignimbrite: Bulletin of Volcanology, v. 54, p. 504–520, doi:10.1007/BF00301396.

Branney, M. J., and P. Kokelaar, 2002, Pyroclastic density currents and the sedimentation of ignimbrites: Geological Society (London) Memoirs 27, 143 p.

Bruhn, C. H. L., and R. G. Walker, 1995, High-resolution stratigraphy and sedimentary evolution of coarse-grained canyon-filling turbidites from the upper Cretaceous transgressive megasequence, Campos Basin, Brazil: Journal of Sedimentary Research, v. 65, p. 426–442.

Fonnesu, F., 2003, 3D seismic images of a low-sinuosity slope channel and related depositional lobe (west Africa deep-offshore): Marine and Petroleum Geology, v. 20, p. 615–629, doi:10.1016/j.marpetgeo.2003.03.006.

Ghiglione, M. C., V. A. Ramos, and E. O. Cristallini, 2002, Estructura y estratos de crecimiento en la faja plegada y corrida de los Andes fueguinos: Revista Geológica de Chile, v. 29, p. 17–41.

Goodwin, R. H., and D. B. Prior, 1989, Geometry and depositional sequences of the Mississippi Canyon, Gulf of Mexico: Journal of Sedimentary Petrology, v. 59, p. 318–329.

Haq, B. U., J. Hardenbol, and P. R. Vail, 1987, Chronology of fluctuating sea levels since the Triassic: Science, v. 235, p. 1156–1167, doi:10.1126/science.235.4793.1156.

Jennette, D. C., K. Fouad, T. Wawrzyniec, D. Dunlap, R. Muñoz, and J. Meneses-Rocha, 2003, Slope and basin-floor reservoirs from the Miocene and Pliocene of the Veracruz Basin, southeastern Mexico: Marine and Petroleum Geology, v. 20, p. 587–600, doi:10.1016/j.marpetgeo.2003.01.001.

Kneller, B. C., and M. J. Branney, 1995, Sustained high-density turbidity currents and the deposition of thick massive sands: Sedimentology, v. 42, p. 607–616, doi:10.1111/j.1365-3091.1995.tb00395.x.

Kranck, K., 1973, Flocculation of suspended sediment in the sea: Nature, v. 246, p. 348–350, doi:10.1038/246348a0.

Malumián, N., 1999, La sedimentación y el volcanismo terciarios en la Patagonia extraandina. La sedimentación en la Patagonia extraandina, *in* Geología Argentina, Instituto de Geología y Recursos Minerales, Anales, v. 29, p. 557–612.

Malumián, N., and E. B. Olivero, 2006, El Grupo Cabo Domingo, Tierra del Fuego: bioestratigrafía, paleoambientes y acontecimientos del Eoceno-Mioceno marino: Revista de la Asociación Geológica Argentina, v. 61, p. 139–160.

Mayall, M., and I. Stewart, 2000, The architecture of turbidite slope channels, *in* P. Weimer, R. M. Slatt, J. Coleman, N. C. Rosen, H. Nelson, A. H. Bouma, M. J. Styzen, and D. T. Lawrence, eds., Deep-water reservoirs of the world: Gulf Coast Section SEPM Foundation, 20th Annual Research Conference, p. 578–586.

Mikkelsen, O., and M. Pejrup, 1998, Comparison of flocculated and dispersed suspended sediment in the Dollard estuary, *in* K. S. Black, D. M. Paterson, and A. Cramp, eds., Sedimentary processes in the intertidal zone: Geological Society (London) Special Publication 139, p. 199–209.

Mutti, E., and F. Ricci Lucchi, 1972, Le torbiditi dell' Appennino Settentrionale: introduzione all'analisi di facies: Memorie della Societa Geologica Italiana, v. 11, p. 161–199.

Nemec, W., 1990, Aspects of sediment movement on steep delta slope, *in* A. Colella and D. B. Prior, eds., Coarse-grained deltas: International Association of Sedimentologists, Special Publication, v. 10, p. 29–73.

Normark, W. R., 1970, Growth patterns of deep-sea fans: AAPG Bulletin, v. 53, p. 2170–2195.

Olivero, E. B., and N. Malumián, 2008, Mesozoic-Cenozoic stratigraphy of the Fuegian Andes, Argentina: Geologica Acta, v. 6, p. 5–18.

Plink-Björklund, P., and R. J. Steel, 2004, Initiation of

turbidity currents: Outcrop evidence for Eocene hyperpycnal flow turbidites: Sedimentary Geology, v. 165, p. 29–52, doi:10.1016/j.sedgeo.2003.10.013.

Ponce, J. J., and E. B. Olivero, 2008, Sediment transfer from shelf to deepwater—Revisiting the delivery mechanisms: AAPG Hedberg Research Conference Guidebook, p. 94.

Ponce, J. J., E. B. Olivero, and D. R. Martinioni, 2005, Estratigrafía y facies sedimentarias del Oligoceno-Mioceno Medio? de la Cuenca Austral de Tierra del Fuego, Argentina: Actas del XV Congreso Geológico Argentino (CD-ROM), Artículo no. 468, 2 p.

Ponce, J. J., E. B. Olivero, and D. R. Martinioni, 2008a, Upper Oligocene–Miocene clinoforms of the foreland Austral Basin of Tierra del Fuego, Argentina: Stratigraphy, depositional sequences and architecture of the foredeep deposits: Journal of South American Earth Sciences, v. 26, p. 36–54, doi:10.1016/j.jsames.2007.12.001.

Ponce, J. J., E. B. Olivero, and D. R. Martinioni, 2008b, Deep-marine hyperpycnal channel-levee complexes in the Miocene of Tierra del Fuego, Argentina: Architectural elements and facies associations, (abs.), *in* J. J. Ponce and E. B. Olivero, conveners, Sediment transfer from shelf to deepwater—Revisiting the delivery mechanisms: Conference Proceedings, AAPG Hedberg Research Conference, Ushuaia-Patagonia, Argentina, 4 p.

Ponce, J. J., E. B. Olivero, and D. R. Martinioni, 2008c, Hiperpicnitas de pie de talud depositacional en clinoformas del Mioceno marino de Cuenca Austral, Tierra del Fuego, Argentina: XII Reunión Argentina de Sedimentología, Buenos Aires, Resúmenes, p. 145.

Posamentier, H. W., 2003, Depositional elements associated with a basin floor channel-levee system: Case study from the Gulf of Mexico: Marine and Petroleum Geology, v. 20, p. 677–690, doi:10.1016/j.marpetgeo.2003.01.002.

Posamentier, H. W., and R. G. Walker, 2006, Deep-water turbidites and submarine fans, *in* H. W. Posamentier and R. G. Walker, eds., Facies models revisited: SEPM Special Publication 84, p. 399–520.

Prather, B. E., 2003, Controls on reservoir distribution, architecture and stratigraphic trapping in slope settings: Marine and Petroleum Geology, v. 20, p. 529–545, doi:10.1016/j.marpetgeo.2003.03.009.

Schieber, J., J. Southard, and K. Thaisen, 2007, Accretion of mudstone beds from migrating floccule ripples: Science, v. 318, p. 1760–1763, doi:10.1126/science.1147001.

Walker, R. G., 1978, Deep water sandstone facies and ancient submarine fans: Models for exploration for stratigraphic traps: AAPG Bulletin, v. 62, p. 932–966.

Zavala, C., J. J. Ponce, M. Arcuri, D. Drittanti, H. Freije, and M. Asensio, 2006, Ancient lacustrine hyperpycnite: A depositional model from a case study in the Rayoso Formation (Cretaceous) of west-central Argentina: Journal of Sedimentary Research, v. 76, p. 41–59, doi:10.2110/jsr.2006.12.

5

Paim, P. S. G., E. L. C. Lavina, U. F. Faccini, A. S. da Silveira, H. Leanza, and R. S. F. d'Avila, 2011, Fluvial-derived turbidites in the Los Molles Formation (Jurassic of the Neuquén Basin): Initiation, transport, and deposition, *in* R. M. Slatt and C. Zavala, eds., Sediment transfer from shelf to deep water—Revisiting the delivery system: AAPG Studies in Geology 61, p. 95–116.

Fluvial-derived Turbidites in the Los Molles Formation (Jurassic of the Neuquén Basin): Initiation, Transport, and Deposition

Paulo Sergio Gomes Paim, Ernesto Luiz Correa Lavina, and Ubiratan Ferrucio Faccini
Universidade do Vale do Rio dos Sinos (UNISINOS)-Programa de Pós-Graduação em Geologia (Brazil), São Leopoldo, Brazil

Ariane Santos da Silveira
Universidade do Vale do Rio dos Sinos (UNISINOS), São Leopoldo, Brazil

Héctor Leanza
Servicio Geológico Minero Argentino-Consejo Nacional de Investigaciones Científicas y Técnicas (SEGEMAR-CONICET) and Universidad Nacional de La Plata, Argentina

Roberto Salvador Francisco d'Avila
Petróleo Brasileiro S. A. (Petrobras), Rio de Janeiro, Brazil

ABSTRACT

This chapter is about an almost uninterrupted exposure of the Jurassic Los Molles Formation in the Arroyo La Jardinera region in the southern Neuquén Basin (western Argentina). It is based on the analysis of high-resolution satellite images linked with extensive field data. Its main goal is to present a conceptual model for the studied turbidite system capable of explaining how sand and gravel were brought onto and removed from the shelf margin and how these coarse-grained sediments were transported along and deposited on the slope and basin plain. The facies tract ascribed to this turbidite system represents high-frequency lowstand system tracts. From proximal to distal, it includes the record of fluvial channels incised into mid- to outer-shelf deposits, turbidite channels scoured on slope strata, turbidite lobes resting on slope rise facies, and poorly developed turbidite lobe fringes enclosed within basin-plain strata. These

Copyright ©2011 by The American Association of Petroleum Geologists.
DOI:10.1306/13271352St613437

gravity flow strata were assigned to short-lived (surges) moderately efficient turbidity currents derived from the almost continuous failure of delta-front deposits (sand and gravel) during ordinary fluvial floods.

INTRODUCTION

Based on the interpretation of high-resolution satellite images and extensive fieldwork, Paim et al. (2008) connected the gravity flow strata exposed in the Arroyo La Jardinera area (southern Neuquén Basin, Argentina) to submarine ramps ("sensu"; Heller and Dickinson, 1985), including multiple point sources and no major conduit or canyon, related to high-frequency lowstand (shelf-margin) deltas. Similarly, Mutti et al. (2003) had ascribed the succession exposed in the area to turbidite systems closely related to lowstand deltas (their type-A mixed systems). The main difference between both ideas is the ultimate depositional setting of the turbidite strata, whether a slope-to-basinal setting (Paim et al., 2008) or a distal delta-front–prodelta context (Mutti el al., 2003). A regional description of the Jurassic turbidites in the Neuquén Basin is presented by Burguess et al. (2000), but unfortunately the Arroyo La Jardinera area is only briefly mentioned.

Paim et al. (2008) associated the gravity flow deposits of the area with three independent systems: the first type of gravity flow system (GF1) was ascribed to gravel- and sand-rich density flows. Its facies tract encompasses fluvial channels incised into mid- to outer-shelf deposits, turbidite channels scoured on slope strata, turbidite lobes resting on slope rise facies, and poorly developed lobe fringe facies enclosed within the basin-plain strata. The second type corresponds to a sand-rich system with abundant stratification, including an oscillatory component, which was ascribed to low-concentration combined density flows (GF2). The third system is related to mass gravity flows (slumps and slides) and derived debris flows (GF3). A full description of the entire set of gravity flow deposits and enclosing facies within the Los Molles Formation exposed in the Arroyo La Jardinera area can be found in Paim et al. (2008).

This chapter focuses on the previously mentioned type-1 gravity flow system (GF1), by far the most important in terms of sand and gravel deposition in the study area. It presents many features that reveal several important issues related to the removal of coarse-grained sediments from the shelf margin (triggering processes) and their subsequent transport to the deep sea through sediment gravity flows. In terms of gravity flow initiation, some pieces of evidence suggest that type-1 gravity flows derive from long-lived hyperpycnal currents, whereas other data evoke short-lived depletive surges. In addition, some features indicate that fluid turbulence and other supporting mechanisms are major, but diverse, factors on suspended-load transport. Last but not least, some data indicate a strong connection of the GF1 deposits with high-frequency accommodation cycles and paleophysiography.

Therefore, our main goal is to discuss and present a conceptual model for the GF1 system capable of explaining how sand and gravel were brought into and removed from the shelf margin and how these coarse-grained sediments were transported along and deposited on the slope and basin plain.

GEOLOGICAL SETTING

The Neuquén Basin, located in central western Argentina and eastern Chile, extends northward along the axis of the Andean Cordillera up to 31°S. Between 34°S and 37°S, the basin is restricted by the cordilleran belt as a narrow north-south elongated strip, whereas southward, it broadens eastward into the extra-Andean domain, where it is known as the Neuquén Embayment. The Neuquén Basin features a triangular shape, covers more than 160,000 km^2 (61,700 mi^2), and extends up to 700 km (434 mi) in a north–south direction (Zavala, 1993). Besides the Andean chain to the west, this basin is bounded by the Pampeano-Sierra Pintada and North Patagonian massifs (Figure 1). The infill of the Neuquén Basin (Figure 2) ranges from late Triassic to Paleocene in age (Ramos, 1998) and exceeds 7000 m (22,900 ft) of marine and continental strata (epiclastics, carbonates, evaporites, and pyroclastics).

During the late Triassic, central western Argentina and eastern Chile underwent extensional stresses related to an arc/trench system developed along the western margin of the South American plate (Legarreta and Uliana, 1991; Gulisano and Gutierrez Pleimling, 1995). At that time, sedimentation was confined to isolated fault-bounded depressions controlled by structural features of the underlying sialic substrate. These depressions were filled with nonmarine facies, strongly influenced by pyroclastic and volcanic rocks, presenting abrupt thickness changes (Gulisano and Gutierrez Pleimling, 1995) and known as the Precuyo Group

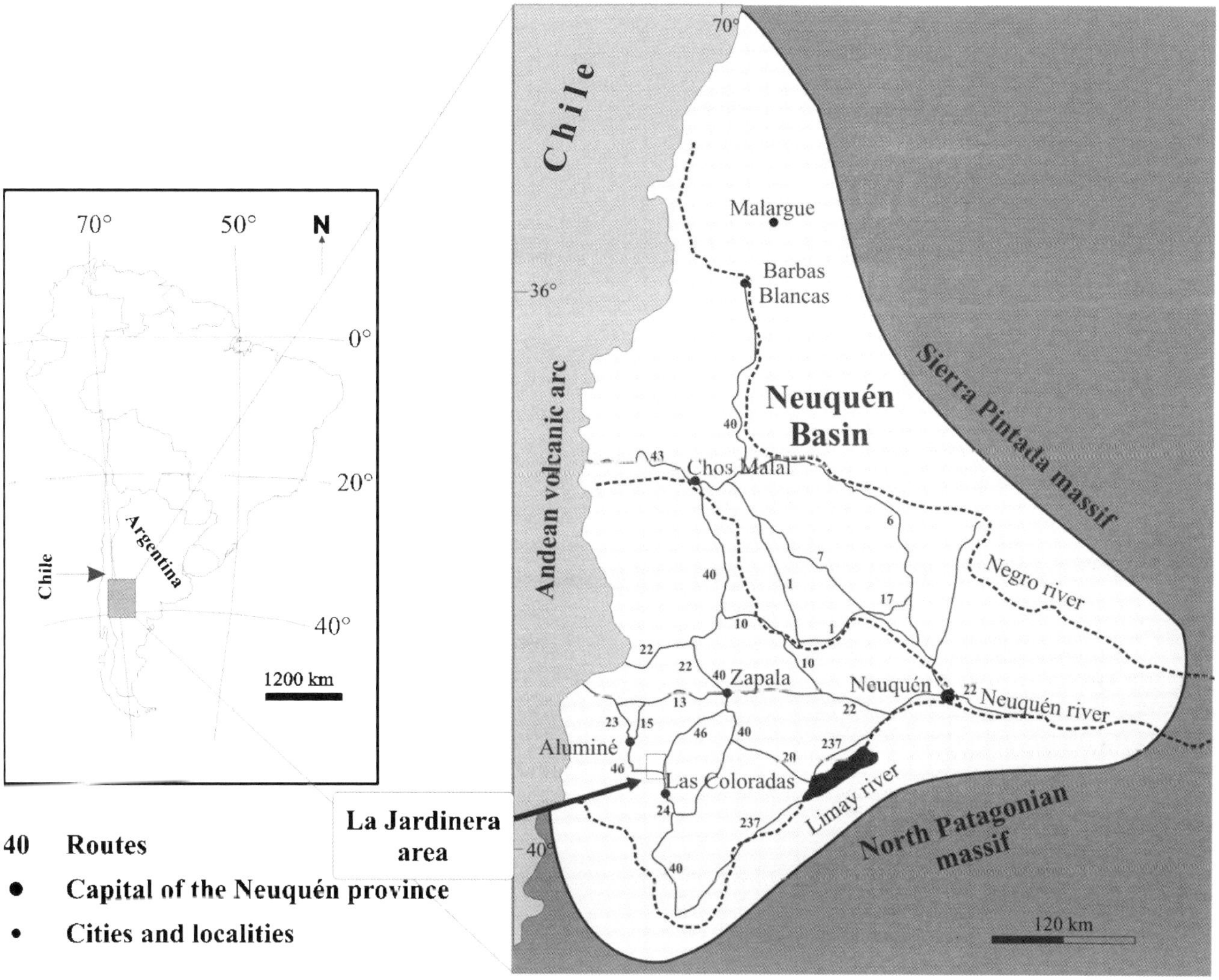

FIGURE 1. The Neuquén Embayment in South America and the location of the study area (Arroyo La Jardinera).

(Lapa Formation in southern Neuquén). These depocenters, sited on the east of the arc/trench system, became ever more unified to ultimately (Pliensbachian) merge into an extensive area of marine sedimentation placed between the volcanic arc to the west and the South American foreland to the east (Vicente, 2005).

From the early Jurassic to part of the middle Jurassic, deposition was chiefly driven by tectonics, a phenomenon that steadily reduced, giving way to a regional subsidence stage from the middle Jurassic to the early Paleogene (Ramos, 1998). Although most of the Neuquén Basin has maintained a nearly steady subsidence rate, records of local episodes of uplift, folding, erosion, and synsedimentary tectonics (Vergani, 2005) exist. Accordingly, sedimentation style and stratigraphic succession are records of those episodes of structural inversion and allied unconformities. The most important of such events produced the Huincul arch, a large east–west transcurrent fault zone in the southern Neuquén Basin associated with extensional stresses caused by the Gondwana fragmentation and ensuing Atlantic Ocean opening (Vergani et al., 1995). Several transgressions and regressions, mainly driven by eustatic sea level changes, occurred before the progressive growth of the Andean chain and cut off any oceanic influence into the basin (Legarreta and Gulisano, 1989). A rise in Andean tectonic activity as from the Miocene made the basin evolution even more complex.

The database used herein comes from the analysis of seismic-scale (10 × 10 × 1.6 km [6.2 × 6.2 × 0.99 mi]) high-resolution satellite images and the examination and logging (sedimentologic and gamma ray) of large natural exposures and road cuts. Bedsets and key stratigraphic surfaces can be followed for distances more than 10 km (6.2 mi). The adopted sequence-stratigraphic approach is similar to the approaches proposed by Van Wagoner et al. (1988, 1990), as sequence

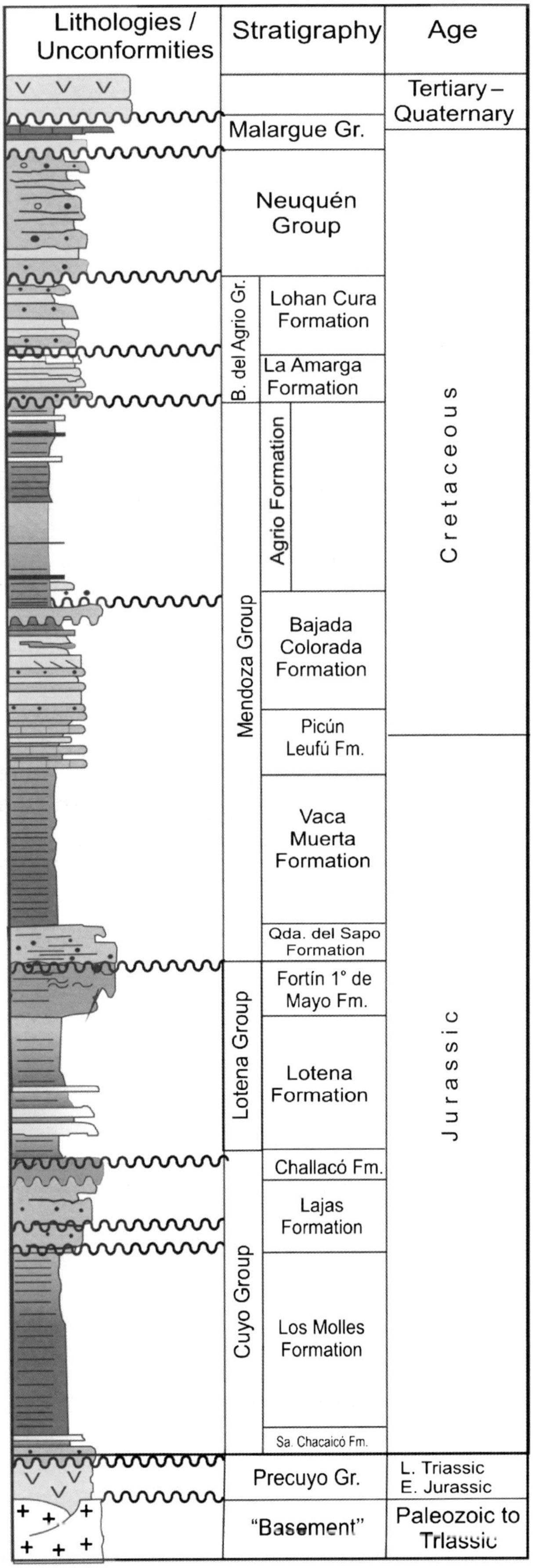

FIGURE 2. Regional stratigraphic column for the southern and central parts of the Neuquén Basin (modified after Leanza and Hugo, 1997). The studied interval comprises part of the Los Molles and Lajas formations. Gr. = Group; B. = Bajada; Fm. = Formation; Qda. = Quebrada; Sa. = Sierra, L. = Late; E. = Early.

boundaries are tied to the end of base-level fall, and by Hunt and Tucker (1992) and Helland-Hansen and Gjelberg (1994), as four system tracts are discriminated (lowstand, transgressive, highstand, and falling stage).

In the study area, a main unconformity (sequence boundary 1 [SB1]) outlines the base of the Cuyo Group and corresponds to the boundary between the Los Molles and Sierra Chacaico formations (Figure 3). The Cuyo Group comprises 1100 m (3600 ft) of deep-marine strata (Los Molles Formation), 550 m (1800 ft) of shelfal to coastal facies of the Lajas Formation (between SB5 and SB8) and, at the top, the red beds of the Challacó Formation (Figure 4). The studied section is bounded by SB2 and SB6 and includes the middle and upper parts of the Los Molles Formation as well as its transition to the Lajas Formation (Figures 3, 4). It was subdivided into four 3d-order depositional sequences (J21–J24) that encompass nearly 6 m.y. (Figure 4).

In the ensuing sections, the database is presented to highlight the main themes used to understand the turbidity currents in terms of initiation, transport, and deposition, and therefore to support the conceptual model proposed to explain the GF1 system.

MAIN FEATURES

Physiography

Based on the analysis of the facies, facies associations, and architectural elements, the Los Molles and Lajas formations have been ascribed to four paleophysiographic elements, that is, genetic units larger than architectural elements and analogous to present-day physiographic units from several marine basins (Paim et al., 2008). They include the slope rise or basin plain and the slope in the Los Molles Formation, the mid or outer shelf, including the shelf margin, at the transition between the Los Molles and the Lajas formations, and the inner shelf or nearshore in the Lajas Formation (Figures 3, 4). These 10- to 100-m (33- to 330-ft)-thick paleophysiographic units include smaller, few to several meters thick, architectural elements with distinct geometries and facies associations (Figure 4).

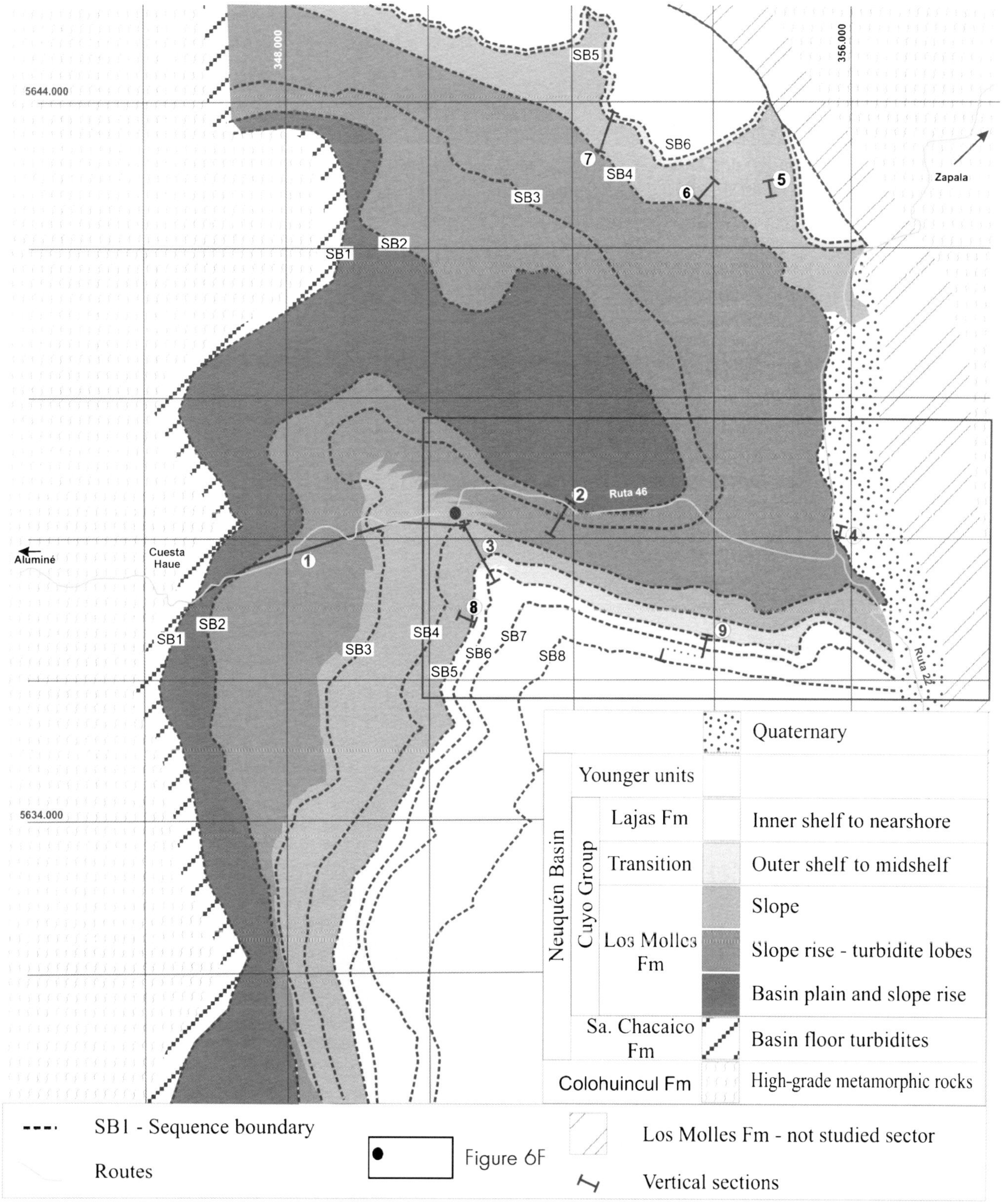

FIGURE 3. Geological units in the study area (Arroyo La Jardinera Region) and paleophysiographic elements and sequence boundaries in the Los Molles and Lajas formations from the interpretation of a high-resolution satellite image (by IKONOS). The acquisition scale of the vertical sections ranges between 1:100 and 1:500. Fm = Formation; Sa. = Sierra.

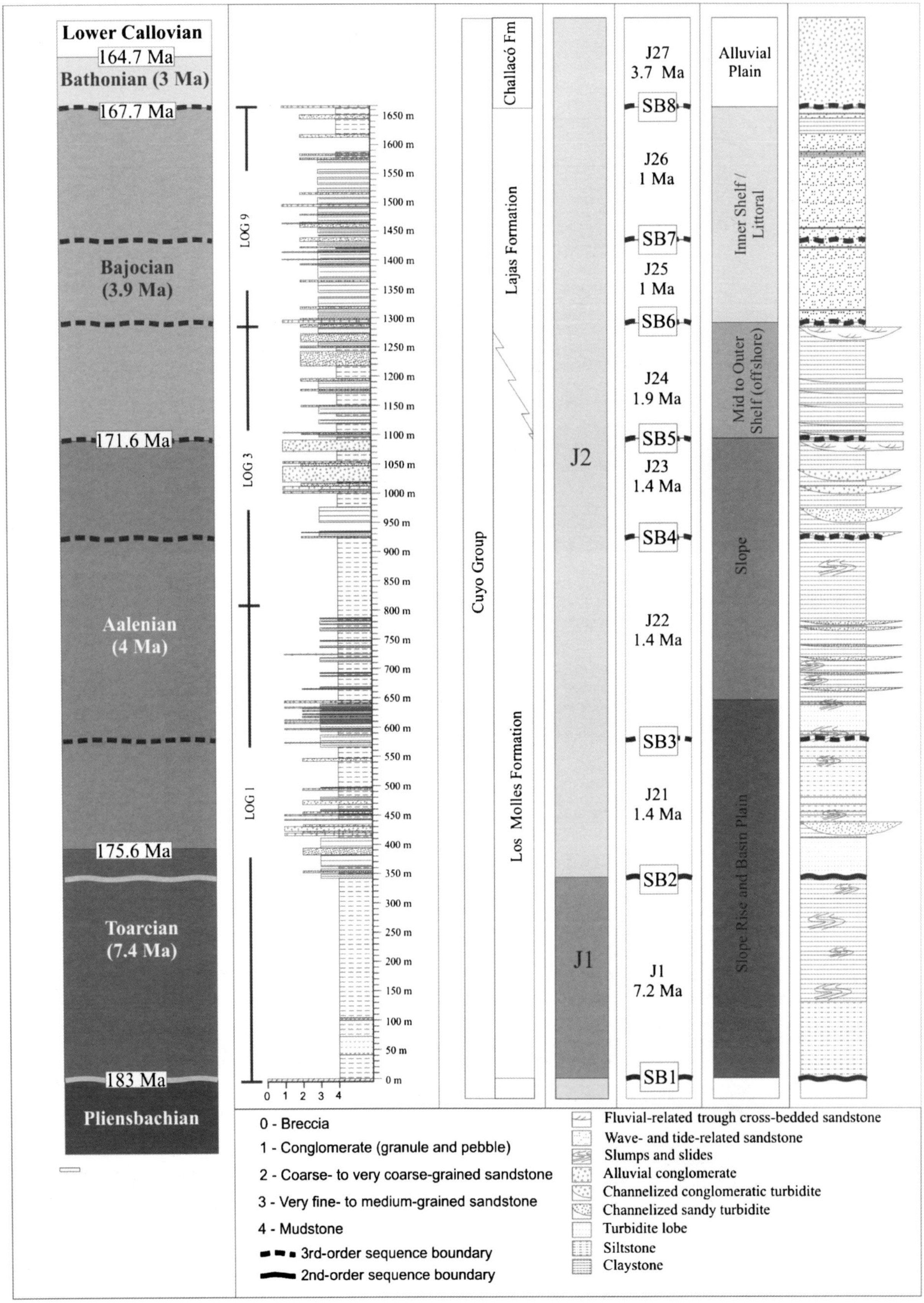

FIGURE 4. Simplified composite vertical section of the study area. Stages obtained from correlation to Gulisano and Gutierrez Pleimling (1995) chart and absolute ages from Gradstein et al. (2004) chart. SB = sequence boundary; Fm = Formation; J21– J27 = 3rd-order depositional sequences; J1, J2 = 2nd-order depositional sequences.

FIGURE 5. Physiography: (A) Fine-grained and (B) heterolithic lobe fringe deposits related to basin plain; (C) lobe sandy (sandstone bedset ranges from 5–15 m [16.4–49.2 ft] in thickness) and (D) slumped beds ascribed to the slope rise; (E) turbidite channel and (F) slide (white arrow) associated with slope strata; and (G) falling-stage wave-rippled sandstone resting on outer-shelf mudstone and scoured by (H) incised channel-fill facies.

A
B
C
inner shelf
SB6
mid to outer shelf
SB5
Incised fluvial channel
slope
Turbidite channel
slope rise
D
E
F
G
H

In the Los Molles Formation, profuse fine-grained deposits (Figure 5A, B), tabular beds of sandstone sets ascribed to turbidite lobes (Figure 5C), fossil-rich (ammonites and small bivalves) layers, and dominantly planktonic radiolarian specimens among the microfauna (Kochhann et al., 2009) characterize the slope rise to basin plain. The distinction between slope rise and basin-plain strata is based on the exclusive occurrence of slumped beds in the former (Figure 5D). Similar faunal content, slumped beds, commonly coupled with slide deposits (Figure 5F), and turbidite channel fills encased by muddy interchannel facies (Figure 5E) define the slope realm. The transition between the Los Molles and Lajas formations consists of mid- to outer-shelf siltstone, sporadically enclosing sharp-based nearshore facies (Figure 5G) scoured by fluvial channels (Figure 5H). Lastly, a sandy interval (Lajas Formation), which includes wave-, storm- and tidal-related structures and some muddy intervals a few meters thick, defines the inner-shelf–nearshore paleophysiographic element.

Basin-plain shale and lobe–lobe fringe turbidites of the lower and mid–Los Molles Formation, respectively, followed by turbidite and incised fluvial channels encased in extensive fine-grained facies of the slope (upper Los Molles Formation) and mid to outer shelf (transition from Los Molles to Lajas formations), in that order, make up a succession that is very similar to that presented by Gulisano and Gutierrez Pleimling (1995) for the same area. This succession is supposed to be an outcrop analog of the tripartite subdivision of the Los Molles Formation defined by Gómez Omil et al. (2002) and Verzi et al. (2005) from seismic data.

In short, the GF1 system is linked to a paleophysiography composed of two gently dipping surfaces (shelf and basin plain) separated from each other by a steeper unstable slope. The thickness of the slope succession (about 400 m [1300 ft] thick; Figure 4) suggests a fairly deep-water setting (hundreds of meters of water depth) for the basin-plain facies. Similar physiography and water depth of the basin-floor turbidites were described by Steel et al. (2000).

Architectural Elements

The basinward tract of architectural elements of the GF1 system encompasses incised fluvial channels (shelf and shelf margin), turbidite channels (slope), and widespread turbidite lobes and lobe fringes (slope rise and basin plain).

The downdip and vertical changes of the GF1 architectural elements (Figures 3, 4) illustrate the basinward transition from incised fluvial channels through turbidite channels to turbidite lobes and lobe fringes (Figure 6). Incised fluvial channels are encased within shelf to shelf-margin strata; turbidite channels are scoured into and surrounded by slope, commonly remobilized fine-grained deposits; and turbidite lobes and lobe fringes are enveloped by fine-grained slope rise and basin-plain facies.

Facies Tract

Plane- and cross-bedded sandstone and both massive and stratified clast-supported conglomerates with sparse wood pieces represent the main fluvial channel-fill facies (Figure 7A–C). A minor tidal influence is suggested by the rare presence of rippled fine grained sandstone with flaser and mud drapes near the base of the channel fills. Together, these pieces of evidence, in addition to the lack of sedimentary features associated with hyperconcentrated flows (sensu; Pierson and Costa, 1987), point to prevailing usual conditions of fluvial transport and deposition within incised braided fluvial channels, even during flood events (i.e., ordinary floods).

In contrast, the turbidite channel fills display a clear basinward reduction in mean grain size from clast-supported through matrix (sandy)-supported conglomerates to sandstones (Figure 7D–G). This down-dip segregation strongly suggests that fluid turbulence was a major factor in the long-distance transport of gravel and sand. This inference is reinforced by profuse flute casts at the base of the turbidite channels (Figure 8E) and rip-up clasts near their bases (Figures 7G, 8C). However, nonturbulent granular flows are indicated by the massive (Figure 7D–G) nongraded nature of most of the turbidite channel-fill beds. Their simple (in opposition to complex) and relatively thin character (0.1–0.5 m [0.33–1.6 ft]; Figure 8A, C) suggests short-lived currents. Jointly, both sets of data point to turbulent surges that developed downcurrent, with ever finer grained basal granular layers as a result of gravity transformation (sensu; Fisher, 1983).

FIGURE 6. **Architectural elements and facies associations: (A, B) Incised stratified fluvial deposits (see arrows for location); (C) turbidite and fluvial channels scoured on slope and shelf-margin facies, respectively; (D, E) massive turbidite channel strata; (F, G) tabular bedsets of massive sandstone related to the turbidite lobe; and (H) thin-bedded turbidites of the lobe fringe. SB = sequence boundary.**

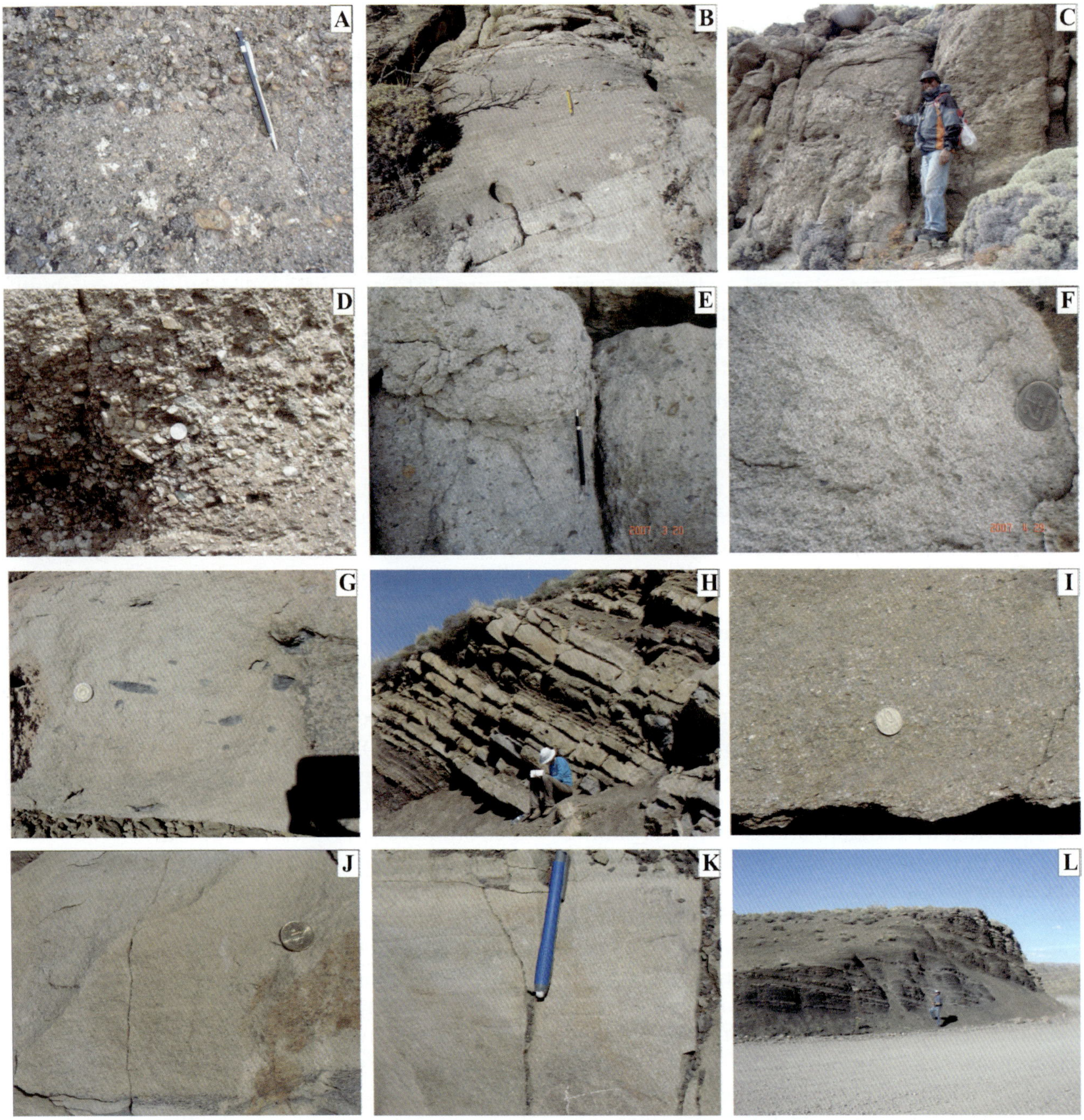

FIGURE 7. Facies tract: Clast-supported conglomerate (A) and plane and trough cross-bedded sandstone (B), sometimes with large pieces of trunks (C), associated with incised fluvial channel fill; clast (D)- and matrix (E)-supported conglomerate and massive sandstone (F), usually with rip-up clasts (G), related to turbidite channel fill; laterally extensive turbidite lobe strata (H) comprising massive coarse (I)- to fine (J)-grained sandstone, occasionally with incipient lamination (K); and heterolithic deposits (L) of the lobe fringe.

Downcurrent turbidite lobe and lobe fringe facies display similar features to those of the turbidite channel fills, except for their extensive geometry (Figures 5C, 6F, 7H, 8B), finer grained nature (Figure 7I–K), less abundant flute casts and rip-up clasts (Figure 8D, E), and more frequent normal grading (Figure 8F) and plane bedding (Figures 7K, 8G). Relative to turbidite channel fill, lobe strata are supposed to represent similar bipartite turbidity currents, but moving a finer grained suspended load in poorly constrained to wholly unconfined flows.

In short, GF1 deposits have been related to short-lived bipartite turbidity currents. Fluid turbulence has acted as the main supporting mechanism in

FIGURE 8. Surgelike density flow suggested by discrete medium-scale sedimentary bodies within turbidite channels (A) and lobes (B); pieces of evidence pointing to the turbulent nature of the long-distance transport, including rip-up clasts in both turbidite channel (C) and lobe strata (D), common to abundant flute casts at the turbidite channel base (E) and normal grading (F) and plane bedding (G) in turbidite lobe strata; and nearby continental sources indicated by common plant debris within both turbidite channel (H) and lobe (I) deposits.

long-distance transport of the coarse-grained suspended load from the shelf margin to the deep sea. However, the massive nongraded nature of most of the turbidites indicates that nonturbulent flow conditions would have left the ultimate mark on the deposits themselves. This non-Newtonian (pseudoplastic) rheology is supposed to represent transient conditions confined to the basal layer of the bipartite density flows and produced just before deposition could occur through frictional freezing.

Paleocurrents

Paleocurrent readings of trough cross-bedding within fluvial beds (incised channels) point to east–northeastward effluents (mean value, 83°; n = 24). Flute casts on turbidite beds show a similar trend (mean value, 67°; n = 60). All together, GF1 paleocurrent data point to northeastward currents (mean value, 72°; number of readings (n) = 84) that are similar to the inferred sediment transport path proposed for the area by Burguess et al. (2000) This mean value is roughly orthogonal (angle of 87°) to that presented by wave-induced cross-bedding (mean value, 159°; n = 126) ascribed to alongshore currents (Figure 9). It is also worth noticing that the sediment dispersal trends of both GF1 and nearshore systems have not changed with time (Figure 10).

The aforementioned paleocurrent data strengthen the idea of some genetic linkage between fluvial effluents and turbidity currents. Both dispersal systems comprise multiple small-scale source points (Figure 6C),

(A)

Depositional systems	Morphological element - sedimentary structure - mean paleocurrent vector		General mean vector
Nearshore, wave and tidal current.	Upper shoreface (trough cross-bedding - 122 readings - mean vector: 159) Flood tidal delta (trough cross-bedding - 4 readings - mean vector: 154)	159	GF1 n = 84
GF1 system tract	Muddy slope rise and plain and lobe fringe (current ripples - 22 readings - mean vector: 63) Turbidite lobe (flutes, current ripples, and rare trough cross-bedding - 9 readings - mean vector: 105) Turbidite channel (flutes and a few trough cross-bedding - 29 readings - mean vector: 59) Incised fluvial channel (trough cross-bedding - 24 readings - mean vector: 83)	72	n = 126

(B)

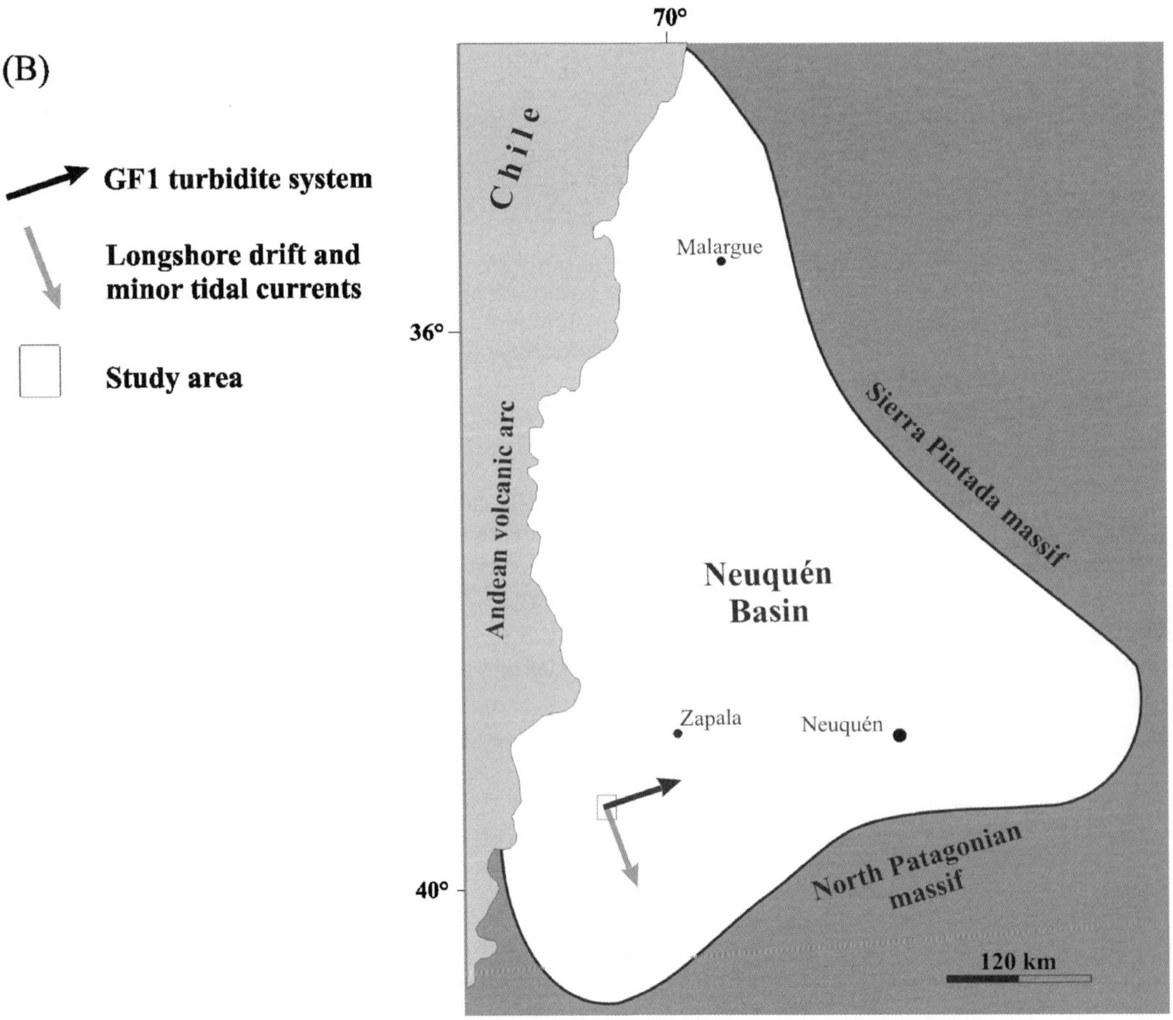

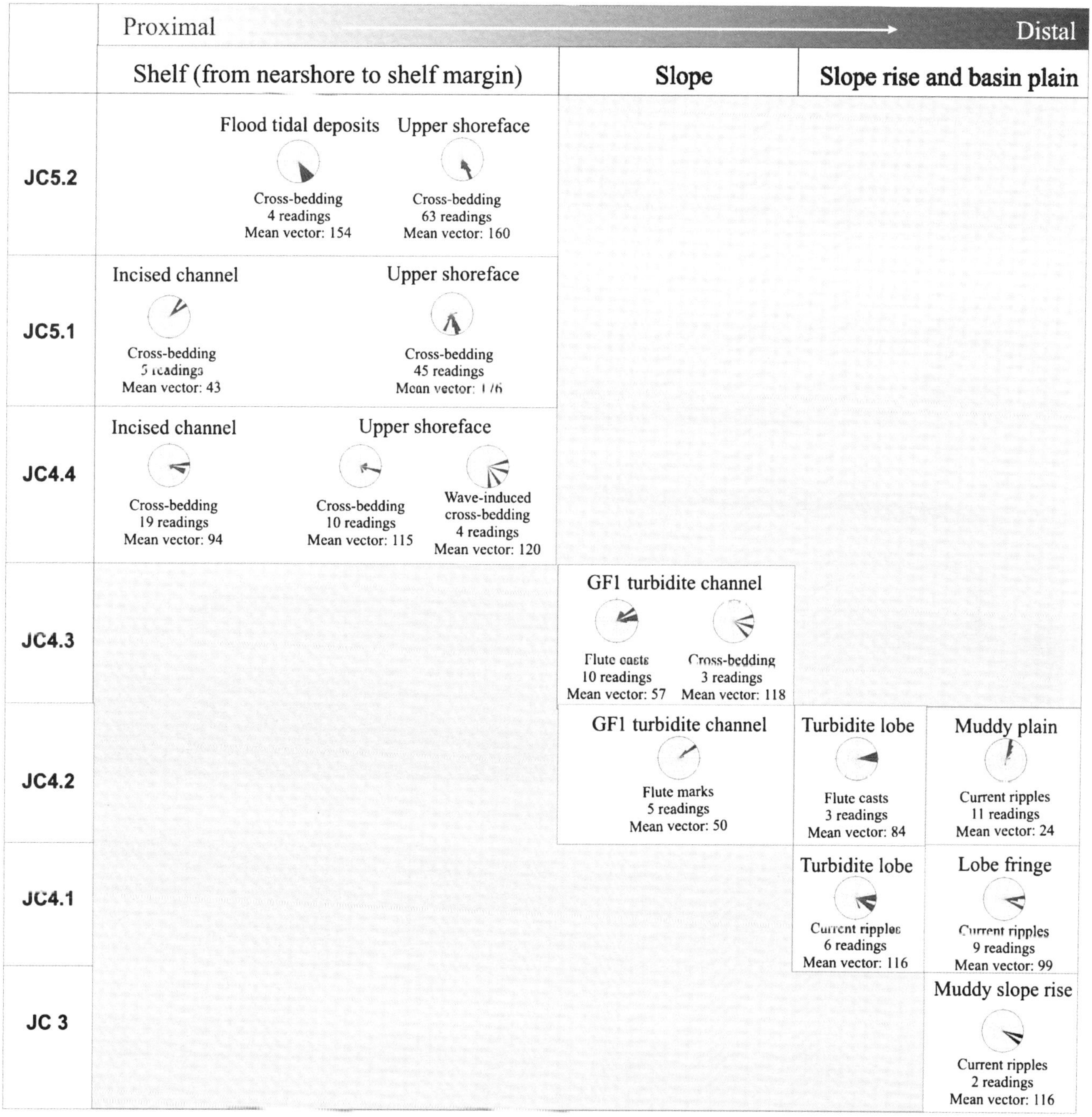

FIGURE 10. Paleocurrents of the first type of gravity flow system (GF1) through time and space. Nearshore, and wave- and tide-induced paleocurrents are also shown for comparison. Notice the steady transport trend of the GF1 strata within distinct paleophysiographic elements, from shelf to basin plain, as well as along time through different 3d-order depositional sequences.

FIGURE 9. The first type of gravity flow system ([GF1] fluvial and turbidity currents) and nearshore, mostly alongshore currents in the study area displaying mean vectors at right angles (A). In the map (B), notice alongshore currents parallel to the presumed local basin margin (north-northwest–south-southeast) and northeasternward toward the basin depocenter GF1 fluvial and turbidity currents. n = number of readings.

A
B
C
D
inner shelf
mid to outer shelf
slope
slope rise
E
F
G

and they have always flowed toward the basin depocenter, whereas alongshore transport is parallel to the orientation of the local basin border (Figures 9, 10).

High-frequency Sequences

The Cuyo Group is subdivided into two 2d-order depositional sequences, named J1 and J2, which record around 7 and 12 m.y., respectively (Figure 4). A key surface (SB2) delineates their boundary and the onset of a thick package of sandy turbidites that abruptly rests upon a thick package mostly composed of deep marine shale (Figures 3, 4). The J1 depositional sequence comprises the base of the Los Molles Formation, and it records a transgressive-regressive succession. The upper depositional sequence (J2) consists of the Los Molles (mid and upper sections), Lajas, and Challacó formations, and records a regressive unit (Figure 4). The latter sequence is subdivided into seven 3d-order depositional sequences (J21–J27), each one recording 1 to 3 m.y. (Figures 3, 4) and comprising up to two paleophysiographic elements (Figure 3). Besides, the analysis of the facies successions within all paleophysiographic elements and sequences indicates a high-frequency (4th- to 5th-order) modulation of the geological record, which is the theme of the next paragraphs (see also Figures 11, 12). The stacking pattern presented by the turbidite strata exposed in the study area has been used by Mutti (1992) as an example of high-frequency cyclicity on the scale of facies sequences.

In the outer- to midshelf strata, transgressive and highstand mudstones are sharply overlain (regressive surfaces of marine erosion) by wave- and tide-related nearshore deposits of the falling-stage system tract (Figures 5G; 11A, D). In turn, these nearshore facies are then scoured by small fluvial channels (Figure 6B, C) related to the later development of the falling-stage system tract and ensuing expansion of the subaerial unconformity. At the start of relative sea level rise, incised channels were then filled by fluvial (Figures 5H; 6A, B; 7A–C; 11B, D) locally tide-influenced facies. The subsequent increase in the rate of relative sea level rise once more drives the onset of the transgressive system tract over the shelf and consequent extensive silt deposition. In short, high-frequency relative sea level oscillations are recorded in outer- to midshelf strata by the falling-stage (regressive surface of marine erosion and succeeding fluvial incision), lowstand (channel-fill), and transgressive (shelfal siltstone above the surfaces of maximum regression) system tracts.

In the slope, by the end of the falling stage, widespread mudstone connected to the previous system tracts (transgressive, highstand, and falling-stage) was scoured by point-sourced turbidity currents (Figures 6D, E; 11D). Through the lowstand stage, the resultant channels were then filled by amalgamated lenses of gravel- to sand-rich upward and downdip fining turbidites (Figures 6C; 7D–F; 11C). Channel abandonment and resuming widespread mud deposition define the beginning of a new transgressive episode. Thus, 4th- to 5th-order relative sea level fluctuations are recorded in the slope by turbidite scour (early lowstand), turbidite channel fill (late lowstand), and widespread mud deposition (transgressive, highstand, and falling-stage system tracts).

On the slope rise and proximal basin plain, mudstone credited to a preceding high frequency cycle (late lowstand, transgressive, highstand, and falling-stage system tracts) are regularly overlain (Figure 11G) by tabular bedsets (Figures 5C; 6F; 11E) composed of sand-rich (Figures 6G; 7I–K) few decimeters thick (Figures 7H; 8B) turbidites that supposedly represent the early lowstand system tract. A new increase in the rate of relative sea level rise leads to the deposition of a mud blanket over the sandy turbidites (Figure 11E, F). This package, which is between few to few tens of meters thick, of fine-grained deposits includes the subsequent late lowstand, transgressive, highstand, and falling-stage system tracts.

DISCUSSION

Short- Versus Long-lived Sediment Gravity Flows

Some features suggest a close link between fluvial discharge and sediment gravity flows. The tripartite basin physiography and high-frequency lowstand context promote appropriate conditions for the initiation of turbidity currents from hyperpycnal flows (e.g., Steel et al., 2000; Plink-Björklund and Steel,

FIGURE 11. High-frequency depositional sequences: (A) sharply based tidal bar deposits (falling stage) and (B) incised fluvial strata (lowstand), both within dominantly muddy shelfal strata; (C) sandy turbidite channel (lowstand) surrounded by muddy slope facies (sandstone bedset ranges from 5–15m [16.4–49.2 ft] in thickness); (D) relative location of (A), (B), and (C); (E) general view of the high-frequency depositional sequences on the slope rise to basin-plain realm (active and abandonment phases of the turbidite lobes); and detailed view of the upper (F) and lower (G) abrupt limits of the lobe bedsets (sandy facies related to the lowstand active phase) with the dominantly muddy deposits (abandonment phases) ascribed to the transgressive and highstand systems tracts.

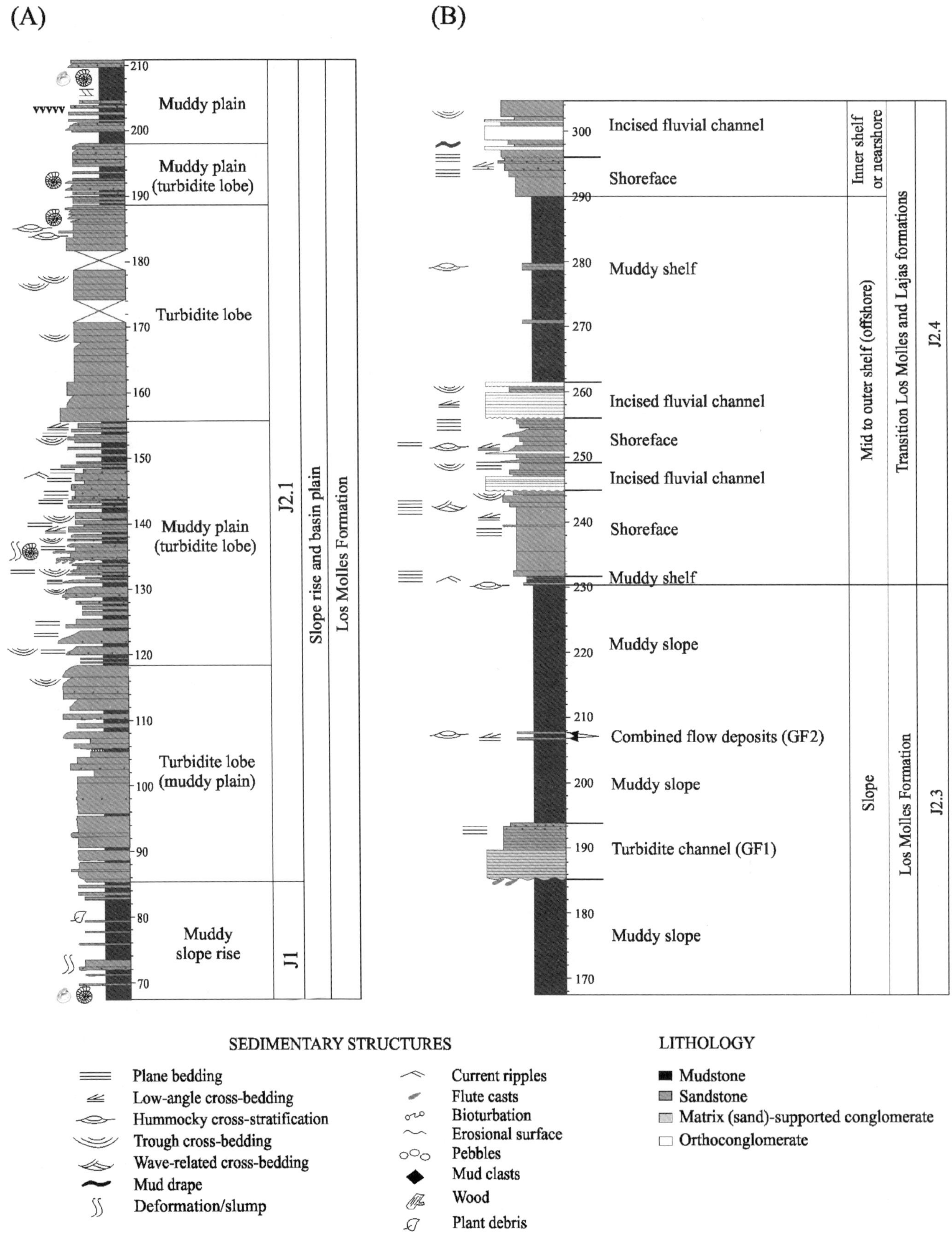

FIGURE 12. Segment of logs 2 (A) and 7 (B). Notice the high-frequency signal modulating the stratigraphic architecture. See log location in Figure 3. Log depths in m. J1 = 2d depositional sequence; J2.1–J2.4 = 3d order depositional sequences.

2004). It is worth noticing that the GF1 facies tract records a downdip change from fluvial facies through gravel- to sand-rich sediment gravity flows. In addition, the fluvial and GF1 systems show parallel paleocurrent trends, a wide range of available grain size (gravel to very fine-grained sand) and prolific plant debris (Figure 8H, I). These data reveal some sort of connection between the fluvial effluents, which delivered sand and gravel to the shelf margin and the GF1 system. However, no support exists in the fluvial deposits to indicate high-density effluents being directly delivered to the slope. On the contrary, the fluvial facies point to braided streams with ordinary gravel and sand bars. The nontorrential character of the fluvial facies associated with the massive and simple (in opposition to complex) nature of the relatively thin gravity flow beds, which lack any internal erosional surface, are evocative of short-lived depletive surges instead of long-lived waxing to waning hyperpycnal currents (Mulder et al., 2003). A brief review of some criteria used elsewhere to recognize hyperpycnal flow deposits in the geological record and their correlation with the GF1 facies is presented in Table 1.

Although these two interpretations (sustained- versus surge-type) seem incompatible at first glance, the specific features that support each one are herein assembled in terms of an indirect fluvial control, that is, surgelike turbidity currents derived from the recurring failure of mouth-bar and delta-front sediments. It is also worth noticing that the clean nature of the sand- and gravel-rich GF1 strata, which display a very fine-grained (or coarser) sandy matrix, indicates that a large part of the fluvial suspended load (silt and clay) was not present in the turbidity currents.

Based on the concepts of Bates (1953) and Wright (1977), and on aforementioned pieces of evidence, it is assumed herein the existence of stratified fluvial effluents with well-defined bed and suspended load that did not enter the sea as inertia flows (truly hyperpycnal flows). Instead, friction-dominated fluvial effluents dumped their bed load (gravel and sand) in the delta front, whereas their suspended load was transported much farther as a hypopycnal plume. This case is similar, but not equal, to the subaerial inflow with an intermediate average density simulated by McLeod et al. (1999). This kind of inflow is strongly influenced by its internal stratification as its sediment-depleted upper layer (richer in finer grained suspended load) becomes a buoyant plume in the receiving basin, whereas its particulate-enriched lower section penetrates the tank as a dense flow. In the case presented herein, this lower sediment-enriched part of the flow could be responsible for the high rate of sediment fallout occurring in the delta front in conjunction with the tractive accumulation of the bed load.

Turbidity Currents Versus Sandy Debris Flows

Massive sandstones are dominant along the entire GF1 system. This fact stresses that other supporting mechanisms instead of fluid turbulence (e.g., fluid escape, hindered settling, and buoyancy) were important factors in the transport of the sediments, as in noncohesive sandy debris flows (Shanmugam, 2000). However, the quite distinct downcurrent grain-size segregation underlines the importance of the fluid turbulence in the long-distance transport of sand and gravel, that is, turbidity currents.

To accommodate such apparently conflicting ideas, bipartite turbidity currents, similar to the density-stratified flows envisaged by many authors (e.g., Kuenen, 1951; Bagnold, 1956; Sanders, 1965; Postma et al., 1988; Mutti et al., 1999), are proposed to equally explain the long (grain-size segregation caused by fluid turbulence)- and the short (ephemeral prefreezing basal inertia flows)-distance transport. The option for using the term "turbidity currents" instead of "noncohesive debris flows" emphasizes the long-distance mechanism of transport and its resultant facies tract instead of short-distance fluid rheology in the density flow basal layer and its associated massive fabric.

Autogenic and Allogenic Control

Autogenic and allogenic processes control the initiation of turbidity currents. As previously mentioned, turbidity current initiation has been related to high rates of sand and gravel deposition at distributary mouth bars during fluvial floods. This condition has induced the recurrent failure of the delta front and the resulting turbidity currents. However, as already presented, age (Gulisano and Gutierrez Pleimling, 1995) and sedimentologic (facies successions and associations) data about the shelf margin, slope, and basin-plain deposits indicate that GF1 surges were associated with high-frequency (4th to 5th order) lowstand stages and closely related short-lived shelf-margin deltas. This idea implies that an allogenic control, that is, relative sea level fall, is a major factor in promoting the right conditions for turbidity current triggering, that is, high rates of coarse-grained sediments (gravel and sand) being rapidly delivered over a relatively steep slope (lowstand shelf-margin deltas).

Table 1. Positive (yes) and negative (no) correlation between features found in the GF1 deposit and some criteria used elsewhere in terms of hyperpycnal flow identification (from Paim et al., 2008).

Criteria of Hyperpycnal Flow Recognition	*Reference*	*Correlation*
Physical connection between fluvial and turbidite channels at the shelf margin	Plink-Björklund and Steel (2004)	Yes, through small incised channels
Alternating massive and stratified intervals within a single bed	Plink-Björklund and Steel (2004); Bhattacharya and MacEachern (2008); Arcuri and Zavala (2008); Bourget et al. (2008); Vesely (2008)	No. Massive, stratified, or massive with stratified top discrete beds
Coarsening- and thickening-upward trend near the bed base and the reverse above		No. Thin, discrete, and uniform or graded beds
Abundant cut-and-fill features in a single bed		No. Absent
Laterally irregular grain-size changes		No. Absent
Sandy nature and abundant meter-scale beds	Piper and Savoye (1993); Mulder et al. (1998); Kneller and Buckee (2000); Mulder and Alexander (2001); Plink-Björklund and Steel (2004); Arcuri and Zavala (2008); Vesely (2008)	No. Gravel and sand as discrete decimeter-scale-thick beds
Systematically accreted shelf margin with associated turbidites	Plink-Björklund and Steel (2004)	Yes
Abundant continental plant material (leaves, phytodetritus, and coal fragments) in the turbidite beds	Mulder and Alexander (2001); Johnson et al. (2001); Plink-Björklund and Steel (2004)	Yes
Turbidite beds with lamination outlined by phytodetritus	Zavala (2008)	Yes, but more common in GF2 beds
Small amount of slump, slide, and debris-flow deposits	Plink-Björklund and Steel (2004)	No. High proportion of mass and debris-flow facies (GF3 facies)
Downcurrent pinch-out of thick beds of sandy turbidite	Plink-Björklund and Steel (2004)	No (only in GF2 strata)
Rounded and aligned mud clasts		Yes (usually aligned)
Siltstone and claystone-graded couplets with syneresis cracks	Bhattacharya and MacEachern (2008)	No, absent
Low to moderate trace fossil diversity, low bioturbation intensity, normal size ichnotaxa, lack of complex behavior strategies, suspension-feeding organisms	Bhattacharya and MacEachern (2008); Carmona et al. (2008)	Yes, usually low trace fossil diversity and bioturbation intensity and lack of complex behavior strategy

GF1 = first type of gravity flow system.

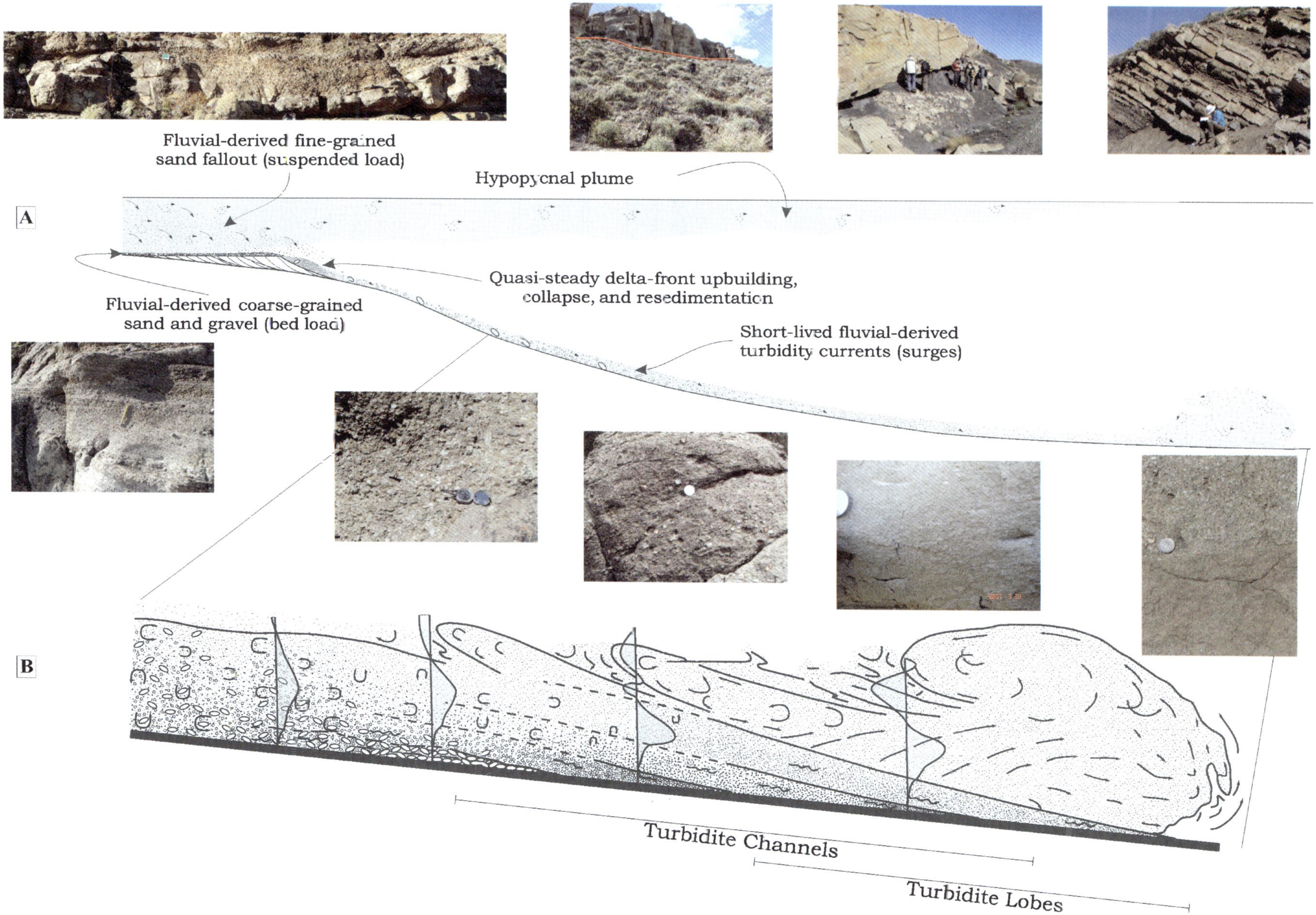

FIGURE 13. Conceptual model of triggering (A), transport with grain segregation (A, B) and deposition (B) of fluvial-derived turbidite surges. The top photographs exemplify the main architectural elements (incised fluvial channels, conglomeratic and sandy turbidite channels, and turbidite lobes), and photographs in the midportion represent the facies tract. Draft shown in panel B was adapted from Postma et al. (1988).

CONCLUDING REMARKS

Fluvial channels at the shelf margin, turbidite channels in the slope, and turbidite lobes and lobe fringes in the slope rise and basin plain represent the GF1 system in terms of architectural elements and physiographic settings. The increase of the gradient at the transition from the outer shelf to the slope enhances gravity flow buildup and entrenchment (turbidite channels in the slope), whereas the opposite alteration at the transition from the slope to the slope rise and basin plain favors large-scale sand deposition through poorly confined channels to widespread turbidite lobes. Multiple point sources related to high-frequency lowstand deltas gave origin to areas of high rates of sedimentation along the shelf margin.

The conceptual model (Figure 13) assumes average fluvial floods producing a quasi-steady delivery of sand and gravel onto continuously aggrading unstable delta-front ramps prograding into ever deeper water. Concurrently, the fluvial suspended load (silt and clay) was dispersed in the basin as hypopycnal plumes. After a brief storage period, debris flows were initiated through episodes of remobilization of the sediments deposited in the distributary mouth bars formed in such unstable areas. Because of flow acceleration along the slope, the gravity flow then builds up, becoming fully turbulent (turbidity current surges) and causing slope erosion. Subsequent deceleration produced a downcurrent grading through consecutive waves of sedimentation. These accumulation pulses were the result of gravity flow transformations driving flow bipartition and ensuing frictional freezing of the lower granular layers composed of ever finer grained sediments (massive strata). The lack of fine-grained particles in the turbidity currents, which results from flow separation at the mouth of the distributary channels, is indicated as the cause of the poor development of the truly tabular lobe beds and extensive lobe fringe facies as flow efficiency is likely to have diminished because of the small amount of clay in the fluid.

ACKNOWLEDGMENTS

This research was supported by Petróleo Brasileiro S. A. (Petrobras) and conducted within the terms of the Agreement for Cooperation and Technical Assistance signed by the Universidade do Vale do Rio dos Sinos (UNISINOS-Brazil) and the Secretaría de Estado de Energía y Minería de la Provincia del Neuquén (Argentina). Special thanks to M. Carminatti, J. E. Faccion, A. J. Andrade Ramos, O. C. Carbone (Petrobras), H. M. Palacios, C. E. Portilla, A. C. Garrido, J. C. Danieli, and Sergio Cocca (Secretaría de Estado de Energía y Minería de la Provincia del Neuquén).

REFERENCES CITED

Arcuri, M., and C. Zavala, 2008, Hyperpycnal shelfal lobes: Some examples from the Lotena and Lajas formations, Neuquén Basin, Argentina (abs.), *in* J. J. Ponce and E. B. Olivero, conveners, Sediment transfer from shelf to deepwater—Revisiting the delivery mechanisms: Conference Proceedings, AAPG Hedberg Research Conference, Ushuaia-Patagonia, Argentina, 4 p.

Bagnold, R. A., 1956, The flow of cohesionless grains in fluids: Philosophical Transactions of the Royal Society of London, Series A: Mathematical and Physical Sciences, v. 249, p. 235–297, doi:10.1098/rsta.1956.0020.

Bates, C. C., 1953, Rational theory of delta formation: AAPG Bulletin, v. 37, p. 2119–2162.

Bhattacharya, J. P., and J. A. MacEachern, 2008, Hyperpycnal rivers and prodeltaic shelves in the Cretaceous seaway of North America (abs.), *in* J. J. Ponce and E. B. Olivero, conveners, Sediment transfer from shelf to deepwater—Revisiting the delivery mechanisms: Conference Proceedings, AAPG Hedberg Research Conference, Ushuaia-Patagonia, Argentina, 4 p.

Bourget, J., S. Zaragosi, T. Mulder, T. Garlan, N. Ellouz-Zimmermann, A. Vantoer, and J.-L. Schneider, 2008, Monsoon-induced hyperpycnal flows recorded in the Gulf of Oman (NW Indian Ocean) (abs.), *in* J. J. Ponce and E. B. Olivero, conveners, Sediment transfer from shelf to deepwater—Revisiting the delivery mechanisms: Conference Proceedings, AAPG Hedberg Research Conference, Ushuaia-Patagonia, Argentina, 2 p.

Burguess, P. M., S. Flint, and S. Johnson, 2000, Sequence stratigraphic interpretation of the turbidítica strata: An example from Jurassic strata of the Neuquén Basin, Argentina: Geological Society of America Bulletin, v. 112, no. 11, p. 1650–1666, doi:10.1130/0016-7606(2000)112<1650:SSIOTS>2.0.CO;2.

Carmona, N., J. J. Ponce, M. I. Lopez-Cabrera, and E. Olivero, 2008, Trace fossil diversity in hyperpycnites: Ethologic implications and comparison to trace fossil in episodic gravity flows (abs.), *in* J. J. Ponce and E. B. Olivero, conveners, Sediment transfer from shelf to deepwater—Revisiting the delivery mechanisms: Conference Proceedings, AAPG Hedberg Research Conference, Ushuaia-Patagonia, Argentina, 2 p.

Fisher, R. V., 1983, Flow transformations in sediment gravity flows: Geology, v. 11, p. 273–274, doi:10.1130/0091-7613(1983)11<273:FTISGF>2.0.CO;2.

Gómez Omil, R. G., J. Schmithalter, A. Cangini, L. Albariño, and A. Corsi, 2002, El Grupo Cuyo en la Dorsal de Huincul: Consideraciones estratigráficas, tectónicas y petroleras, Cuenca Neuquina. Actas 5° Congreso de

Exploración y Desarrollo de Hidrocarburos, Mar del Plata Actas (CD-ROM).

Gradstein, F. M., J. G. Ogg, and A. G. Smith, 2004, A geologic time scale: Cambridge, Cambridge University Press, 589 p.

Gulisano, C. A., and A. R. Gutierrez Pleimling, 1995, Field guide: The Jurassic of the Neuquén Basin, Neuquén Province, Secretaria de Minería de la Nación (Publicación no. 158) y Asociación Geológica Argentina, Serie E (2), 111 p.

Helland-Hansen, W., and J. G. Gjelberg, 1994, Conceptual basis and variability in sequence stratigraphy: A different perspective: Sedimentary Geology, v. 92, p. 31–52, doi:10.1016/0037-0738(94)90053-1.

Heller, P. L., and W. R. Dickinson, 1985, Submarine ramp facies model for delta-fed, sand rich turbidite systems: AAPG Bulletin, v. 69, no. 6, p. 960–976.

Hunt, D., and M. E. Tucker, 1992, Stranded parasequences and the forced regressive wedge systems tract: Deposition during base-level fall: Sedimentary Geology, v. 81, p. 1–9, doi:10.1016/0037-0738(92)90052-S.

Johnson, K., C. K. Paull, J. P. Barry, and F. P. Chavez, 2001, A decal record of underflows from a coastal river into the deep sea: Geology, v. 29, p. 1019–1022, doi:10.1130/0091-7613(2001)029<1019:ADROUF>2.0.CO;2.

Kneller, B., and C. Buckee, 2000, The structure and fluid mechanics of turbidity currents: A review of some recent studies and their geological implications: Sedimentology, v. 47, no. 1, p. 62–94, doi:10.1046/j.1365-3091.2000.047s1062.x.

Kochhann, K. G. D., S. Baecker-Fauth, G. Fauth, A. S. Silveira, 2009, Aalenian (middle Jurassic) radiolarian fauna from Los Molles Formation, Argentina, and its paleobiogeographic affinities, *in* Congresso Brasileiro de Paleontologia, 2009, Livro de Resumos, v. 1, p. 39.

Kuenen, Ph. H., 1951, Properties of turbidity currents of high density, *in* J. L. Hough, ed., Turbidity currents and the transportation of coarse sediments to deepwater: A symposium: SEPM Special Publication 2, p. 14–33.

Leanza, H. A., and C. A. Hugo, 1997, Hoja Geológica 3969-III, Picún Leufú, provincias del Neuquén y Río Negro, Programa Nacional de Cartas Geológicas de la República Argentina, escala 1: 250.000: Boletín Instituto de Geología y Recursos Naturales, v. 218, p. 1–135.

Legarreta, L., and C. A. Gulisano, 1989, Análisis estratigráfico secuencial de la Cuenca Neuquina (Triásico superior-Terciario inferior, Argentina), *in* G. Chebli and L. Spalletti, eds., Cuencas sedimentarias Argentinas, Instituto Miguel Lillo, Universidad Nacional de Tucumán, San Miguel de Tucumán, Serie Correlación Geológica 6, p. 221–243.

Legarreta, L., and M. A. Uliana, 1991, Jurassic/Cretaceous marine oscillations and geometry of a back-arc basin fill, central Argentine Andes, *in* D. I. M. Mc Donald, ed., Sedimentation, tectonics and eustasy: Oxford, International Association of Sedimentologists Special Publication 12, p. 429–450.

McLeod, P., S. Carey, and R. S. J. Sparks, 1999, Behavior of particle-laden flows into the ocean: Experimental simulation and geological implications: Sedimentology, v. 46, p. 523–536, doi:10.1046/j.1365-3091.1999.00229.x.

Mulder, T., and J. Alexander, 2001, The physical character of subaqueous sedimentary density flows and their deposits: Sedimentology, v. 48, p. 269–299, doi:10.1046/j.1365-3091.2001.00360.x.

Mulder, T., J. P. M. Syvitski, and K. I. Skene, 1998, Modeling of erosion and deposition of turbidity currents generated at river mouths: Journal of Sedimentary Research, v. 68, p. 124–137.

Mulder, T., J. P. M Syvitski, S. Nigeon, J. C. Faugeres, and B. Savoye, 2003, Marine hyperpycnal flows: Initiation, behavior and related deposits—A review: Marine and Petroleum Geology, v. 20, p. 861–882, doi:10.1016/j.marpetgeo.2003.01.003.

Mutti, E., 1992, Turbidite sandstones San Donato Milanese: Parma, Italy, Azienda Generale Italiana Petroli-Instituto di Geologia, Università di Parma, p. 275.

Mutti, E., R. Tinterri, E. Remacha, N. Mavilla, S. Angela, and L. Fava, 1999, An introduction to the analysis of ancient turbidite basins from an outcrop perspective, AAPG Continuing Education Course Note Series 39, p. 61.

Mutti, E., R. Tinterri, G. Benevelli, D. Di Biase, and G. Cavanna, 2003, Detaic, mixed and turbidite sedimentation of ancient foreland basins: Marine and Petroleum Geology, v. 20, p. 733–755, doi:10.1016/j.marpetgeo.2003.09.001.

Paim, P. S. G., A. S. Silveira, E. L. C. Lavina, U. F. Faccini, H. A. Leanza, J. M. M. Teixeira de Oliveira, and R. S. F. d'Ávila, 2008, High resolution stratigraphy and gravity flow deposits in the Los Molles Formation (Cuyo Group-Jurassic) at La Jardinera region, Neuquén Basin, Argentina: Revista de la Asociación Geológica Argentina, v. 63, no. 4, p. 728–753.

Pierson, T. C., and J. E. Costa, 1987, A rheologic classification of subaerial sediment-water flows, *in* J. E. Costa and G. F. Wieczorek, eds., Debris flows/avalanches: Process, recognition, and mitigation: Geological Society of America Reviews in Engineering Geology, v. VII, p. 1–12.

Piper, D. J. W., and B. Savoye, 1993, Processes of Late Quaternary turbidity current flow and deposition on the Var fan, north-west Mediterranean Sea: Sedimentology, v. 40, p. 557–582, doi:10.1111/j.1365-3091.1993.tb01350.x

Plink-Björklund, P., and R. J. Steel, 2004, Initiation of turbidity currents: Outcrop evidence for Eocene hyperpycnal flow turbidites: Sedimentary Geology, v. 165, p. 29–52, doi:10.1016/j.sedgeo.2003.10.013.

Postma, G., W. Nemec, and K. L. Kleinspehn, 1988, Large floating clasts in turbidites: A mechanism for their emplacement: Sedimentary Geology, v. 58, p. 47–61, doi:10.1016/0037-0738(88)90005-X.

Ramos, V. A., 1998, Estructura del sector occidental de la Faja Plegada y Corrida del Agrio, Cuenca Neuquina, Argentina, Actas 10° Congreso Latinoamericano Geología y 6° Congreso Nacional Geología Económica 2, Buenos Aires, Argentina, p. 105–110.

ARTHUR LAKES LIBRARY
COLORADO SCHOOL OF MINES
GOLDEN, CO 80401

Sanders, J. E., 1965, Primary sedimentary structures formed by turbidity currents and related sedimentation mechanisms, *in* G. V. Middleton, ed., Primary sedimentary structures and their hydrodynamic interpretations: SEPM Special Publication 12, p. 192–219.

Shanmugam, G., 2000, 50 Years of the turbidite paradigm (1950s–1990s): Deep-water processes and facies models—A critical perspective: Marine and Petroleum Geology, v. 17, p. 285–342, doi:10.1016/S0264-8172(99)00011-2.

Steel, R. J., J. Crabaugh, M. Schellpeper, D. Mellere, P. Plink-Björklund, J. Deibert, and T. Loeseth, 2000, Deltas vs. river on the shelf edge: Their relative contributions to the growth of the shelf-margin and basin-floor fans (Barreminan to Eocene, Spitsbergen), *in* P. Weimer, ed., Deep-water reservoirs of the world: Gulf Coast Section-Society of Sedimentary Geology (GC-SEPM) Research Conference (CD-ROM), p. 981–1009.

Van Wagoner, J. C., H. W. Posamentier, R. M. Mitchum Jr., P. R. Vail, J. F. Sarg, T. S. Loutit, and J. Hardenbol, 1988, An overview of the fundamentals of sequence stratigraphy and key definitions, *in* C. K. Wilgus, B. S. Hastings, C. G. St. C. Kendall, H. W. Posamentier, C. A. Ross, and J. C. Van Wagoner, eds., Sea level changes: An integrated approach: SEPM Special Publication 42, p. 40–45.

Van Wagoner, J. C., R. M. Mitchum Jr., K. M. Campion, and V. D. Rahmanian, 1990, Siliciclastic sequence stratigraphy in well logs, cores, and outcrops: Concepts for high-resolution correlation of time and facies: AAPG Methods in Exploration Series 50, 10, p. 2150–2175.

Vergani, G. D., 2005, Control estructural de la sedimentación jurásica (Grupo Cuyo) en la Dorsal de Huincul, Cuenca Neuquina, Argentina, Modelo de falla lístrica rampa-plano, invertida: Boletín de Informaciones Petroleras, v. 1, no. 1, p. 32–42.

Vergani, G. D., A. J. Tankard, H. J. Belotti, and H. J. Welsink, 1995, Tectonic evolution and paleogeography of the Neuquén Basin, Argentina, *in* A. J. Tankard, R. Suárez Soruco, and H. J. Welsink, eds., Memoir petroleum basins of South America: AAPG Bulletin, v. 62, p. 383–402.

Verzi, H., M. F. Raggio, and M. Suarez, 2005, Volume interpretation of a turbidite system, Los Molles Formation, Neuquén Basin, Argentina, *in* D. Soubies, M. Arteaga, and F. Fantín, eds., La Sísmica de Reflexión, más allá de la Imagen Estructural, Buenos Aires, Instituto Argentino del Petroleo y del Gas, 2005, p. 219–226.

Vesely, F. F., 2008, Subaqueous sandstones deposited by melt water-fed density flows in a late Paleozoic glaciomarine succession, eastern Paraná Basin, Brazil (abs.), *in* J. J. Ponce and E. B. Olivero, conveners, Sediment transfer from shelf to deepwater—Revisiting the delivery mechanisms: Conference Proceedings, AAPG Hedberg Research Conference, Ushuaia-Patagonia, Argentina, 1 p.

Vicente, J. C., 2005, Dynamic paleogeography of the Jurassic Andean Basin: Pattern of regression and general considerations of the main features: Revista de la Asociación Geológica Argentina, v. 61, no. 3, p. 408–437.

Wright, L. D., 1977, Sediment transport and deposition at river mouths: A synthesis: Geological Society of America Bulletin, v. 88, p. 857–868, doi:10.1130/0016-7606(1977)88<857:STADAR>2.0.CO;2.

Zavala, C., 1993, Estratigrafía y análisis de facies de la Formación Lajas (Jurásico medio) en el sector sur occidental de la Cuenca Neuquina. Provincia del Neuquén, República Argentina, Tesis Doctoral Departamento de Geología, Universidad Nacional del Sur, Bahía Blanca, Buenos Aires, Argentina, 260 p.

Zavala, C., 2008, Toward a genetic facies tract for the analysis of hyperpycnal deposits (abs.), *in* J. J. Ponce and E. B. Olivero, conveners, Sediment transfer from shelf to deepwater—Revisiting the delivery mechanisms: Conference Proceedings, AAPG Hedberg Research Conference, Ushuaia-Patagonia, Argentina, 3 p.

6

Campos, C., O. Guzmán, A. Pilloud, R. Uzcátegui, A. Cabrera, and M. Toro, 2011, Gravelly and sandy facies of Rio Guache Formation, Venezuelan Andes: Evidence for transformation of gravity flows in deep-marine water, *in* R. M. Slatt and C. Zavala, eds., Sediment transfer from shelf to deep water—Revisiting the delivery system: AAPG Studies in Geology 61, p. 117–128.

Gravelly and Sandy Facies of Rio Guache Formation, Venezuelan Andes: Evidence for Transformation of Gravity Flows in Deep-marine Water

Corina Campos, Oswaldo Guzmán, Andrés Pilloud, Redescal Uzcátegui, and Ana Cabrera

Universidad Simón Bolívar, Edificio de Física y Electrónica II (FE-II), Departamento de Ciencias de La Tierra, Valle Sartenejas, Baruta, Miranda, Venezuela

Manuel Toro

Core Laboratories, Houston, Texas, U.S.A.

ABSTRACT

The Rio Guache Formation represents deep-marine deposits accumulated within a diachronic foreland basin setting formed as a result of the Paleogene oblique collision between the transpressional front of the Caribbean plate and the Cretaceous to Paleogene passive margin of northern South America.

This study attempted to provide the first comprehensive sedimentologic analysis of the Rio Guache Formation in the Guaramacal area. Based on the study of outcrops and samples, eight facies have been identified as composing the Rio Guache Formation, five of them gravelly and three sandy. The facies analysis and the interpretation of the processes of transport and deposition for each of the different facies indicate that they are the products of the progressive transformation of gravity flows within the basin. The distinct assemblages of facies, the high percentage of coarse-grained sediments, the poor textural sorting, and the minor development of sedimentary structures indicate that these gravity flows were short-lived and poorly efficient. They could be the product of episodic mass flows and sliding induced by the high tectonic activity in western Venezuela during the late Paleogene. No direct evidence of a hyperpycnal origin for these coarse-grained strata exists.

Copyright ©2011 by The American Association of Petroleum Geologists.

DOI:10.1306/13271353St613440

INTRODUCTION

The Paleocene to Miocene strata of northern South America represent dominantly deep-marine deposits that accumulated within a diachronic foreland basin setting in response to the eastward migration of a fold and thrust belt. This foreland basin was formed as a result of the Paleocene to Holocene oblique collision between the transpressional front of the Caribbean plate and the Cretaceous to Paleogene passive margin of northern South America (Speed, 1985; Lugo and Mann, 1995; Colletta et al., 1997; Avé Lallemant and Sisson, 2005; Ostos et al., 2005; Pindell et al., 2005).

Most of these deep-marine strata were deposited in a diachronic foredeep basin. These rocks were incorporated and highly deformed into the Caribbean fold and thrust belt. Therefore, the study of these rocks helps to understand the geologic evolution and the hydrocarbon potential of the passive margin of northern South America.

One of these units, the Rio Guache Formation, has been recognized in the southeastern flank of the Venezuelan Andes (Figure 1), specifically at Lara, Portuguesa, and Trujillo states. This formation is composed of conglomerates, sandstones, siltstones, claystones, and exotic blocks (metamorphic and limestones), which have been interpreted as having been deposited on a slope by gravity flows and turbidity currents. Its age is uncertain, having been placed between the Maastrichtian and middle Eocene by Von Der Osten and Zozaya (1957), Ramírez (1968), and Blin (1989). Nevertheless, these authors have not conducted detailed sedimentologic studies that would allow an understanding of the processes of sedimentation of the Rio Guache Formation. The present work is the first sedimentologic study of the Rio Guache Formation in the Guaramacal area, in the state of Trujillo (Figure 1).

We present a detailed facies analysis, an interpretation of deposition and transport processes for each facies, and finally, a depositional model for the Rio Guache Formation. These results provide insights into understanding the diachronic foreland basins of northern South America and the role of these strata in the petroleum systems of the area.

REGIONAL SETTING

The oblique collision between the Caribbean and South American plates has controlled the tectonic evolution of northern Venezuela since the start of the Cenozoic (Mann et al., 1990). The southern limit of the Caribbean plate is a very complex deformation zone that starts in Colombia and includes the Bocono, San Sebastian, and El Pilar strike-slip faults (Hung, 2005). The topography of this area is defined by five mountainous systems: the Venezuelan Andes, the Serrania del Interior Central, the Cordillera de la Costa, the Serrania del Interior Oriental, and the Cordillera de Araya-Paria. The Venezuelan Andes are located in western Venezuela and represent one of the northern culminations of the South American Andes (Figure 1). These mountains have an N50°E orientation; they are approximately 400 km (250 mi) in length and 80 km (50 mi) in width. The Venezuelan Andes are divided almost symmetrically along their entire length by the Bocono fault, and its genesis is still a subject of debate. Its structure has been interpreted as a positive flower megastructure (Stephan, 1982), a thrust fold belt produced by a subduction complex (Colletta et al., 1997), and by overthrusting toward the northwestern structures with a root in the continental crust and involved basement (Audemard, 1997). The trace of the Bocono fault has been used to divide the Venezuelan Andes into North Andean and South Andean flanks. The study area is located in the northern part of the South Andean flank (Figure 1). In this zone, the Caribbean front deformation overthrust the Eocene to Paleozoic autochthonous rocks (Stephan et al., 1990; Colletta et al., 1997). The geodynamic evolution of western Venezuela has been divided into six major episodes (Figure 2), each associated with at least one tectonostratigraphic sequence: (1) Prerift phase in Paleozoic time; (2) rifting and drifting phases during the Jurassic; (3) passive margin period during the Cretaceous; (4) transitional phase to a compressive regime during the late Cretaceous–early Paleocene, with compression resulting from the collision and obduction of the Pacific volcanic arc with the South American plate; (5) foreland basin in front of the volcanic arc and emplacement of the Lara nappes during the late Paleocene–middle Eocene; (6) modification of the foreland basin by the collision of the Panama arc against the western South American plate, which caused the uplifting of the Serrania de Perija, Macizo de Santander, and Venezuelan Andes during Eocene–Pleistocene time, creating the present configuration (Parnaud et al., 1995).

In the study area, only the prerift phase and the foreland basin phase, during the late Paleogene, have been recognized (Figure 1). The rocks are in fault contact. The prerift phase is represented by Paleozoic

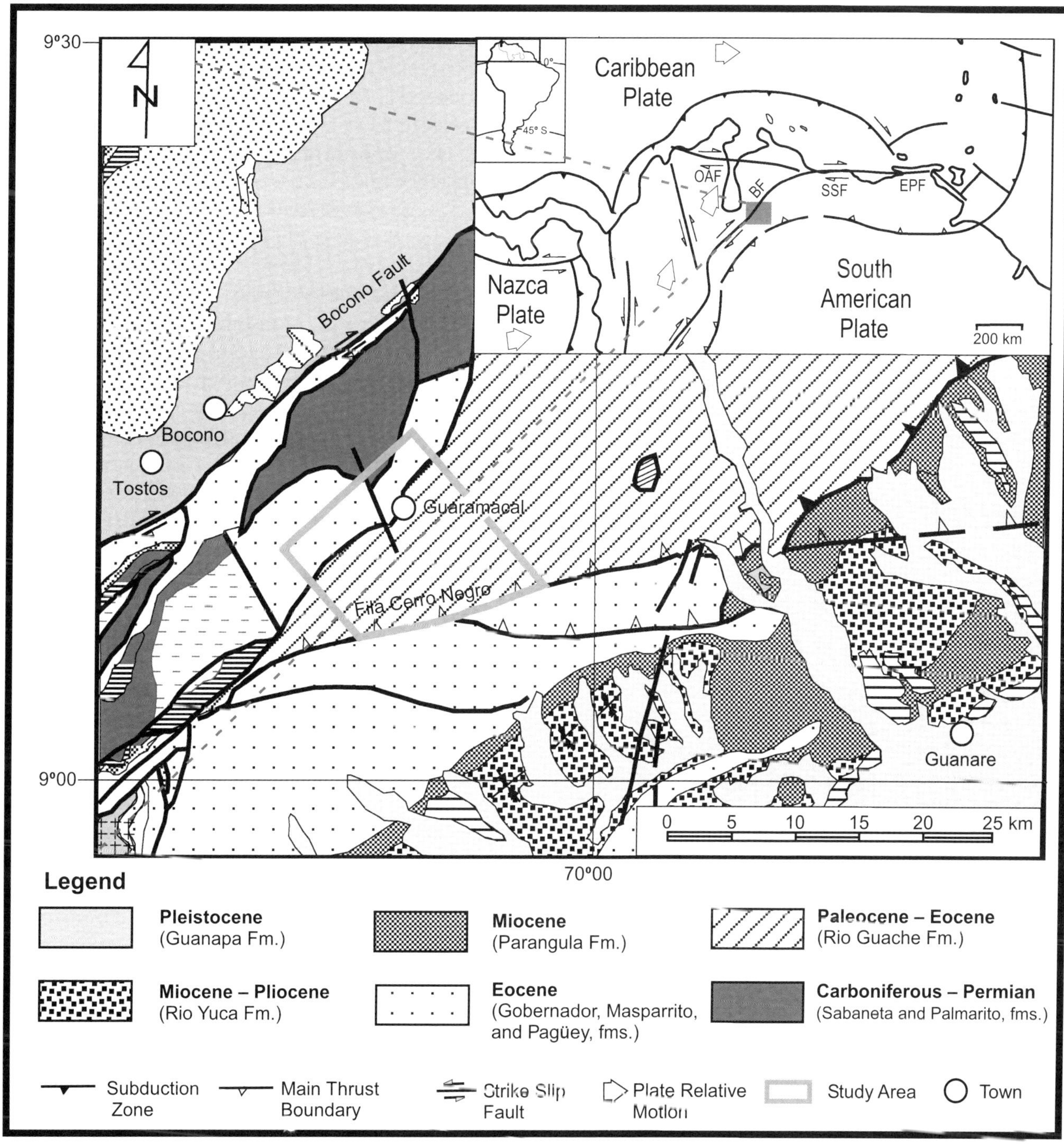

FIGURE 1. Location of the study area and simplified geologic map of northwestern Venezuela. Modified from Stephan (1982) and simplified after Bellizia et al. (1976). OAF = Oca-Ancon fault; SSF = San Sebastian fault; BF = Bocono fault; EPF = El Pilar fault; Fm. = formation.

volcanic and marine deposits of the Sabaneta and Palmarito (Arnold, 1966). The foreland basin phase is represented by transitional and deep-marine deposits of the Gobernador, Pagüey, and Rio Guache formations, with the sedimentology of the latter being the main focus of this study.

The tectonic complexity present in the South Andean flank has impeded the building of a stratigraphic column for the Rio Guache Formation. However, a synthesized column with a thickness of 4830 m (15,846 ft) for the Rio Guache Formation was proposed by Blin (1989).

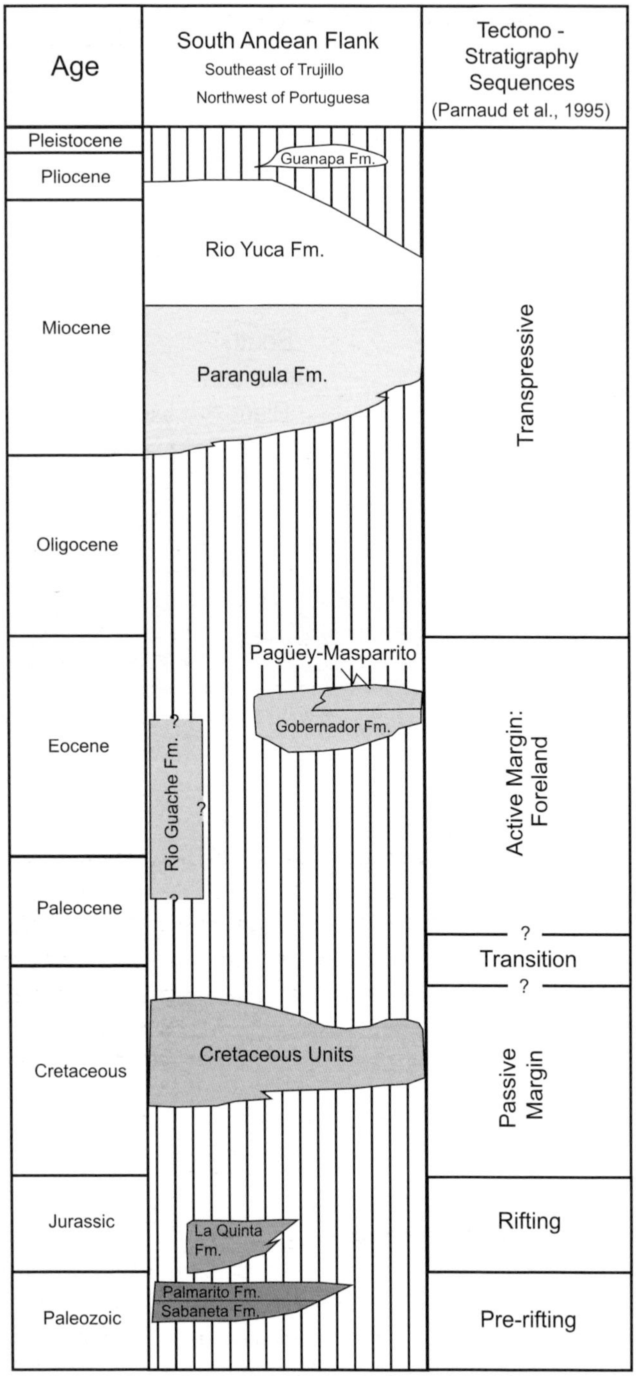

FIGURE 2. **Chronostratigraphic chart of southeast of Trujillo state and northwest of Portuguesa state, South Andean flank. The main tectono-stratigraphic sequences are identified. Modified from Kiser (1989) and Parnaud et al. (1995). Fm. = formation.**

FACIES ANALYSIS

The data used in this chapter come mainly from observations and study of 188 outcrops and 44 petrographic analyses of different facies. Based on the study of these localities and samples, eight facies have been identified, out of which five are gravelly and three are sandy. These facies were identified by their lithotypes and sedimentary structures. In addition, the processes of transport and deposition of each of the different facies were interpreted to determine the progressive transformations that the gravity flows underwent while being transported into the basin. A summary of facies description is provided in Table 1.

Facies Df1

This facies consists of massive gravelly mudrocks and mud-supported conglomerates. The clasts in these rocks have angular to rounded shapes, low sphericity, and range in size from granules to boulders (<3 m [9.8 ft] in diameter). Poor grain sorting and large clasts floating in the middle and top of the beds are characteristics of this facies (Figure 3A). The clasts are floating in a muddy matrix; in some cases, this matrix can be sandy and slightly calcareous. The beds of this facies are about 1 to 10 m (3.3–33 ft) thick and are generally associated with rocks that belong to facies Df2 and facies Gl.

The clasts found are (1) Cretaceous clasts, such as black micritic limestones with *Hedbergella* species, *Heterohelix* species, and fragments of *Inoceramus* species (probably of La Luna Formation), micritic limestone with rudist and echinoderm fragments (probably of Maraca, Aguardiente, or Lisure formations), and micritic limestones with fragments of mollusk and bivalves (similar rocks to Guayacán Member); (2) Paleocene–Eocene clasts such as micritic limestone with fragments of oncoids, *Nummulites* species, red algae, bivalves, gastropods, and bryozoans. In addition, we found other clasts such as gneiss, slate, schist, quartzite, plutonic and volcanic rock fragments, glauconitic and calcareous sandstones, mudstones, and others.

Interpretation

The muddy matrix and the poor grain sorting with large clasts floating in the middle and top of the beds indicate that these rocks are the result of cohesive debris flows, which were deposited en masse (Lowe, 1982; Mutti, 1992; Mulder and Alexander, 2001; Gani, 2004). The lack of significant basal scour can indicate the phenomenon of hydroplaning (Mulder and Alexander, 2001; Sohn et al., 2002).

Facies Df2

This facies consists of massive clast-supported conglomerates with a muddy matrix. The clasts in these rocks have angular to rounded shapes and variable

Table 1. Sedimentary facies identified in the Eocene deposits of Rio Guache Formation.

Facies	*Thickness*	*Lithology*	*Sedimentary Structures*
Df1	1–10 m (3.2–32 ft)	Gravelly mudrocks and mud-supported conglomerates	Massive
Df2	1–5 m (3.2–16 ft)	Clast-supported conglomerates with muddy matrix	Massive
Gs	1–5 m (3.2–16 ft)	Sand-supported conglomerates to grain-supported conglomerates with sandy matrix	Erosive bases, massive, in some cases, with crude grading
Gc	1–4 m (3.2–13 ft)	Clast-supported conglomerates with calcite cement	Massive
Gl	1–50 cm (0.4–19.7 in.)	Conglomerates and microconglomerate lenses	Erosive bases, massive, in some cases, with normal and inverse grading
S1	20 cm (7.8 in.)–4 m (13.1 ft)	Fine to coarse sandstones	Erosive bases, massive, in some cases, with normal grading and parallel lamination
S2	30 cm (11.8 in.)–1.7 m (5.6 ft)	Very fine to medium sandstones	Massive or with ripples, parallel, cross- and convolute lamination
T	1–30 cm (0.39–11.8 in.)	Fine to very fine sandstones and siltstones interbedded with claystones	Massive, in some cases, with parallel lamination, cross-lamination, ripples, and climbing ripples. Trace fossils like Paleodictyon

sphericity. The predominant grain sizes are granules to boulders (<50 cm [<20 in.] in diameter), and the grain sorting is highly variable. The clasts are similar to those described for facies Df1. The conglomerates are supported by clasts, and the matrix comprises from 10 to 40% of the rock (Figure 3B). The beds of this facies have a thickness less than 5 m (16 ft). This facies is generally associated with rocks that belong to facies Df1.

Interpretation

The muddy matrix, the poor grain sorting, and the lack of significant basal scours suggest that these rocks were deposited en masse from cohesive debris flows (Lowe, 1982; Mutti, 1992; Mulder and Alexander, 2001; Gani, 2004), but in contrast with the facies Df1, the sizes of clasts in these conglomerates are smaller. This can imply that the debris flows were unable to transport big clasts. For this reason, facies Df2 can be a result of little dilution of a parental debris flow (Mulder and Alexander, 2001; Sohn et al., 2002), without transforming into a hyperconcentrated flow.

Facies Gs

This facies consists of sand- to grain-supported conglomerates with a sandy matrix (Figure 3C). The clasts show a higher degree of rounding and a lower sphericity compared with the previous facies. The predominant grain sizes are granules to cobbles and are similar to those described in facies Df1. The beds of this facies are less than 5 m (16 ft) thick. These rocks are massive or crudely graded, and they can show basal scours and a little imbrication in grain-supported conglomerates. This facies is generally associated with rocks that belong to facies S1.

Interpretation

The sandy matrix, the basal scours, and the crude gradation indicate that this facies is the result of hyperconcentrated flows. The main mechanism for deposition of this flow is frictional freezing because of grain-to-grain interaction and inability of the progressive increase of water into the flow, losing the capacity to carry larger clasts (Mulder and Alexander, 2001; Sohn et al., 2002).

Facies Gc

This facies consists of massive clast-supported conglomerates with calcite cement (Figure 3D). In this facies, the muddy matrix is almost absent, and it can be mistaken for pseudomatrix. The clasts in these rocks have angular to subangular shapes, low sphericity, and range in size from cobbles to boulders (<50 cm [<20 in.] in diameter). The clasts are similar to those described for facies Df1, and the beds are irregular and have thicknesses less than 4 m (13 ft).

FIGURE 3. Facies of Rio Guache Formation in the Guaramacal area, northwestern Venezuela. (A) Facies Df1, mud-supported conglomerate. The clasts are floating in a muddy matrix. It is the product of cohesive debris flow. (B) Facies Df2, clast-supported conglomerate with muddy matrix. Deposit of a cohesive debris flow. (C) Facies Gs, grain-supported conglomerate with sandy matrix. Deposit of hyperconcentrated density flows. (D) Facies Gc, clast-supported conglomerate with calcite cement (see arrow), it is the product of residual deposits due to the progressive loss of strength of the cohesive debris flow.

Interpretation

The size of the clasts, the irregular form of the beds, and the poor grain sorting provide evidence that these rocks are residual deposits caused by the progressive dilution of a parental debris flow (Mutti, 1992) and inability to transport big clasts. The coarsest clasts were segregated at the base of the flow.

Facies Gl

This facies consists of conglomerates and microconglomerate lenses with erosive bases (Figure 4A). These conglomerates are clast supported with a muddy and slightly sandy matrix. The clasts in these rocks have subangular to rounded shapes, low sphericity, and range in size from granules to cobbles. Elongate clasts have their long axis orientated along the same direction. The clasts of this facies are similar to those described in facies Df1, and the beds are less than 50 cm (<20 in.) thick and 5 m (16 ft) in length. These rocks can be either massive or show normal or inverse gradation. This facies is generally associated with pelitic rocks, facies Df1, and facies Df2.

Interpretation

The erosive bases are interpreted as an initial erosional event caused by a turbulent flow. The erosion was immediately followed by deposition from a hyperconcentrated flow (Sohn et al., 2002). The main mechanism for deposition of this hyperconcentrated flow was frictional freezing (Mulder and Alexander, 2001; Gani, 2004). The preferential orientation of the long axis of the clasts suggests their ability to move independently within the flow.

Facies S1

The sandstones of this facies are fine to coarse grained. This facies can show erosive bases with abundant pebbles and cobbles. However, it is common to find clasts randomly distributed in the bed. These clasts have subangular to subrounded shapes and low

FIGURE 4. Facies of Rio Guache Formation in the Guaramacal area, northwestern Venezuela. (A) Facies Gl, microconglomerate lenses with erosive bases (see arrow). They are the product of hyperconcentrated density flows. (B) Facies S1, fine to coarse sandstone with erosive base. This facies is associated with rocks belonging to T and S2 facies. (C) Facies S2, massive and tabular sandstones. These sandstones were deposited by processes of frictional freezing in concentrated density flows. (D) Facies T, fine sandstone and siltstones interbedded with claystones. The sandstones show ripples and cross-lamination.

sphericity, and are similar to those described in facies Df1. The sandstones are lithic and moderately to poorly sorted. In general, these sandstones are massive but can show normal gradation and parallel lamination either in the middle or at the top of the beds (Figure 4B). The beds have thicknesses between 20 cm (7.9 in.) and 4 m (13.1 ft). This facies is generally associated with rocks that belong to facies Gs, facies S2, and facies T.

Interpretation

The size of the sediments, the erosive bases, the normal gradation, and the parallel lamination in the middle and top of the beds indicate that these sandstones were deposited by processes of friction and traction fallout from concentrated density flows (Mulder and Alexander, 2001). However, we argue that these flows suffered some stratification. The basal scours with pebbles and cobbles indicate that the base of the flow had an intense near-bed transport of coarse particles. This part of the flow may behave like a hyperconcentrated flow (Mulder and Alexander, 2001). In contrast, the parallel lamination in the middle and top of the beds indicates that the upper part of the flow was turbulent because of a lower sediment concentration (Mulder and Alexander, 2001).

Facies S2

This facies encompasses medium to very fine sandstones. The sandstones are lithic to sublithic and are moderate to well sorted. The grains have angular to subangular shapes and have variable sphericity. These rocks can be tabular (Figure 4C) and massive, or they might show parallel lamination, cross-lamination, ripples, and convolute lamination either in the middle or at the top of the beds. The beds of this facies have thicknesses between 30 cm (11.8 in.) and 1.7 m (5.6 ft). This facies is generally associated with pelitic rocks and with rocks that belong to facies S1 and facies T.

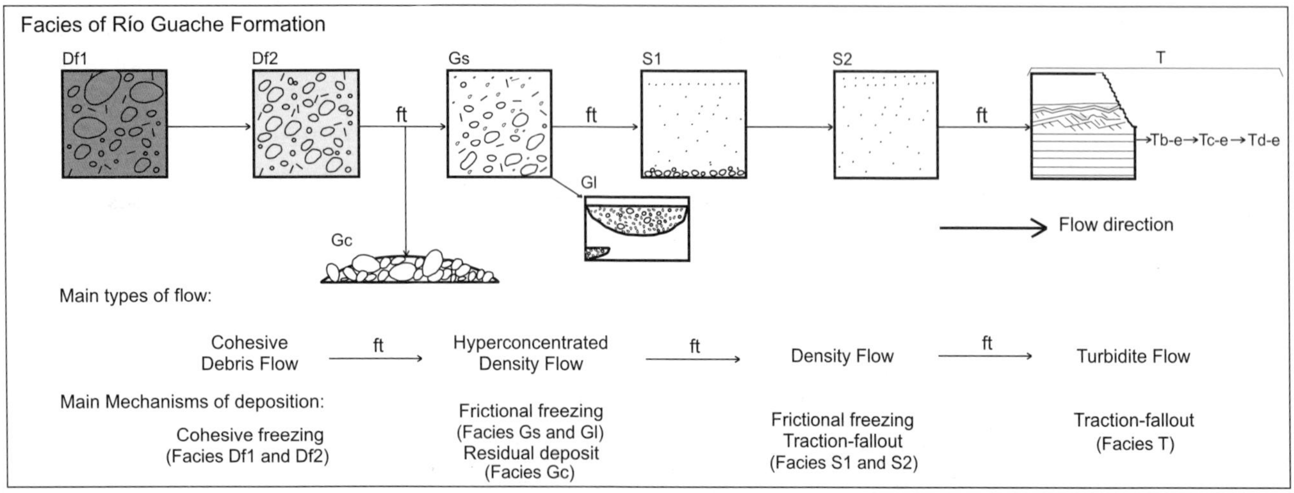

FIGURE 5. Facies of Río Guache Formation with the main types of flows and mechanisms of deposition. Modified from Mutti (1992). ft = flow transformation; Tb–e = Bouma sequences.

Interpretation

The size of the sediments, the thicknesses of the beds, and the lack of sedimentary structures indicate that these sandstones were deposited by processes of frictional freezing in concentrated density flows (Mulder and Alexander, 2001). However, the sedimentary structures in the middle and top of the beds indicate that the upper part of this flow was turbulent because of a lower sediment concentration (Mulder and Alexander, 2001).

Facies T

This facies consists of fine to very fine sandstones and siltstones interbedded with claystones. The sand/clay ratio is about 10 to 50%. The sandstones and siltstones can be either massive or show parallel lamination, cross-lamination, ripples, and climbing ripples either in the middle or at the top of the beds (Figure 4D). The coarser beds have thicknesses between 1 and 30 cm (0.39–11.8 in.), and can have trace fossils as *Paleodictyon* on the basal surface of the beds. The claystones are gray to dark gray, have pyrite and micas as accessory minerals, are massive, or might show parallel lamination and do not show evidence of bioturbation. The pelitic rocks have thicknesses between 10 cm (3.9 in.) and 3 m (9.8 ft).

Interpretation

This facies represents the classical turbidite deposit defined by Bouma (1962). It can show the Bouma sequences Tb-e, Tc-e, and Td-e. The poor development of internal depositional structures indicates that these sandstones were deposited rapidly by processes of traction and fallout (Lowe, 1982; Mutti, 1992) from periodic short-term surgelike turbidite flows (Mulder and Alexander, 2001). The trace fossils *Paleodictyon* on the basal surface of the beds are indicative of deep bathyal and abyssal zones, where low-energy conditions generally prevail (Boggs, 1995). Furthermore, the pelitic rocks were deposited by fallout in quiet water.

FACIES TRACTS AND DEPOSITIONAL MODEL

The interpretation of the facies tracts and their depositional model is based on the models given by Mutti (1992) and Mutti et al. (1999). These authors propose that different facies occupy predictable positions within a succession. Such succession of facies is named by Mutti "facies tract." In this model, the facies are related genetically and represent the evolution of submarine gravity flows during basinward travel.

The group of facies observed in the outcrops allows us to infer that they are the result of the progressive downstream transformation of high-sediment-concentration flows (Figure 5). These flows are short-lived and could be the product of episodic mass flows and sliding. These processes could have been induced by the high tectonic activity in western Venezuela during the late Paleogene, which was related to the emplacement of the Caribbean fold and thrust belt over the South American plate (Speed, 1985; Stephan et al., 1990; Lugo and Mann, 1995; Colletta et al., 1997; Avé Lallemant and Sisson, 2005; Pindell et al., 2005; Ostos et al., 2005; among others). In addition, the trace fossils *Paleodictyon* identified in facies T are indicative of deep bathyal and abyssal conditions, which could have existed in the foredeep of the

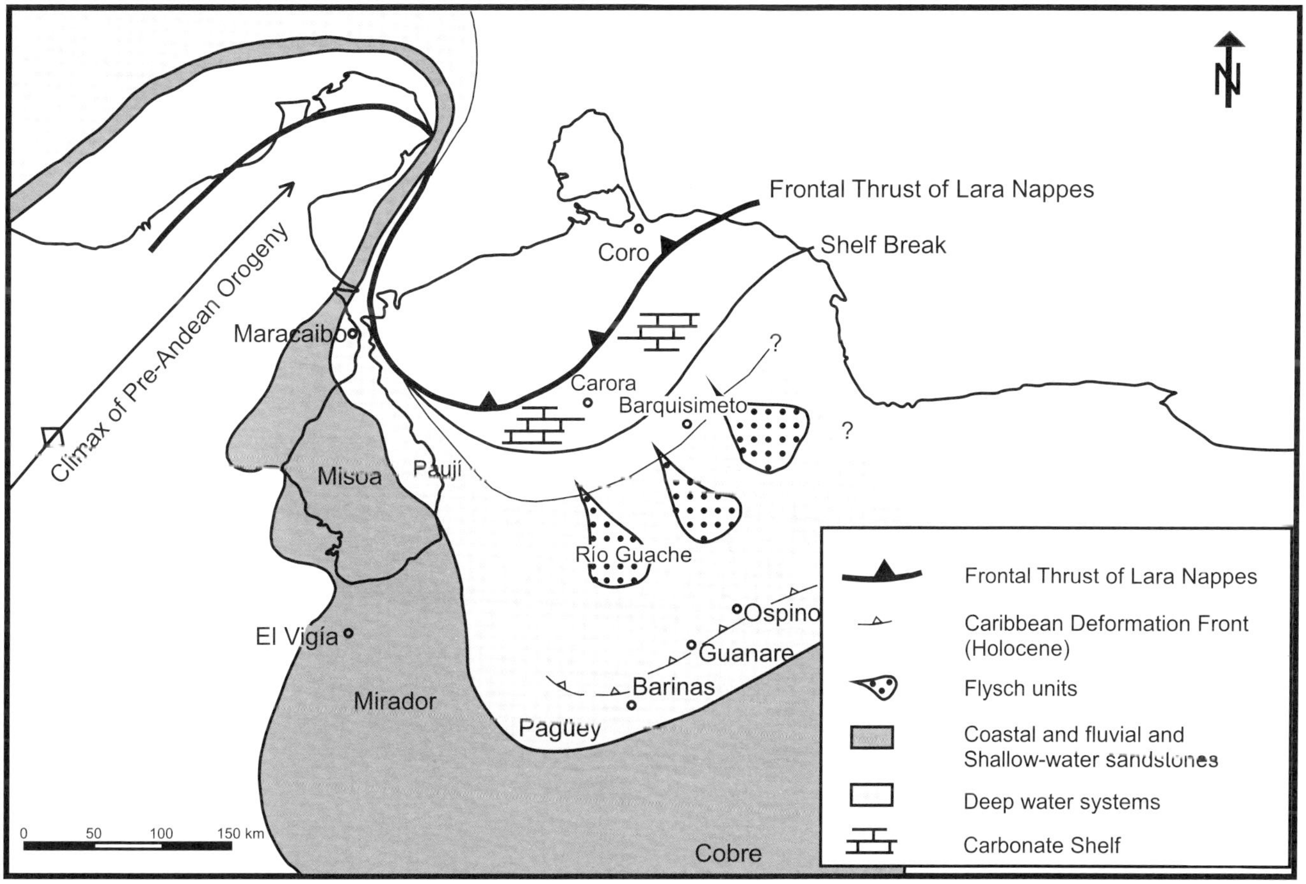

FIGURE 6. Paleogeographic scheme of the middle Eocene, western Venezuela. The frontal thrust of the Caribbean fold and thrust belt is located in the actual position of Falcon and Lara states. The flysch units could be the product of episodic mass flows and sliding derived from the north and northwest of Venezuela. Based on Villamil (1999).

Paleogene foreland basin formed in front of the Caribbean fold and thrust belt, as suggested by Parnaud et al. (1995), Villamil (1999), Pindell et al. (2005), and others.

However, the clasts identified in conglomerate facies Df1, Df2, Gs, Gc, and Gl, and the lithic fragments in sandstone facies S1, S2, and T, comprise three groups: (1) Cretaceous clasts, (2) Paleocene–Eocene clasts, and (3) metamorphic and igneous clasts. This variety of clasts could indicate at least three simple sources of sediments or a complex source of sediments that involves all these types of rocks, such as the Caribbean fold and thrust belt. The paleocurrent measurements in current ripples indicate that the source of the Rio Guache Formation was located in the northwest of our study area, correlating with the position of the Caribbean fold and thrust belt as described by Villamil (1999) for the middle Eocene (Figure 6).

Cretaceous clasts in the Rio Guache Formation were deposited in western Venezuela during the passive margin phase. The Paleocene–Eocene limestone clasts might belong to the carbonate shelf, which was developed in the outer Caribbean thrust belt during the Paleocene–Eocene. The Cretaceous, Paleocene–Eocene, metamorphic and igneous clasts probably were transported from northwest to southeast as a result of the advance toward the south-southeast of the Caribbean thrust belt and finally were deposited by gravity flows in the foredeep of the foreland basin.

Our interpretation of the facies tracts and their depositional model is mainly based on the closeness of the source area. The most proximal facies within this set of facies correspond to gravelly mudrocks, massive mud-supported conglomerates (facies Df1), and clast-supported conglomerates with muddy matrix (facies Df2). They were deposited en masse from cohesive debris flows (Figure 5). In some cases, these cohesive debris flows underwent a downslope transformation because of a progressive mixing with subaqueous fluids, becoming the cohesive debris flows in hyperconcentrated density flows (Hampton, 1972; Mutti, 1992; Mulder and Alexander, 2001; Sohn et al., 2002). Such flows were deposited by frictional freezing caused

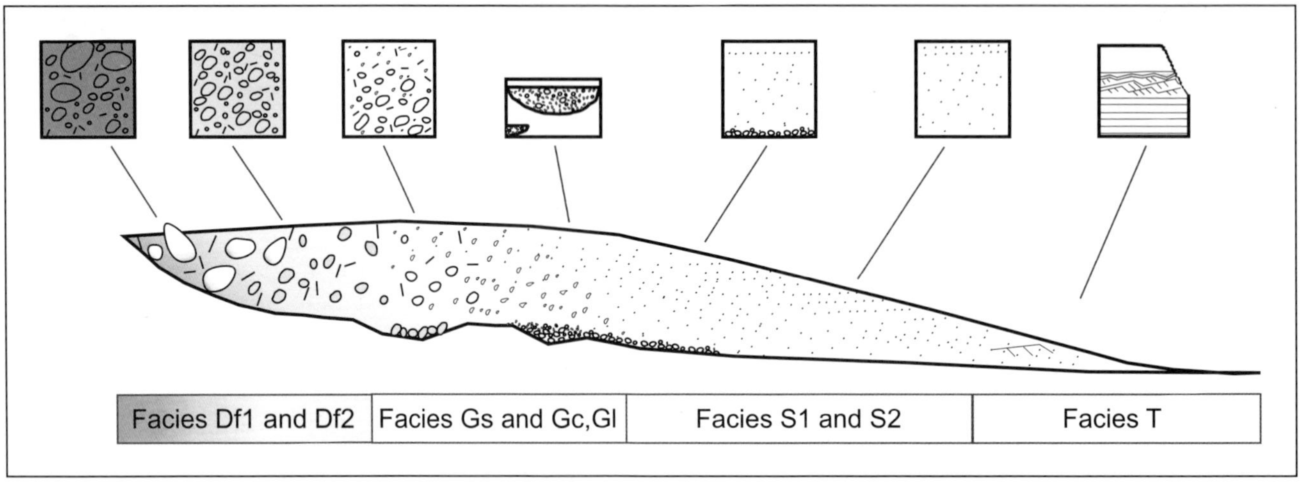

FIGURE 7. Conceptual model of the lateral relations between facies deposited by poorly efficient gravity flows. These facies show a high percentage of coarser grained sediments, poor textural sorting, and little development of sedimentary structures.

by grain interaction (Mulder and Alexander, 2001). These deposits are represented by sand-supported to grain-supported conglomerates with sandy matrix (facies Gs), as well as conglomerate and microconglomerate lenses (facies Gl).

The progressive loss of strength of the flow results in the segregation of coarser clasts at the base of the flow, forming residual deposits such as clast-supported conglomerates with later-formed calcite cement (facies Gc). The transformation of hyperconcentrated flows in density flows generates the deposition by friction and traction-fallout processes of this proximal-middle facies (Figure 5). Theses facies are represented by fine- to coarse-grained sandstones with basal scours (facies S1) and medium to very fine sandstones (facies S2).

Finally, the density flows had a basinward transformation because of a progressive mixing with water and dilution of particles in the flow, becoming turbulent flows. The turbulent flows underwent high rates of deceleration and were deposited by traction-plus-fallout processes (Figure 5). The distal facies is represented by very fine-grained sandstones and siltstones interbedded with claystones (facies T); this facies is known also as classical turbidite deposits but with minimal development of sedimentary structures because of the rapid deceleration (Mutti, 1992).

The high percentage of coarse-grained sediments, the poor textural sorting, and the minimal development of sedimentary structures in the previously defined facies provide evidence that these gravity flows were poorly efficient (Figure 7). The efficiency of the flow is defined as its ability to carry its sediment load basinward and segregate its populations into distinct facies types with distance (Mutti, 1992; Mutti et al., 1994, 1999). In addition, the very short lifetime of the flows, the lack of freshwater fossils and continental organic matter, and few sedimentary structures such as climbing ripples suggest that the Rio Guache Formation deposits are the product of episodic gravity flows. Furthermore, we do not have evidence for deposits of quasi-steady hyperpycnal flows in the study area.

CONCLUSIONS

The Paleogene rocks of northern South America represent dominantly deep-marine deposits that accumulated within a diachronic foreland basin setting in response to the eastward migration of the Caribbean fold and thrust belt. One of these deep-marine deposits is the Rio Guache Formation. The origin of these deposits is related to tectonic instability of this basin; such instability caused episodic mass flows and sliding, whereas flexural subsidence related to thrusting played the major factor in creating space for sedimentation.

The facies study presented in this work allows us to infer that these facies are the result of the progressive transformation of gravity flows while transported inside the basin, resulting in the formation of a facies tract. The proximal facies in this tract correspond to massive mud-supported and clast-supported conglomerates with muddy matrix, sand-supported to grain-supported conglomerates with sandy matrix, and conglomerate and microconglomerate lenses. The middle facies are represented by fine to coarse sandstones with basal scours and medium to very fine sandstones. The distal facies are represented by very fine sandstones and siltstones interbedded with claystones (classical turbidites).

The gravity flows that generated these facies were poorly efficient with a short time; this is evidenced by the high percentage of coarse-grained sediments, the poor textural sorting, and the minimal development of sedimentary structures. In addition, the lack of freshwater fossils and terrestrial organic matter, as well as the few sedimentary structures such as climbing ripples, suggest that the Rio Guache Formation deposits are the result of episodic gravity flows. No direct evidence for deposits of quasi-steady hyperpycnal flows in the study area exists.

ACKNOWLEDGMENTS

We thank Repsol-YPF, especially Alejandro Franco, Laszlo Benkovics, and Ángela Echanove, for financial support. We also thank Servicios Geocinteg C. A. for the collaboration, Max Furrer for his assistance in the biostratigraphic analysis, and A. Alexis, R. Mirian, H. Carelis, Q. Jhon, and O. Christian for their assistance in the field. Carmen Contreras, Carlos Zavala, and the anonymous reviewers provided helpful comments.

REFERENCES CITED

Arnold, H. C., 1966, Upper Paleozoic Sabaneta-Palmarito sequence of Mérida Andes, Venezuela: AAPG Bulletin, v. 50, p. 2366–2387.

Audemard, F. A., 1997, Holocene and historical earthquakes on the Bocono fault system, southern Venezuelan Andes: Trench confirmation: Journal of Geodynamics, v. 24, p. 155–167, doi:10.1016/S0264-3707(96)00037-3.

Avé Lallemant, H. G., and V. B. Sisson, 2005, Prologue, *in* H. Avé Lallemant and V. B. Sisson, eds., Caribbean–South American plate interactions, Venezuela: Geological Society of America Special Paper 394, p. 1–5.

Bellizia, A., W. Pimentel, and R. Bajo, 1976, Mapa geológico estructural de Venezuela. Hoja NC-1-DC: Caracas, Venezuela, Ministerio de Minas e Hidrocarburos, Foninves, scale 1:500,000, 30 sheets.

Blin, B., 1989, Le front de la chaîne caraïbe vénézuélienne entre la serranía de Portuguesa et la région de Tiznados (surface et subsurface): Ph.D. dissertation, Université de Bretagne Occidentale, Brest, France, 359 p.

Boggs Jr., S., 1995, Principles of sedimentology and stratigraphy: Columbus, Ohio, Merrill Publishing Co., 774 p.

Bouma, A. H., 1962, Sedimentology of some flysch deposits: A graphic approach to facies interpretation: Amsterdam, Netherlands, Elsevier, 168 p.

Colletta, B., F. Roure, B. De Toni, D. Loureiro, H. Passalacqua, and Y. Gou, 1997, Tectonic inheritance, crustal architecture, and contrasting structural styles in the Venezuela Andes: Tectonics, v. 16, p. 777–794, doi:10.1029/97TC01659.

Gani, M., 2004, From turbid to lucid: A straightforward approach to sediment gravity flows and their deposits: The Sedimentary Record, SEPM, v. 2, p. 4–8.

Hampton, M. A., 1972, The role of subaqueous debris flow in the generating turbidity current: Journal of Sedimentary Petrology, v. 42, p. 775–793.

Hung, E., 2005, Thrust belt interpretation of the Serranía del Interior and Maturín subbasin, eastern Venezuela, *in* H. Avé Lallemant and V. B. Sisson, eds., Caribbean–South American plate interactions, Venezuela: Geological Society of America Special Paper 394, p. 251–270.

Kiser, G. D., 1989, Relaciones estratigráficas de la cuenca Apure/Los Llanos y áreas adyacentes. Venezuela Suroeste y Colombia Orienta: Caracas, Venezuela, Monografías Sociedad Venezolana de Geólogos, 71 p.

Lowe, D. R., 1982, Sediment gravity flows: II. Depositional models with special reference to the deposits of high-density turbidity currents: Journal of Sedimentary Petrology, v. 52, p. 279–297.

Lugo, J., and P. Mann, 1995, Jurassic–Eocene tectonic evolution of Maracaibo basin, Venezuela, *in* A. J. Tankard, R. S. Suárez, and H. J. Welsink, eds., Petroleum basins of South America: AAPG Memoir 62, p. 699–725.

Mann, P., C. Schubert, and K. Burke, 1990, Review of Caribbean neotectonics, *in* G. Dengo and J. E. Case, eds., The Caribbean region: The Geology of North America: Geological Society of America, v. H, p. 307–338.

Mulder, T., and J. Alexander, 2001, The physical character of subaqueous sedimentary density flows and their deposits: Sedimentology, v. 48, p. 269–299, doi:10.1046/j.1365-3091.2001.00360.x.

Mutti, E., 1992, Turbidite sandstones: Milan, Italy, Agip, Instituto di Geologia, Università di Parma, San Donato Milanese, 275 p.

Mutti, E., G. Davoli, S. Mora, and L. Papani, 1994, Internal stacking patterns of ancient turbidite systems from collisional basins, *in* P. Weimer, A. H. Bouma, and B. Perkins, eds., Submarine fans and turbidite systems, SEPM 15th Research Conference, p. 257–268.

Mutti, E., R. Tinterri, E. Remacha, N. Mavilla, S. Angella, and L. Fava, 1999, An introduction to the analysis of ancient turbidite basins from an outcrop perspective: AAPG Continuing Education Course Note Series 39, 95 p.

Ostos, M., F. Yoris, and H. G. Avé Lallemant, 2005, Overview of the southeast Caribbean–South American plate boundary zone, *in* H. Avé Lallemant and V. B. Sisson, eds., Caribbean–South American plate interactions, Venezuela: Geological Society of America Special Paper 394, p. 53–89.

Parnaud, F., Y. Gou, J. Pascual, M. A. Capello, I. Truskowski, and H. Passalacqua, 1995, Stratigraphic synthesis of western Venezuela, *in* A. J. Tankard, R. S. Suárez, and H. J. Welsink, eds., Petroleum basins of South America: AAPG Memoir 62, p. 681–698.

Pindell, J. L., L. Kennan, W. V. Maresch, K.-P. Stanek, G. Draper, and R. Higgs, 2005, Plate-kinematics and crustal dynamics of circum–Caribbean arc–continent interactions: Tectonic controls on basin development in Proto-Caribbean margins, *in* H. Avé Lallemant and V. B. Sisson, eds., Caribbean–South American plate interactions, Venezuela: Geological Society of America Special Paper 394, p. 7–52.

Ramírez, C., 1968, Definición de la Formación Río Guache: Boletín Geología, v. 9, p. 565–567.

Sohn, Y. K., M. Y. Choe, and H. R. Jo, 2002, Transition from debris flow to hyperconcentrated flow in a submarine channel (the Cretaceous Cerro Toro Formation, southern Chile): Terra Nova, v. 14, p. 405–415, doi:10.1046/j.1365-3121.2002.00440.x.

Speed, R. C., 1985, Cenozoic collision of the lesser Antilles arc and continental South America: The origin of the El Pilar fault: Tectonics, v. 4, p. 41–69, doi:10.1029/TC004i001p00041.

Stephan, J.-F., 1982, Evolution géodinamique du domaine Caraïbe, Andes et chaîne Caraibe sur la transversale de Barquisimeto (Vénézuéla): Ph.D. dissertation, Université Pierre et Marie Curie, Paris, France, 512 p.

Stephan, J.-F., et al., 1990, Paleogeodynamic maps of the Caribbean: 14 Steps from Lias to present: Bulletin de la Societe Geologique de France, v. 6, p. 915–919.

Villamil, T., 1999, Campanean–Miocene tectonostratigraphy: Depocenter evolution and basin development of Colombia and western Venezuela: Paleogeography, Paleoclimatology, Paleoecology, v. 153, p. 239–275, doi:10.1016/S0031-0182(99)00075-9.

Von Der Osten, E., and D. Zozaya, 1957, Geología de la parte suroeste del Edo. Lara, región de Quibor (carta 2.308): Boletín Geología, v. 4, p. 3–52.

7

Henriksen, S., A. Pontén, N. Janbu, and B. Paasch, 2011, The importance of sediment supply and sequence-stacking pattern in creating hyperpycnal flows, *in* R. M. Slatt and C. Zavala, eds., Sediment transfer from shelf to deep water—Revisiting the delivery system: AAPG Studies in Geology 61, p. 129–152.

The Importance of Sediment Supply and Sequence-stacking Pattern in Creating Hyperpycnal Flows

Sverre Henriksen, Anna Pontén, Nils Janbu, and Britta Paasch
Statoil Research Centre, Rotvoll, Trondheim, Norway

ABSTRACT

The Eocene Central Basin on Spitsbergen, Neogene offshore eastern Venezuela, and offshore northern Sakhalin are high-sediment supply systems that do not obey standard rules of sequence stratigraphy. These studies show that high sedimentation rates are capable of filling accommodation space created during transgression and highstand. The thick successions of coastal plain deposits in these basins consist of fluvial and estuarine deposits and are interpreted to result from rapid increases and cutoffs in fluvial sediment supply generated by tectonic pulses and climatic cycles.

We suggest that the areal positioning of fluvial input points along a margin will give different shelf geometries. A shelf may be fluvial dominated or wave or tide influenced at the same time, which together with multiple fluvial input points creates lateral variations in the depositional architecture. The results from any studied two-dimensional section will therefore be very dependent on which part of the margin is studied, and changes in depositional architecture might be incorrectly interpreted as basinwide sea level changes.

Prolonged or periodically high sedimentation rates result in progradation of the fluvial-deltaic system toward the shelf edge. With high sedimentation rates, the sediments reaching the shelf edge are prone to collapse, creating failure back to the contemporary shoreline. The collapse scars will act as conduits, and the fluvial system will be capable of overriding the shelf edge and transporting deposits to the deep water by hyperpycnal flows.

INTRODUCTION

One field study from the Central Tertiary Basin (CTB) on Spitsbergen and two seismic studies from (1) Neogene offshore eastern Venezuela and (2) northeastern offshore Sakhalin, Russia, respectively, show basins with thick fluvial-deltaic successions. In addition, one field example from a modern delta is used to illustrate process changes across high-sediment supply systems.

Copyright ©2011 by The American Association of Petroleum Geologists.
DOI:10.1306/13271354St613441

The sedimentary infill of these basins displays high-frequency cyclicity related to major tectonic movements and changes of onshore drainage systems. Considering the volume of fluvial input to these basins, we classify them as supply-driven as opposed to eustasy-driven systems (e.g., Porobski and Steel, 2006). The massive fluvial input to the basins must have been highly efficient and probably involved hyperconcentrated flows generated by episodic floods. These floods overran the shelf edge, allowing hyperpycnal flows to enter deeper parts of the receiving basins. For the Venezuela and Sakhalin case studies, this seems to occur even during relative highstand in sea level. This scenario follows the original definition of hyperpycnal flows as fluvial input to a basin where the density of the inflow has greater density than the receiving basin (Bates, 1953). The importance of hyperconcentrated fluvial input to sedimentary basins was demonstrated by Mutti et al. (1996) and has been discussed in several publications (e.g., Mulder and Syvitski, 1995; Mulder et al., 2003; Bhattacharya and MacEachern, 2009).

This further suggests that the established sequence-stratigraphic models and work procedures introduced by Vail et al. (1977) and Posamentier and Allen (1999) do not necessarily apply to these systems. The aims of this study were therefore to (1) document the importance of the fluviodeltaic input to these basins, (2) demonstrate the effect of a high sediment supply on sequence-stacking patterns, and (3) discuss the likelihood of producing hyperpycnal flows when flood-dominated rivers meet the marine basin. Key questions that arise are the following: (1) How efficient are pure hyperpycnal flows (direct river input) for bringing sediments onto the slope or basin floor? (2) Is fall in sea level below the shelf edge necessary for incision of the shelf and efficient bypass of sediments to the deep-marine environment?

SPITSBERGEN

The CTB on the island of Spitsbergen is a small foreland basin formed by active folding and thrusting during the latest Paleocene–Eocene (Figure 1) (Steel et al., 1985). The basin was asymmetrically filled from rivers draining the rising foreland in the west (Figure 1). The sedimentation rates were generally higher than the accommodation created, and a series of seismic-scale clinoforms characterizes the basin infill (e.g., Steel et al., 1981; Helland-Hansen, 1992). The basin filled relatively rapidly, and each clinoform is inferred to represent approximately 300 ka, which is equal to 4th-order sequences (Van Wagoner et al., 1990). The clinoforms along the Van Keulenfjorden transect have been numbered 1 to 20 by Steel and Olsen (2002), and we adopt this terminology.

Most of the clinoform successions in the CTB are shale prone throughout or are sand prone only as far out as their inner reaches and shale prone from outer shelf to basin floor (e.g., Plink-Björklund et al., 2001; Mellere et al., 2002). However, clinoforms that are sand prone across the entire shelf exist, bringing sand onto the slope. On a few occasions, clinoforms that display sandy deposits far beyond the slope and onto the basin floor are seen (Figure 1) (Mellere et al., 2002).

SEDIMENTOLOGY AND FACIES TRACTS

Coastal Plain or Shelf Edge

The sedimentology and facies tracts of the CTB along the Van Keulen Fjord have been described and interpreted in several publications (Mellere et al., 2002; Muto and Steel, 2002; Plink-Björklund and Steel, 2004; Pontén and Plink-Björklund, 2009). It is beyond the purpose of this chapter to review or reinterpret these publications. We therefore give a brief outline of the most commonly observed sedimentary units. Where appropriate, we supplement these previous works with new observations from our own work on clinoform 15 in the area.

All previous studies have described a thick coastal-plain succession consisting of inferred fluvial lowstand and transgressive estuarine deposits, interpreted to consist of a series of incised valley fills (e.g., Plink-Björklund, 2005). The unconformities are of regional extent and show a relief of up to 16 m (52 ft). Considering the size of the CTB on Spitsbergen, this is a significant incision. The overall geometry of the shelf edge is recognized as a distinct geometry in outcrop (Figure 1). Typical features observed at the shelf edge are distributary channels 25 to 100 m (82–328 ft) wide and 5 m (16.4 ft) deep (Figure 2) (Plink-Björklund and Steel, 2005). The incisions on the shelf edge can be followed downdip into slope channels with a main transport direction toward the northeast (Figure 2) (Pontén and Plink-Björklund, 2009). The incisions are inferred to be formed by fluvial erosion surfaces with strong tidal influence and development of estuaries in their most seaward extent (Plink-Björklund, 2005). Thick successions of plane-parallel and climbing-ripple laminated beds at the seaward end of fluvial distributary channels are interpreted to represent mouth bars deposited by sustained hyperpycnal flows (Pontén and Plink-Björklund, 2009).

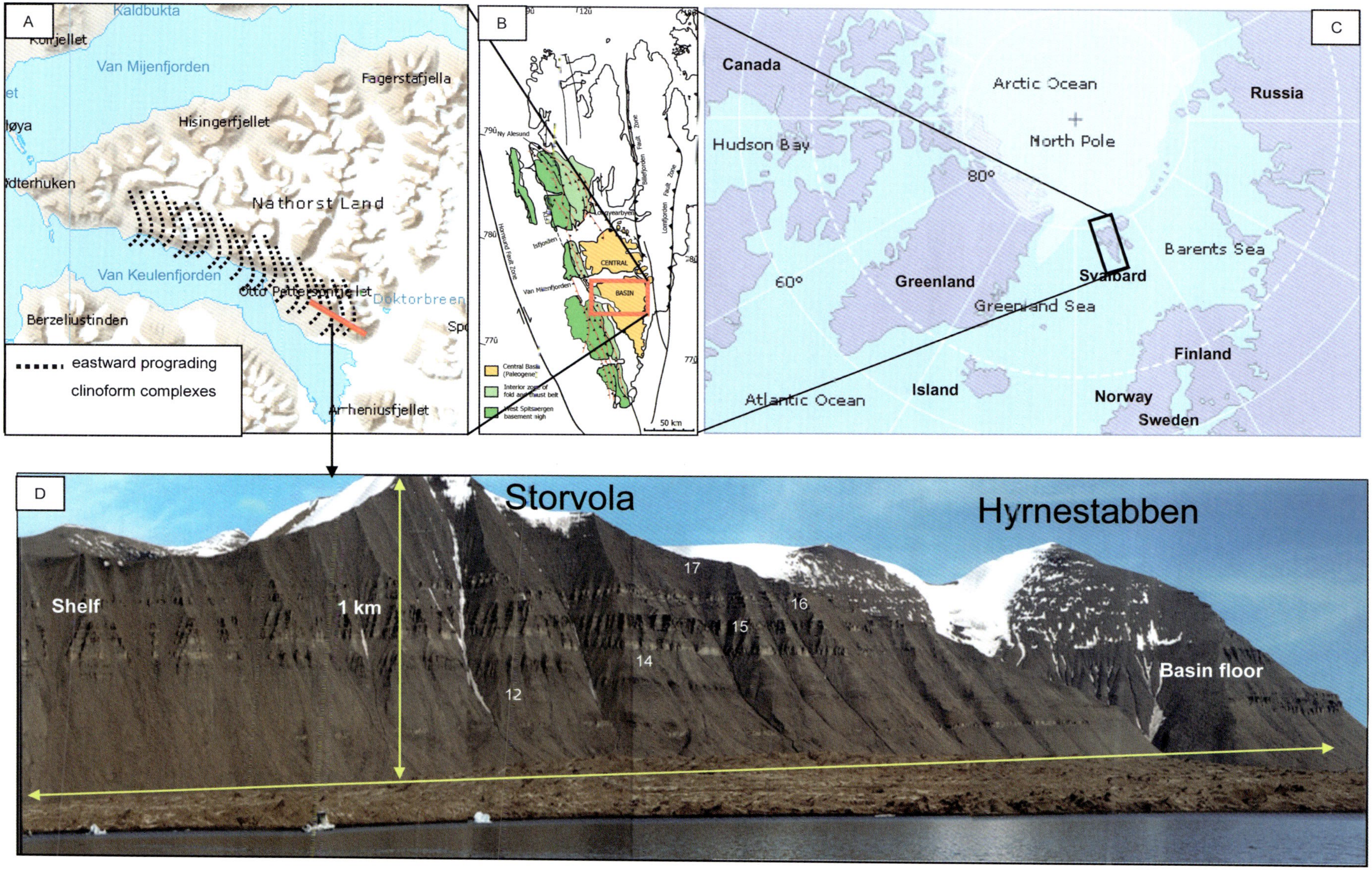

FIGURE 1. The study area along the Van Keulen Fjord. Panel A is located on the the Island of Spitsbergen. Panel B, which is part of the Svalbard archipelago, is situated in the Norwegian Arctic (C). The Eocene Central Basin of Spitsbergen (B) was a foreland or piggyback basin, created in response to strike-slip movement and transgression along the Hornsund Fault Zone. Sediment was supplied by rivers from the West Spitsbergen Orogenic Belt (B). The Eocene Central Basin was filled with southeastward prograding shelf-margin clinoforms and associated basin-floor fans identified in outcrops from Storvola and Hyrnestabben Mountains at Svalbard (D). Numbers refer to established stratigraphy and nomenclature of clinoforms in CTB (Steel et al., 2000; Plink-Björklund and Steel, 2002). The curved dotted lines along the Van Keulen transect (A) represent clinoforms outbuilding from west toward east. CTB = Central Tertiary Basin; SEDL = Svartfjella-Eidemsbukta-Daumannsodden Lineament.

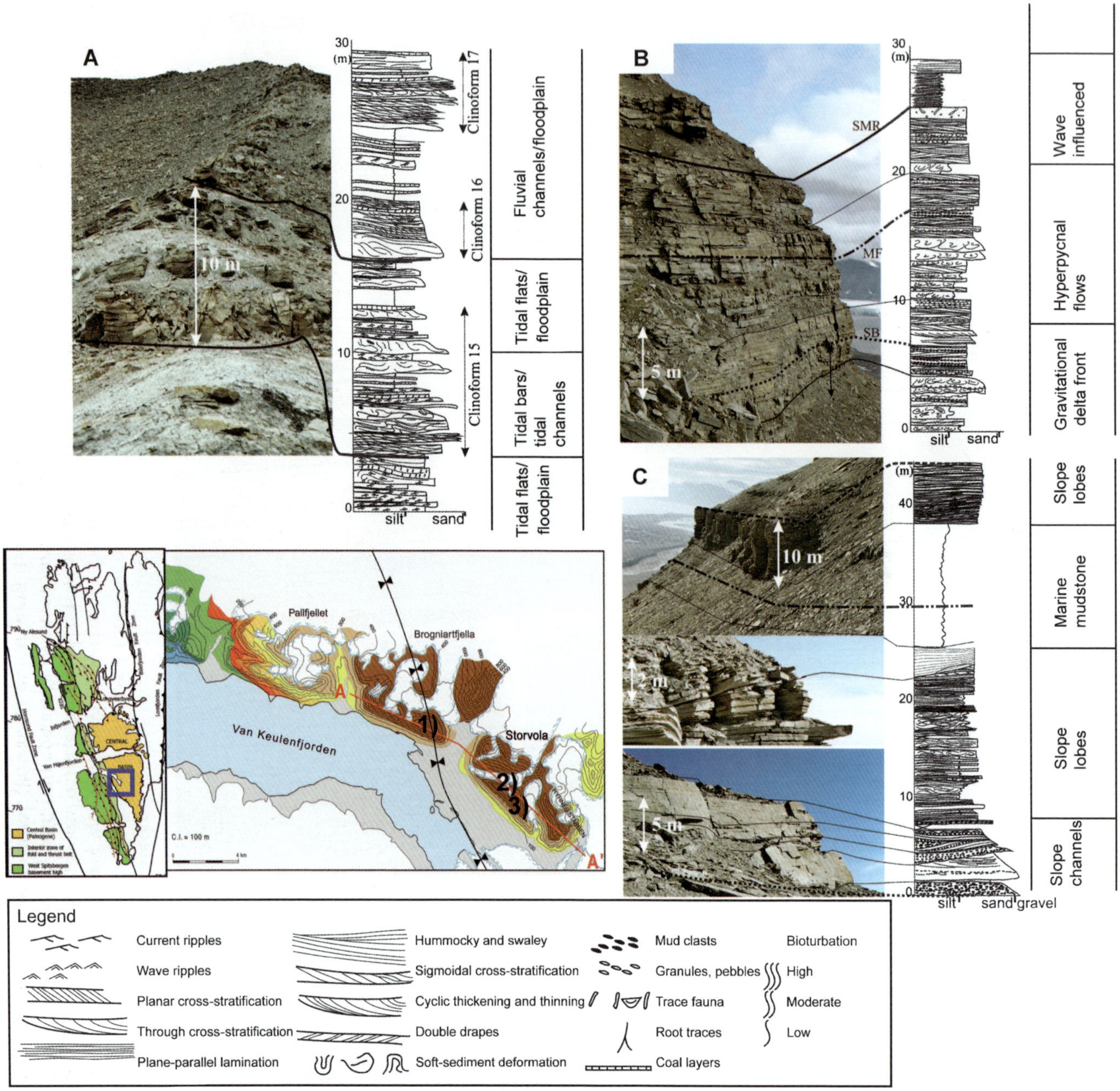

FIGURE 2. **Stratal geometries and interpreted sedimentologic logs at: (A) coastal plain, (B) shelf edge, and (C) upper slope. Field localities marked on inserted map. Legend for logs applies to all logged sections in subsequent figures. Modified from Pontén and Plink-Björklund, 2009. SB = sequence boundary; MF = marine flooding surface; SMR = surface of marine ravinement; SEDL = Svartfjella-Eidemsbukta-Daumannsodden Lineament.**

Whereas sedimentary structures suggesting hyperpycnal flows are documented from both clinoforms 14 (Petter and Steel, 2006) and 15 (Pontén and Plink-Björklund, 2009), no documentation of such features in the younger clinoform successions exists (Uroza and Steel, 2008). The younger clinoforms represent a late stage of the basin fill, and because of the reduced basin relief, they generally also lack development of a distinct shelf edge. The lateral equivalent to the younger clinoforms is probably represented by the outcrops along the northern part of the Nathorst Land Peninsula, the Van Mijen transect (Figure 3) (Olsen, 2008). Although a high sediment supply and a narrow shelf (Uroza and Steel, 2008) may initiate turbidite deposits at delta-front slopes, these sediments will be closely associated with the shoreface successions (Figure 3). These inferred turbidite deposits are notable for displaying, in places, beds with coarsening- or fining-upward grain size (Figure 3). One explanation for such grain-size variations within one bed is deposition

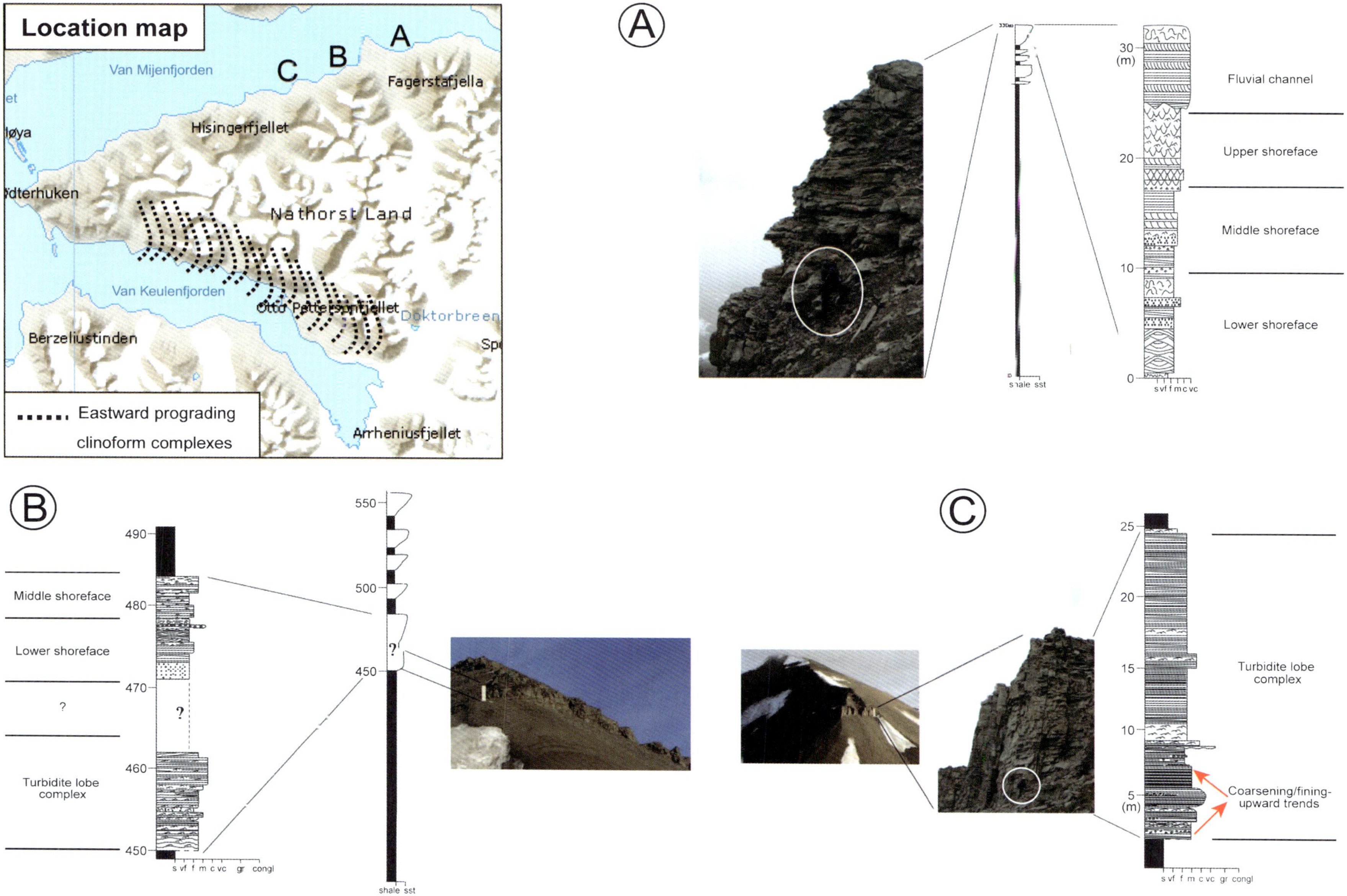

FIGURE 3. Field localities along the Van Mijen Fjord on the north side of Nathorst Land. The transect typically consists of prograding shoreface successions. In places the shoreface is incised by fluvial channels; relative sea level falls are indicated (A). Person encircled for scale. Turbidite deposits are found at several locations but are associated with delta slope and shoreface successions (B, C). Person encircled for additional scale. Note that the turbidite sandstones never fully detach from the shelfal depositional systems. Note also that individual beds display coarsening and fining-upward trends. See text for discussion (from Olsen, 2008; reprinted with permission from William Helland-Hansen, University of Bergen, Norway). Legend for logs is found in Figure 2. sst = sandstone; s = sand; vf = very fine; f = fine; m = medium; c = coarse; vc = very coarse; gr = gravel; congl = conglomerate.

by waxing and waning flows set up by a hyperpycnal delivery system (e.g., Mulder et al., 2003).

DEEP MARINE

Previous Work

Where the deltas reach the shelf edge, they contribute to building the shelf edge and slope in a basinward direction. In these cases, the delta-front deposits may constitute sheetlike turbidite deposits and chutes on the slope, both inferred to originate from sustained turbidite flow (Figure 2) (Plink-Björklund and Steel, 2005). Well-documented examples of sandy turbidite systems that, dependent on the margin type, may or may not reach the basin floor are shown by Mellere et al. (2002), Johannessen and Steel (2005), Plink-Björklund and Steel (2005), and Porobski and Steel (2006). Basin-floor fans seem to be more common in the southern part of the basin. In addition, clinoforms feeding basin-floor fans seem to be more closely spaced in the western part of the basin than in the eastern part (Plink-Björklund, 2005; Johannessen and Steel, 2005).

Compensating lobes of turbidite basin-floor fans belonging to clinoforms 14 and 15 are well exposed on the south side of the Hyrnestabben Mountain (Figure 1) (e.g., Crabaugh and Steel, 2004; Johannessen and Steel, 2005; Petter and Steel, 2006). These basin-floor fans show two main transport directions. For the lower part of the fan succession, transport direction is toward the east; and for the upper part of the fan complex, transport direction is toward the northeast (Crabaugh and Steel, 2004). The basin-floor fan of clinoform 15 is 1 to 5 m (3.3–16.4 ft) thick and displays thin-bedded turbidites of fine- to very fine-grained sand. Locally, the unit displays 2- to 3-m (6.6–9.8 ft)-thick channel geometries consisting of amalgamated fine- to medium-grained sand.

New Observations

Sandy Deep-marine Channels

New field investigations from the northeast side of Hyrnestabben show that the basin-floor fans of clinoform 15 correlate to a stacked deep-marine channel complex (Figure 4). The channels show numerous internal erosion surfaces that stack laterally and vertically (Figure 4, 5A). Apart from multiple internal erosion surfaces, the channel succession is characterized by abundant plant fragments and coal clasts (Figure 5C). Soft-sediment deformation and water-escape structures are relatively frequent (Figure 5B). Bioturbation is minor in the channels, but some trace fossils of small *Skolithos* occur in the muddy heteroliths below and between channels (Figure 5D). Less amalgamated channel geometries occur at the same stratigraphic level across the valley from the channels of Figure 4 (Figure 6). These channels show a basal section of thinly bedded strata displaying multiple sets of climbing-ripple lamination, succeeded by more massive sandstone with discrete accretion surfaces building a dune or bar form (Figure 6). Near the top of the same channel, thin-bedded strata display sets of climbing ripples (Figure 6). The massive sandstone is fine to medium grained, whereas the climbing-ripple successions consist of fine- to very fine-grained sandstone. The climbing ripples are indicative of waning flow and net deposition, whereas the massive and slightly coarser grained sandstone suggests waxing flow and an element of bypass. The base of the upper climbing-ripple package displays groove and flute casts with paleocurrent directions at approximately 90° to the climbing-ripple sets (Figure 6).

This suggests (anti-) helicoidal current in the channels and is a phenomenon that has been attributed to sustained flow (hyperpycnal flows) in sinuous channels that last for several days or weeks (Elliot, 2000; Mutti et al, 2003). Morphological expression of sinuosity is hard to recognize in outcrop, but an increasing number of documented examples of lateral accretion deposits exist, representing point-bar growth, that are currently thought to be diagnostic of sinuous channel forms (e.g., Elliot, 2000; Peakall et al., 2007). Sinuous channel lateral migration and point-bar growth appear to be driven by sustained flow of fluvial-sourced low-density turbidity currents (probably hyperpycnal), although variations in energy conditions in some outcrop examples show switches in erosive and depositional phases of activity (Mutti et al., 2003; Peakall et al., 2007; Wynn et al., 2007; Kane et al., 2008).

Fossil plant material and coal fragments are abundant throughout the channel succession and point to a direct river-fed origin for these deposits (Figures 5, 6). The strongly amalgamated sandstones with lateral accretion surfaces in the channels suggest sustained flows in turbidite channels (Figures 5, 6). This is in accordance with well-documented findings of sustained flow turbidites at several locations throughout the basin (Plink-Björklund and Steel, 2005).

A puzzling aspect of the inferred slope channels on the east side of Hyrnestabben is that they seem to be located downdip and to some extent also stack laterally to sheetlike turbidite deposits on the western flank of Hyrnestabben. Considering that the CTB probably was a highly asymmetric basin and that

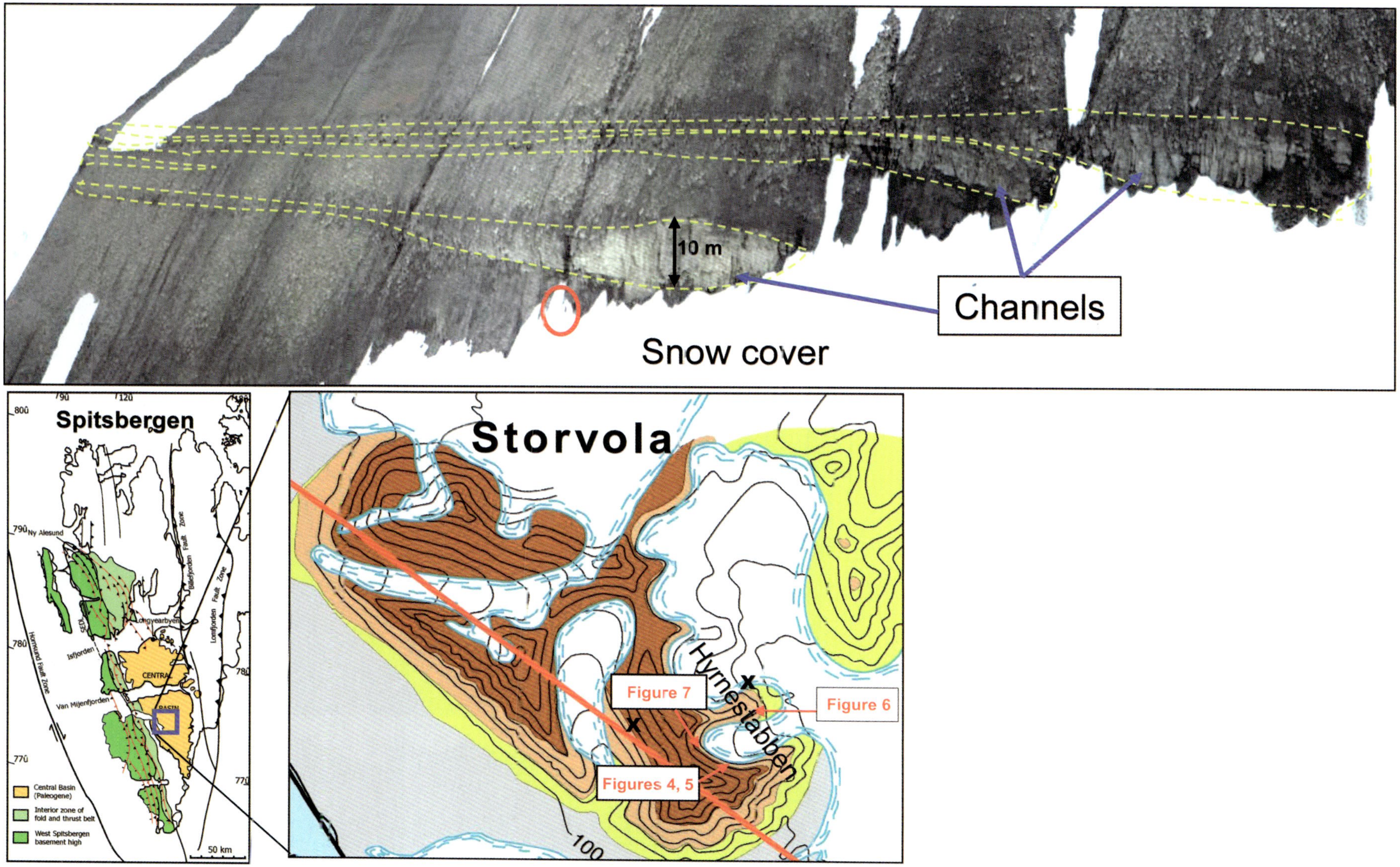

FIGURE 4. Photomosaic from northeast side of Hyrnestabben Mountain. Note the well-exposed laterally stacked channel bodies. Person encircled for scale. Field locality and logged section marked with figure numbers on inserted map. The X marks indicate additional field localities used for correlation of depositional systems. SEDL = Svartfjella-Eidemsbukta-Daumannsodden Lineament.

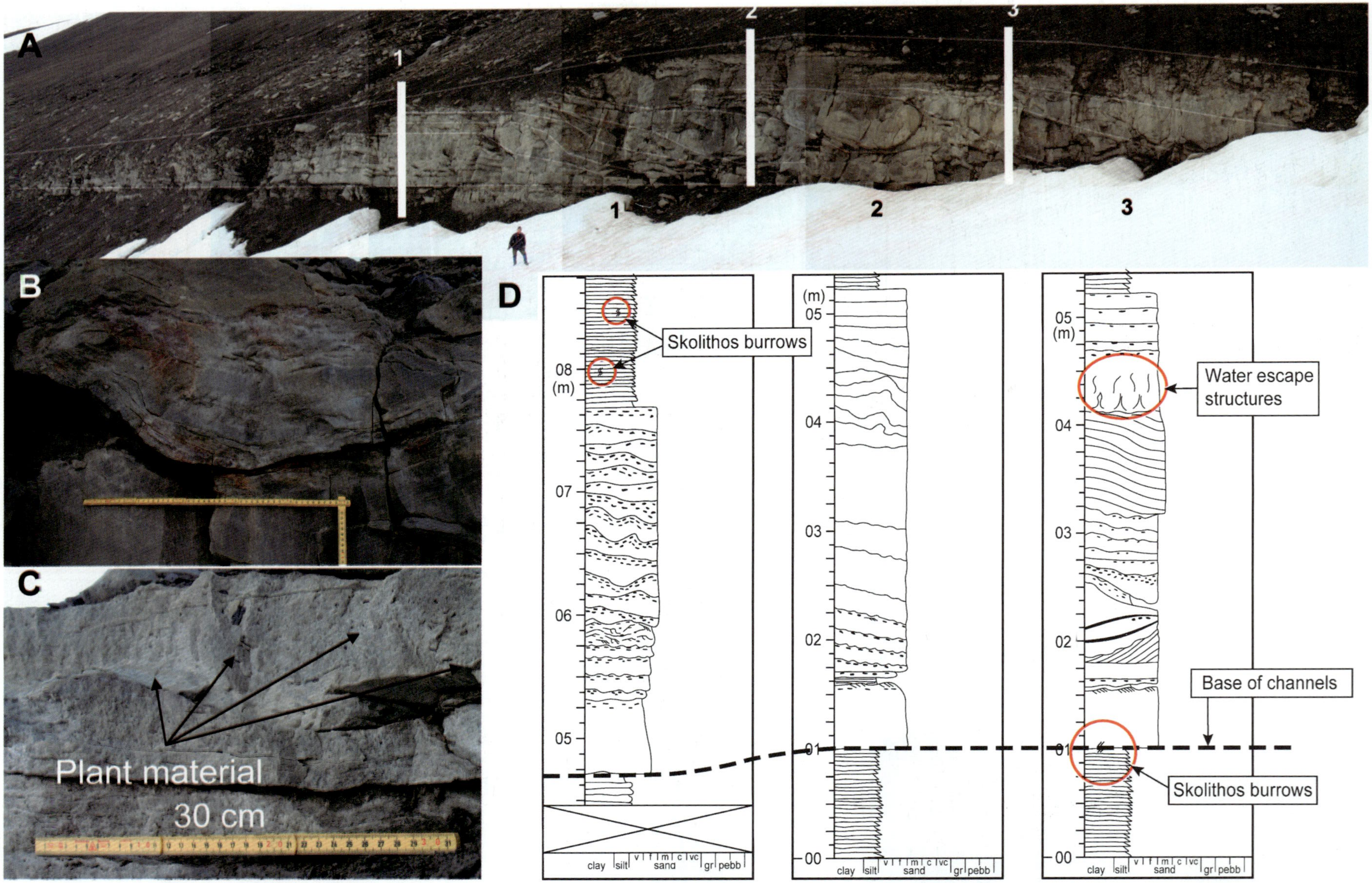

FIGURE 5. (A) Inferred slope channels at the east side of the Hyrnestabben Mountain. Note the highly amalgamated sandstones that also stack laterally. Logged sections and photographs display abundant plant material (B, D), some soft-sediment deformation and water-escape structures (C). Person on photograph gives scale of cliff face. See Figure 4 for location of outcrop. v = very fine; f = fine; m = medium; c = coarse; vc = very coarse; gr = gravel; pebb = pebbles.

FIGURE 6. (A) Inferred dune forms or lateral accretion packages at the base of a slope channel. Note that the facies changes upward and that the upper part of the channel is dominated by climbing ripples (B) with measured current direction approximately 90° to the flow direction suggested by the groove casts (C) at the base of beds. (C) Plant material and coal clasts are abundant but seem to be concentrated in certain beds. One-meter bar for scale. See Figure 4 for location of photograph. Yellow scale bar on photographs gives scale in centimeters.

clinoform 15 represents a late stage of basin filling, it is uncertain if true basin-floor-fan conditions were reached during this stage of the basin fill. In a gradually narrower basin, it is likely that the turbidity currents were diverted along the axis of the basin toward the lowest topography. The changing transport direction and lateral stacking of facies may very well be ascribed to the changing basin physiography during this late stage of basin infill.

Background Mud Sedimentation

Few, if any, of the previous studies in the CTB have dealt with the thick successions of shale between the sandy units. This bias is also common in the literature for other basins, although the mud is an essential component of the sediments delivered to the basin (Bhattacharya and MacEachern, 2009). A part of this imbalance is caused by mechanical weathering and poor exposure of the shaly intervals. Investigation of outcrop localities in newly deglaciated areas in the CTB has, however, made possible more detailed work on one of the shaly intervals.

The stratigraphic unit immediately above the deep-marine channels of clinoform 15 has been interpreted as an overall shaly or heterolithic mud succession in other parts of the basin (Figure 2) (Pontén and Plink-Björklund, 2009). It appears that the muddy intervals are very heterolithic, comprising alternating beds of 1- to 10-cm (0.39- to 39-in.)-thick mud and very fine sand in the field localities on the east side of Hyrnestabben (Figure 7). Some of the sandy beds are sharp based and fine upward like classical surge-type turbidites consisting of C and D parts of a Bouma sequence (Figure 7) (Bouma, 1962). However, other beds typically show coarsening-up to fining-upward trends, indicating waxing and waning of sustained flows (Figure 7) (Bhattacharya and MacEachern, 2009). Sedimentary structures are dominated by planar-parallel stratification and ripple-scale cross-lamination (Figure 7).

The muddy intervals are generally massive with thin stringers of silt and very fine sand. The content of sand increases upward before the next succession of mud is deposited (Figure 7). Plant fragments are overall present in the sandy intervals. Disruption of the lamination by *Skolithos* burrows occurs throughout the section but is not pervasive (Figure 7). *Phycosiphon* burrows are present on bedding planes and suggest a stressed marine environment. No exact bioturbation index is given, but on a scale from low to moderate and high, the sediments within the succession show low to moderate bioturbation.

Minor unconformities of approximately 0.5-m (1.6-ft) relief locally occur in the muddy heteroliths (Figure 7). Lateral to the main cut, the unconformities lose relief, and the succession appears to be conformable (Figure 7). The bases of the cuts are lined with deformed clay, and no signs of bypass currents exist (e.g., coarse-grained lag) (Figure 7). The limited extent of these disconformities suggests that they are formed by localized processes, such as slumps or slides as opposed to regional unconformities.

Interpretation

The scarce occurrence of bioturbation in the muddy intervals points to rapid deposition, where only limited biota had time to establish. We thus suggest that the mud layers represent rapid deposition of flocculated clay from a river inflow supersaturated with fine-grained sediments instead of slow deposition from suspension fallout. The abundant plant fragments support an interpretation of these deposits as linked and continuous deposition from source to sink instead of pelagic settling of clay from suspension.

The localized disconformities and clay smearing may be taken as an additional argument for high sedimentation rates and semicontinuous supply of sediments. Rapid deposition of flocculated clay will give very soft sediment layers that, if subject to overloading and disturbances, are prone to create slide planes.

Similar thin-bedded shale intervals are found on the east side of the Storvola Mountain (Pontén and Plink-Björklund, 2009). This succession is recognized by an upward-increasing bioturbation index dominated by *Phycosiphon*. The depositional environment is interpreted as slope lobes originating from hyperpycnal flows (Pontén and Plink-Björklund, 2009). Stratigraphically, they are in the same position as the muddy heterolith on the east side of Hyrnestabben. These deposits are interpreted to originate from hyperpycnal flows (Pontén and Plink-Björklund, 2009). Although no direct correlation can be worked out, it is likely that our study area represents a distal part of this depositional system.

SAKHALIN

Observations

Early Tertiary transtension provided accommodation for Oligocene to Pliocene deltaic progradation of the paleo-Amur River in the northwestern Pacific (Lindquist, 2000; Davies et al., 2005). During the Miocene to Pliocene, the tectonic regime resulted in uplift of the deltaic sediments now exposed on Sakhalin. As a result, the post-Pliocene sediments from the Amur River catchment were diverted northward (Davies et al., 2005). Sedimentation rates were extremely high (500–800 m [1640–2625 ft]/m.y.) and sufficient to override the basin subsidence, creating a highly prograding system (Figure 8). Apart from Wong et al. (2003) and Davies et al. (2005), no previous publications describing the seismic or sedimentologic facies of the Tertiary succession of the Sakhalin area exist. The database available for this study does not give ground proof of processes, but the high-quality two-dimensional (2-D) seismic data provide new insight for understanding one of the world's largest river systems and its delta.

With an estimated average seismic velocity of 2000 m (6560 ft)/s, the clinoforms typically display two scales of progradation: a few tens of meters and approximately 200 meters (656 ft) (Figure 8A). This is in accordance with the suggested scales of deltaic or shoreline and shelf-edge progradation, respectively (Steel and Olsen, 2002). The deltas are stacked on top of each other and show multiple episodes of advance and retreat over the shelf (Figure 8A). Where the delta clinoforms reach the shelf edge, they become the building blocks in further shelf-edge progradation, contributing to building the shelf edge basinward and upward (Figure 8A).

FIGURE 7. (A) Exposure of muddy heterolithic interval immediately above the stacked channel complex (person for scale). (B) The section is recognized by alternating sand and mud bedsets and the disconformity with clay smearing (hammer for scale). (C) Drawing of main features of the outcrop; note the rhythmic occurrence of very fine sand and mud. (D) Logged section confirms the strikingly rhythmic deposition and suggests continuous and possibly seasonal variations in sediment supply to the basin. See Figure 4 for location of outcrop and text for discussion. Numbers at left-hand side of log refer to meters of section logged.

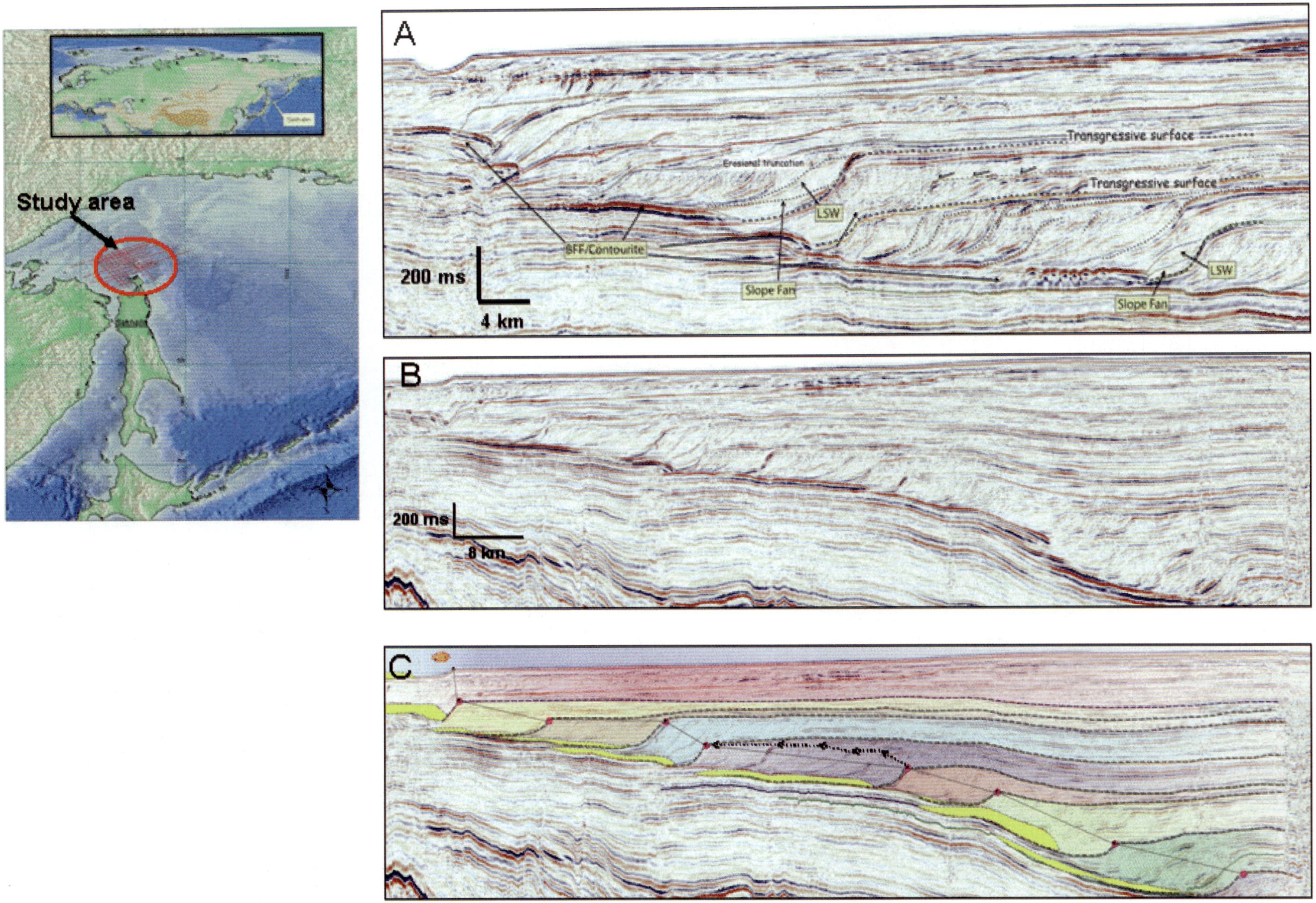

FIGURE 8. (A) Seismic transect from Northeast Sakhalin in northwest Pacific. Note multiple stacked clinothems colored for illustration purposes (B,C). The inferred basin-floor fans (BFF) are colored yellow. Lowstand wedges (LSW) at clinoform slopes are marked off. Note also that the shelf-edge trajectory is overall rising throughout progradation of the system. Seismic, reprinted with permission from TNGS-Nopec.

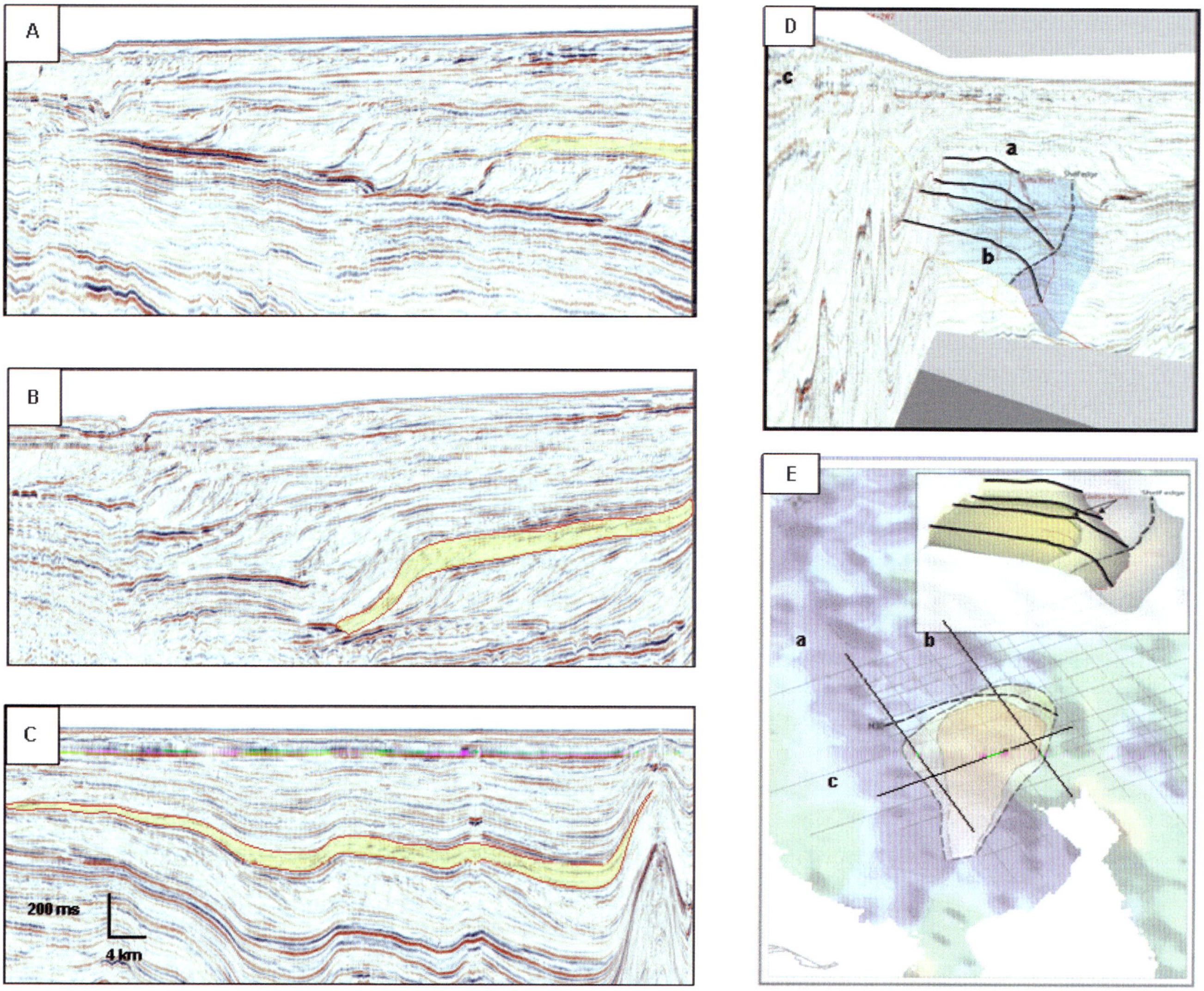

FIGURE 9. A selection of three seismic lines displaying dip sections (A, B) and one strike section (C) across the paleo-Amur Delta. The lines clearly illustrate the along-strike variability of deltas on the shelf. (D) The three-dimensional (3-D) display of the four seismic lines, illustrating the shelf (light blue) and the variable degree of delta progradation over the shelf. Note that the delta overrides the shelf edge at section b, whereas the delta is situated on the shelf in section a. Note also that where the deltas reach and override the shelf-edge break, deposition of basin-floor-fan geometries at the toe of clinoform slopes exists. (E) Map view of the delta's relation to the shelf edge and schematic illustration of how several deltas stack on top of each other. Seismic, courtesy of TNGS-Nopec.

Although the shelf-edge trajectory is dominantly ascending, basin-floor fans are observed at several of the clinoform toes (Figure 9). Our findings also show that these clinothems have a considerable degree of variability along strike (Figure 8A). Different cross sections show that the sequences display different shelf-margin types depending on where the cross section is picked (Figure 9). Where the deltas reach the shelf edge, basin-floor fans are also most commonly found at the toe of clinoform slopes (Figure 9). Some fan features are, however, slightly displaced in relation to the location of the deltas and seem to onlap the slope in an upslope prograding manner (Figure 9).

Interpretation

The multiple scales of clinoforms stacked on top of each other suggest rapid changes in relative sea level, and that the sedimentary input is sufficient to fill in the accommodation created by the mid-Cenozoic strike-slip movements. The massive input of sediments into the basin suggests large-scale uplift or reorganization of the source area, possibly combined with climatic deterioration on the Northern Hemisphere. The extreme basinward progradation of proximal facies was also noted by Davies et al. (2005), who attributed this to a forced regression. The fact that the inferred basin-floor fans also are found at the toe of clinoform sets with ascending trajectories implies that the North Sakhalin Basin was a supply-driven system during most of the Neogene. A high sediment supply is thus evident from both stacking pattern and calculated volumes (Lindquist, 2000). However, the along-strike variability within the sequences shows that only parts of the delta reach the shelf edge. The part of the delta reaching the shelf edge is probably the area of highest fluvial discharge and is thus dominated by high-energy fluvial processes. This is confirmed by field studies by Davies et al. (2005), who suggest a braided fluvial delta system in the Neogene of the Amur River. The areas lateral to the delta point were most likely reworked by waves, tides, and shelfal currents, similar to the older delta successions in the area (Davies et al., 2005).

Deltaic deposition (and overriding) at the shelf edge suggests a link between the fluvial system and the mass wasting processes on the slope. The inferred basin-floor fans with upslope prograding geometries bear close resemblance to contourite drifts (e.g., Henriksen et al., 2005). This was also pointed out by Wong et al. (2003), who interpreted these geometries to result from counterclockwise contour currents, which may have been established as early as the Oligocene (Davies et al., 2005).

VENEZUELA

Observations

The paleo-Orinoco delta, offshore eastern Venezuela, represents a high-supply system where sedimentation rates were capable of filling in the huge accommodation created by upper Miocene or lower Pliocene Canyon incision (DiCroce et al., 1999; Sydow et al., 2003). Sedimentation rates through the Neogene were enormous with 5 to 6 m (16.4–19.7 ft)/ka, reaching 8 m (26.2 ft)/ka in some depocenters (Wood, 2000). The high sedimentation rates resulted in a prograding system, with an overall rising shelf-edge trajectory trend (Figure 10) (Sydow et al., 2003). Thick successions of sediments with a subparallel seismic expression frequently intercepted by smaller channel geometries accumulated in the topset segment of clinoforms (Figures 10, 11). The channelized facies are more prevalent in the landward (western) areas, whereas the continuous stripy facies become increasingly more important in the seaward part of the topsets (Figures 10, 11). Based on their stratigraphic position and correlation to in-house well data, the topset segments are interpreted to consist of deltaic and marginal marine sediments (Figures 10, 11). The shallow-marine sediments display 30- to 60-ms two-way traveltime (TWT) thick blocky to upward-decreasing gamma values separated by abrupt increases in gamma value (Figure 11). These are interpreted as stacked parasequences in a shoreface position. The fluvial facies is characterized by thicker (100-ms TWT) upward-decreasing-and-increasing gamma values interpreted to reflect progradation and retrogradation of the delta system, with maximum progradation being represented by the coarsest part of the gamma log.

Close to the shelf edge, clinoform geometries of a few tens of milliseconds TWT contribute to building the shelf-edge clinoforms (200- to 300-ms TWT) seaward (Figure 11). In the case of an ascending shelf-edge trajectory, the paleoslopes show a distinct curved geometry and abrupt seaward termination of topset strata (Figure 10). In the case of a descending shelf-edge trajectory, the slope profiles appear straighter, and the bounding unconformity truncates the topset segment of the previous sequence set (Figure 10). High-amplitude features with mounded geometries appear at the toe of clinoform slopes and are associated with all types of trajectory sets (Figure 10). It must however be noted that obvious along-strike variabilities of the stacking pattern exist, and trajectory sets may appear descending or ascending only a few kilometers apart.

Interpretation

The thick succession of strata in the topset segment signifies a high sediment input in balance with high subsidence rates in the basin. This was also noted by Wood (2000) and Sydow et al. (2003), who identified a vertical stacking of deltaic sediments at the shelf edge. Furthermore, the high-frequency shifts of facies in the topset segment suggest rapid advances and retreats of the proto-Orinoco delta over the shelf. Considering the high sedimentation rates, the shelf accommodation was filled at all times, and consequently little space for accumulation of sediments was created during transgressions (Wood, 2000). The high-amplitude mounds at the base of clinoform slopes are interpreted to consist of basin-floor fans transported over the paleoshelf edge by mass wasting. These may well represent slump scars and mass transport complexes, respectively.

The fact that transport of sediments to the basin floor seems to be associated with clinoforms that show both descending and ascending trajectories suggests that sedimentation rates were sufficiently high to override the shelf edge, even during highstands of relative sea level. The curved slopes occur when shelf-edge

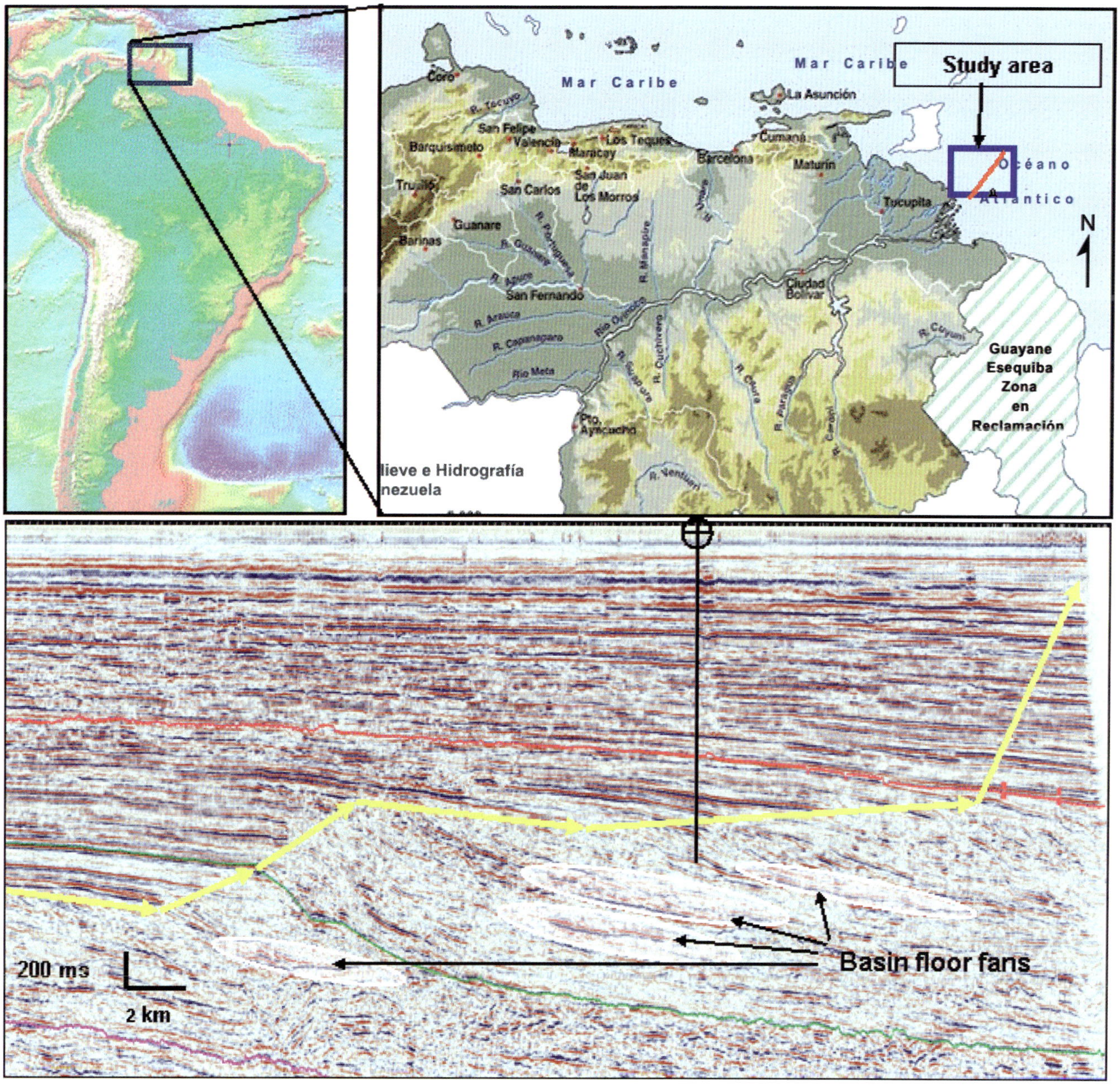

FIGURE 10. Seismic section from offshore eastern Venezuela (paleo-Orinoco delta) illustrating the sequence-stacking pattern resulting from the high sedimentation rates during the Pliocene–Pleistocene. Yellow arrows indicate the shelf-edge trajectory trend. Note that inferred basin-floor-fan features occur also under an ascending trajectory condition (highstand systems tract). Well symbol indicates location of random well used for illustrating facies correlation to well. See text for discussion. Modified from Henriksen et al. (2011).

trajectories are ascending, and the chaotic mounds associated with these strongly suggest slumping of sediments piled up at or near the shelf edge. The slumping will create a topographic low that may focus fluvial channels on the shelf. The initiation of the basin-floor fans may thus not be related to the formation of an incised valley, whereas the incision may be a secondary effect of slope failure.

The erosion associated with descending trajectories is most likely caused by fluvial incision and a classical lowstand situation, where bypass of sediments from the shelf to the basin floor through incised valleys is inferred.

DISCUSSION

Supply-driven Versus Eustasy-driven Systems: Highstand or Lowstand Delivery System?

A cyclic stacking pattern developed at different hierarchical orders is a striking aspect of all three case

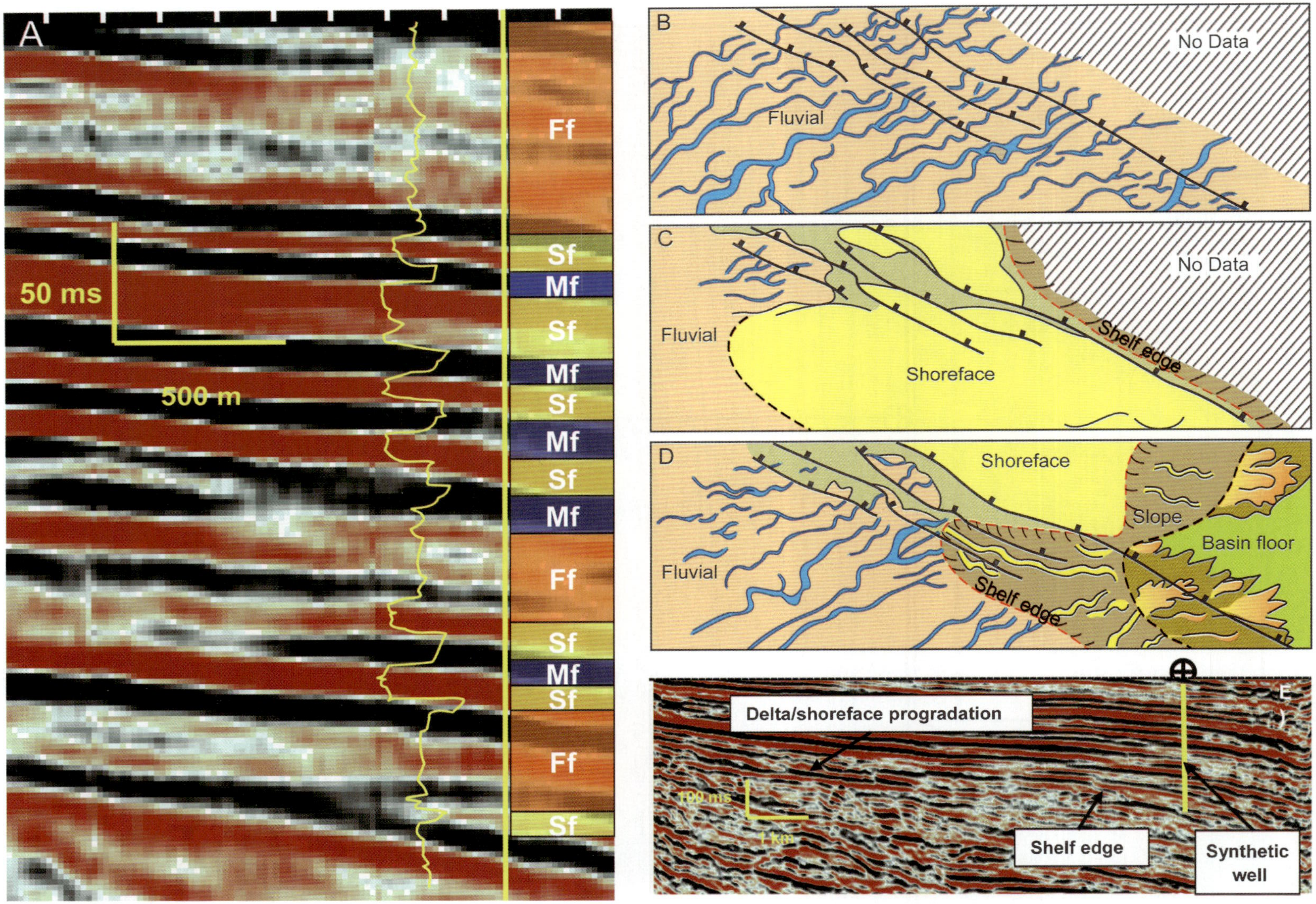

FIGURE 11. (A) Correlation of fluvial-seismic facies (Ff) and shallow-marine seismic facies (Sf) to random well (see Figure 10 for a typical placement of such a well). The Sf display 30 to 60 ms TWT (30–60 m [98–196 ft]) thick blocky to upward-decreasing gamma values, separated by abrupt increases in gamma value representative of marine shale facies (Mf). The Ff displays a different gamma ray pattern; it is characterized by thicker (ca. 100 ms/100 m [328 ft]) upward-decreasing-and-increasing gamma values. Instantaneous amplitude maps of seismic facies recognized in this study are also shown. (B) The Ff with typical disorganized character. (C) Shallow-marine seismic facies. Bright colors represent continuous shoreface sandstones. (D) Fluvial facies in the topset segment abruptly cut by an erosive shelf edge. Note the concave shape of the shelf edge indicated by the stippled line. (E) Seismic section across the shoreface (delta) progradation close to the shelf edge. The white lines on the maps indicate the outline of faults cutting through the succession. Modified from Helle (2008). TWT = two-way time.

studies. These stacking patterns are apparently very similar to those that are thought to be characteristic of sequence-stratigraphic models, all implying a relative fall in sea level to push the deltas to the shelf edge and further dislocation of sediments from the shelf to the deep-marine parts of the basin (Mitchum et al., 1977; Miall, 1991; Posamentier et al., 1991; Posamentier et al., 1992; Schumm, 1993; Bullimore et al., 2005). Indeed, even for the highest sedimentation rates, this is most likely to happen during a fall in relative sea level. However, considering that the three case studies represent high-supply systems, the understanding of other observations must be viewed in the context of the extraordinary basin conditions. In the offshore Sakhalin and Venezuela studies, it is obvious that different margin types are simply a result of along-strike variability of the same sequence (Figure 9). Similar relationships seem to apply for the CTB, where studies from other parts of the basin show that basin-floor fans are nonexistent only 20 km (12.4 mi) from the Van Keulen transect (Figure 3) (Helland-Hansen, 1990; Plink-Björklund and Steel, 2002; Olsen, 2008; Stene, 2009). This addresses the problem of applying rigid sequence-stratigraphic models for deltaic systems as pointed out by Wood (2000) and Anderson (2005).

The fluvial system feeding the shelf-edge deltas in the CTB is considered to have been widespread with numerous channels similar to a braid delta (Muto and Steel, 2002). A modern braid delta with a high sediment supply may therefore serve as a useful analog for the stacking pattern and distribution of depositional environments observed in the CTB (Figure 12). The modern delta clearly shows the existence of multiple depositional environments within a limited geographic area. We will therefore launch the possibility that the inferred margin types observed on Spitsbergen (e.g., Mellere et al., 2002; Plink-Björklund and Steel, 2004) also are the result of three-dimensional (3-D) variability or along-strike variability of depositional systems (Figures 13, 14). A logical consequence of this is that a variety of processes may be active at the same time, and that lateral facies changes also must be viewed in the context of along-strike variability (Yoshida et al., 2007; Pontén and Plink-Björklund, 2009).

For a river to go hyperpycnal and support sustained flow, it needs to be chocked with mud (Bhattacharya and MacEachern, 2009). During a relative sea level rise, a high-supply system will continue to shed sediments to the shelf and slope, but most of the sand will be stored on the shelf. The turbidite system will, however, not shut down but be expressed as a heterolithic succession of alternating mud and fine sands on the slope and basin floor.

The thick accumulations of the shelf and oversteepening of the slope are prone to collapse, and in the case of a hyperpycnal delivery system, the sediment discharge and current velocity may be sufficient to incise the shelf, even during a highstand situation (Figure 14). Indeed, Mulder et al. (2003) documented that a single flood may cause a river to incise 3 m (9.8 ft) as far as 2 km (1.2 mi) inland of the river mouth without implying sea level fall at all. As suggested by this study, the presence of hyperpycnal muds at the slope may in itself be a cause of slope failure.

We will here invoke the possibility that the +15-m (+49-ft) incision on the shelf of clinoform 15 in the CTB is not necessarily caused by a base level fall as suggested by Mellere et al. (2002) and Pontén and Plink-Björklund (2009). If a hyperconcentrated or hyperpycnal river plunges directly into a newly formed slump scar and submarine canyon, the sediment delivery to the deep marine may be efficient as long as the river stays hyperpycnal (Porobski and Steel, 2006). In the case of a collapse of an oversteepened shelf edge, there will be an ignitive phase of the turbidity current that may be a result of, instead of a cause of, hyperpycnal flow (Figure 14). Several studies (e.g., Hiscott et al., 1997; Mulder et al., 2001) suggest hyperpycnal flows as an important mechanism for formation and maintenance of submarine canyons in areas of high sediment supply, either as frequent slide-induced surges or by sustained hyperpycnal flows (Mulder et al., 2003; Carvajal and Steel, 2006). In this case, the hyperpycnal flows confined within the incised shelf will set up a sediment fairway that will adjust its gradient toward the lowest point (the basin floor). This will imply a stratigraphic base level different from mean sea level. Such a model will only be meaningful if a sustained flow from source to sink in the basin exists (Mutti et al., 2009).

The Paleogene of Spitsbergen was deposited during greenhouse conditions, and the archipelago experienced temperate and humid climates (Schweitzer, 1980). In this scenario, eustatic changes are considered minor. However, like today, the Svalbard region was situated at a high northern latitude, between 65°N and 75°N in the Eocene (Irwing, 1977). Although generally warm climates are inferred, clasts commonly present in the Tertiary shales are interpreted to be ice-rafted dropstones (Dalland, 1977), indicating large shifts in temperatures between summer and winter months. Higher frequency stacking patterns and catastrophic

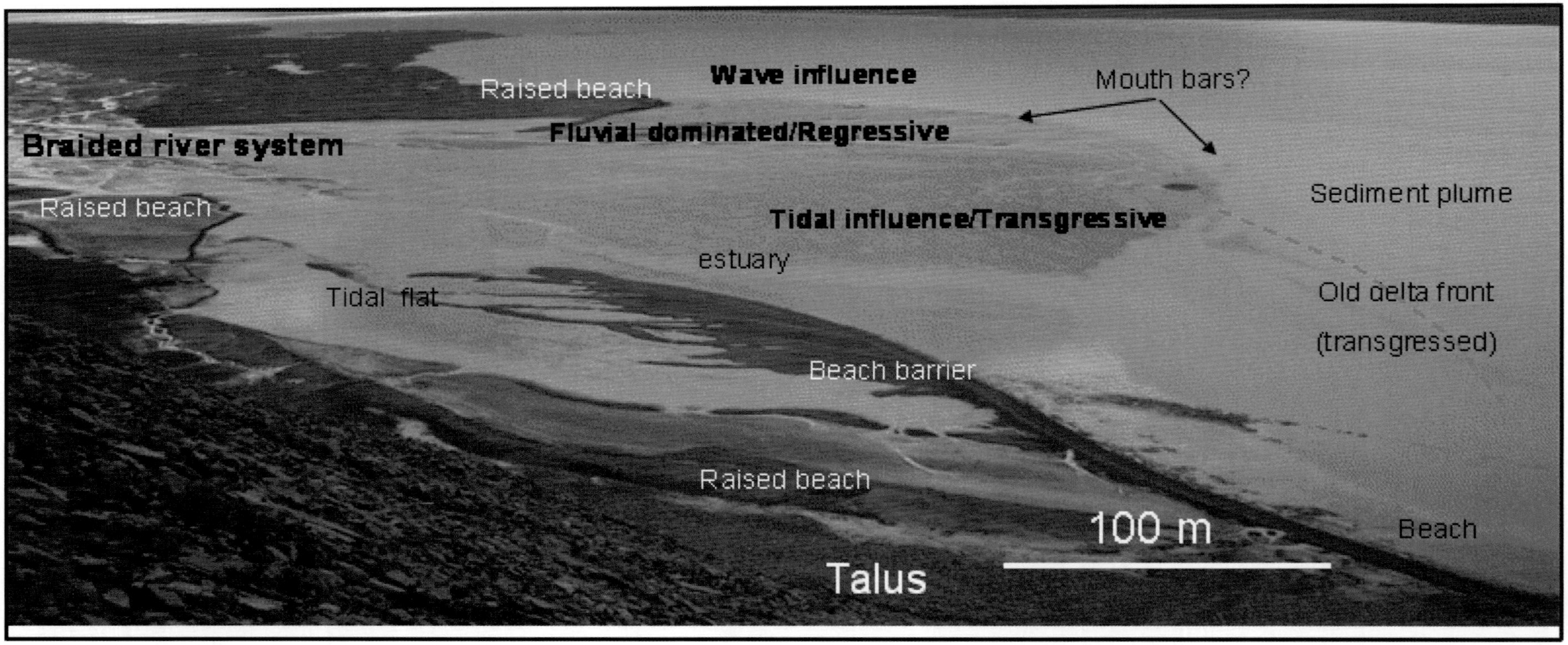

FIGURE 12. Modern delta in front of a braided river system in Van Keulen Fjord, Spitsbergen. A number of depositional environments and processes coexist and are active at the same time. Note also that a vertical stacking of facies on the modern delta plain exists.

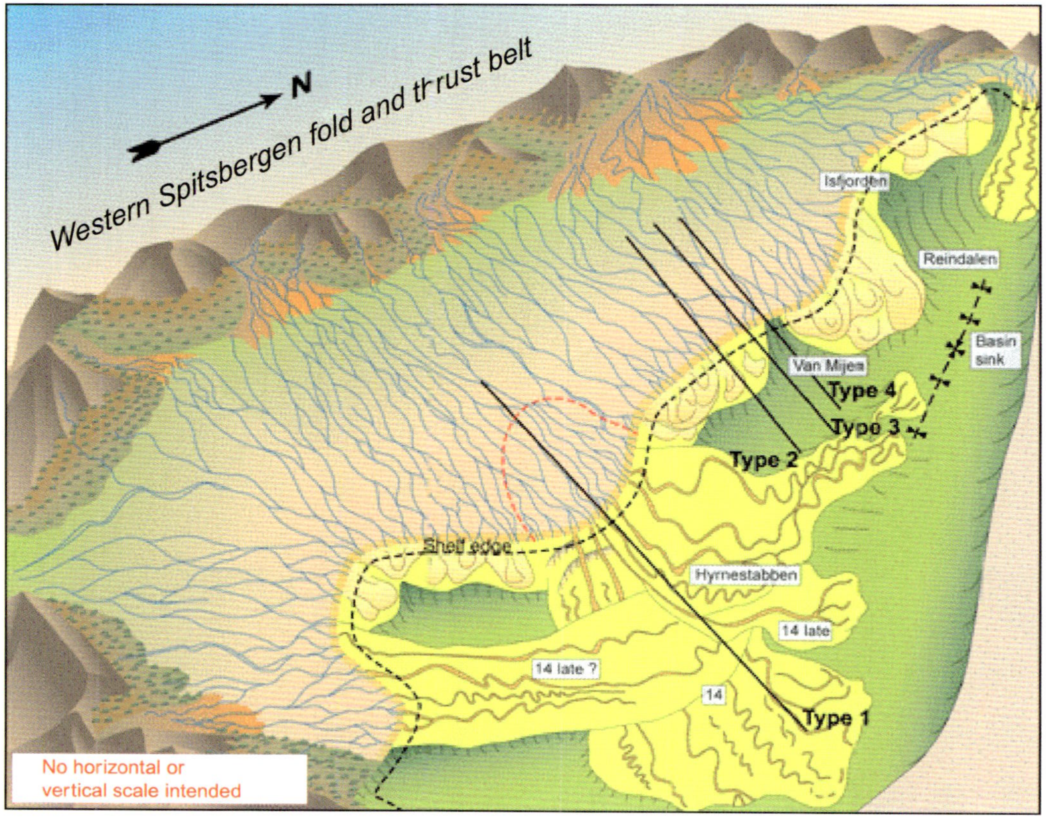

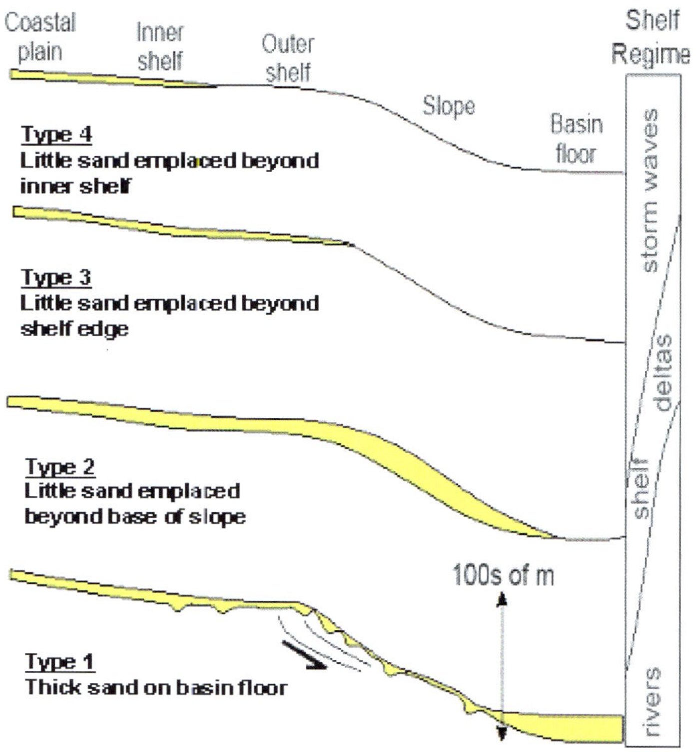

FIGURE 13. Depositional model for the Eocene Battfjellet Formation in Central Tertiary Basin on Spitsbergen, Svalbard. Margin types 1 to 4 are redrawn from Mellere et al. (2002). Note that the different margin types coexist and represent along-strike variability of depositional systems along the margin. Red dotted line indicates area subject to collapse caused by high sedimentation rates (see text for discussion). Note also that all four margin types from Mellere et al. (2002) can occur at the same time, dependent on the orientation of the cross section (see text for discussion).

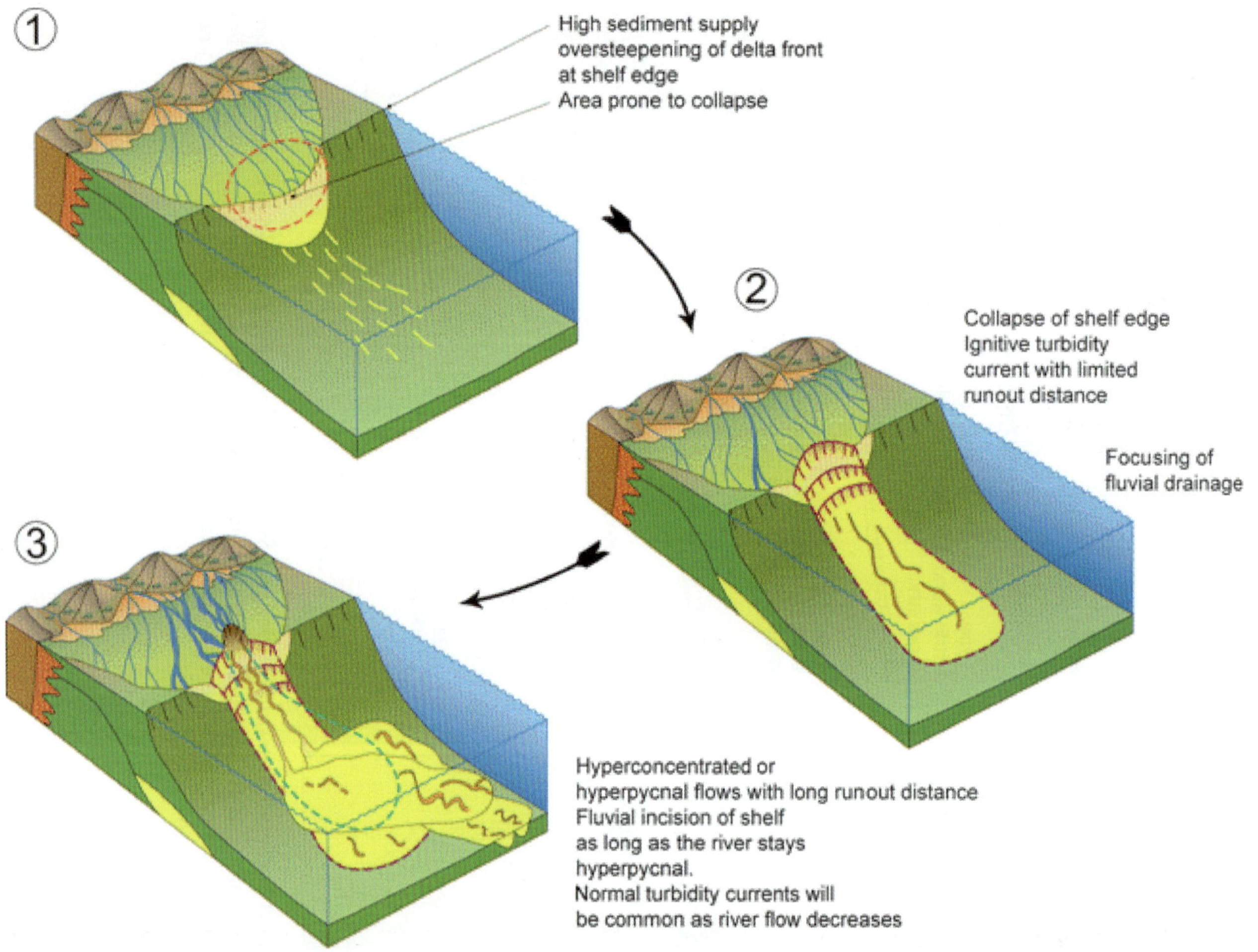

FIGURE 14. Block diagrams displaying the proposed sequence development of a high-supply system like the Central Tertiary Basin (CTB), Svalbard. Note that no large relative sea level changes are inferred as a genetic cause of the sequence. Instead, the observed stacking patterns and facies result from variations in generally high sediment supply. This suggests a model for formation of incised valleys in a supply-driven system. Note that both ignitive and hyperpycnal turbidity currents may coexist, and that there probably are transitional flows between the different stages.

events of sedimentary input within these stages (Plink-Björklund, 2005) may thus be interpreted to represent forestepping-backstepping episodes of fluvial deposition controlled by tectonic pulses interacting with climatic variations (Spielhagen and Tripati, 2009). The rhythmic expression of the muddy heteroliths in this study may well represent seasonal variations in runoff and temperature.

The Neogene of Sakhalin and offshore Venezuela are clearly under the influence of ice house earth oscillations of sea level, and a high-frequency and high-amplitude variation is to be expected. This also explains the rapid transgressive and regressive events inferred for both of these systems.

The relationship between the sediment supply, amplitude of relative sea level change, climate, and tectonic movements in fully marine basins like the Sakhalin and offshore Venezuela remains unclear. In a foreland basin like the CTB, these relations may be even more complex, and a tripartite variation diagram is proposed (Figure 15).

SUMMARY AND CONCLUSIONS

- In a eustatically driven system, basin-floor fans will be deposited during periods of descending shelf-edge trajectory styles (lowstand systems tract). The three case studies represent exceptions to these general rules and seem to be closely related to tectonic and climatic influence.
- In supply-driven systems, basin-floor fans may be deposited during periods of ascending trajectory styles (transgressive and highstand systems tract).
- Variability of sequence-stacking patterns and hyperpycnal flow organization on supply-driven margins is likely a result of 3-D variability along strike.

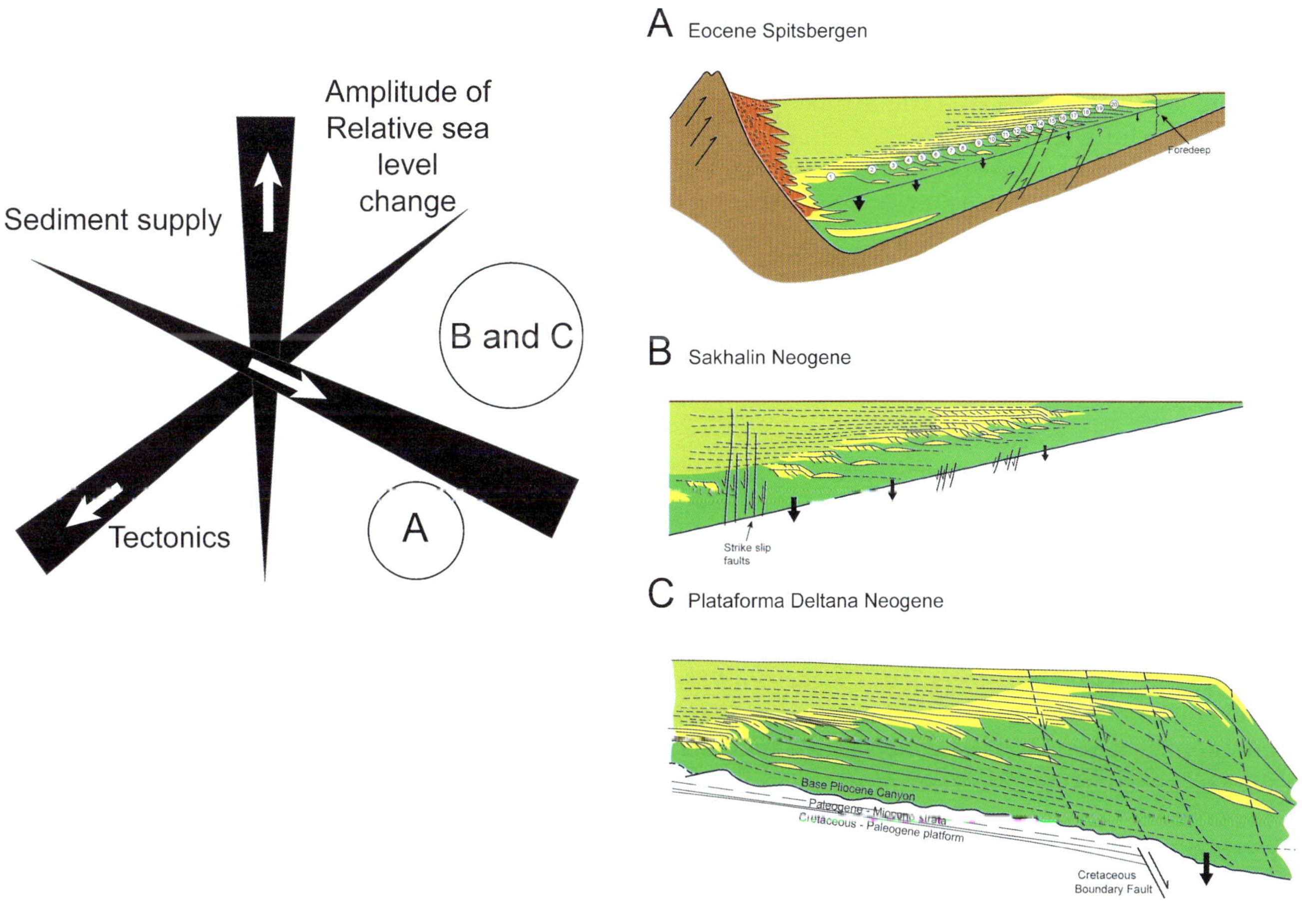

FIGURE 15. Development of sequence-stacking patterns in relation to variable sediment supply, amplitude of relative sea level change, and tectonic movements. Panels A, B, and C in the ternary diagram represent Svalbard, Sakhalin, and Plataforma Deltana, respectively. Note that all systems are high-supply systems, but that they represent variable tectonic settings. In the Svalbard case (A), the tectonic influence is largest for the early stage of the basin fill, and although sedimentation rates are high, the amplitude of sea level change is low. This will give rise to higher clinoform relief in the early versus the late stage of the basin fill. See text for discussion.

- Any river system capable of carrying sediment loads large enough to overcome the highstand accommodation and with sufficient discharge to displace seawater is likely to produce a hyperpycnal flow.
- Hyperconcentrated (catastrophic) river floods are unlikely to be focused river systems that incise the shelf. Incision and focusing of flows may occur because of initial slumping at the shelf edge. Headwall slumping and the momentum of the sediment load of the river are capable of creating incisions, easily misinterpreted as incised valleys.
- A large part of the sediment load in a hyperconcentrated river is mud, and these sediments are deposited as heterolithic successions of alternating clay and very fine sand.
- Muddy intervals void of bioturbation likely result from rapid deposition of flocculated clay. Local slumps with clay smearing at the base suggest collapse of low-competence hyperpycnal muds.

The scenario outlined does not contradict the observations presented in previous sequence-stratigraphic models for CTB; however, it highlights the importance of along-strike variability and sediment supply in sequence-stratigraphic interpretation. In a broader perspective, the models challenge our ideas about the timing of sedimentation on continental margins and the ocean floor. This, in turn, could impact the interpretation of the ancient rock record in terms of the role of eustatic sea level, climate, and sediment supply (e.g., Haq et al., 1987; Massari et al., 2007) because the relationship between deep-marine sedimentation and relative sea level change might be much more complex than commonly assumed (Anderson, 2005).

Because of easier access in outcrops combined with focus on reservoirs in the hydrocarbon industry, the literature is heavily weighted toward studies of sandstone successions. Evaluation of the shaly successions (e.g., Bhattacharya and MacEachern, 2009), as well as systematic studies of trace fossils (e.g., MacEachern et al., 2005), may be a key step to a more correct stratigraphic model in any basin.

ACKNOWLEDGMENTS

We thank Statoil ASA for permission to publish this work. TNGS-Nopec generously gave access to and allowed publication of the high-quality seismic from offshore Sakhalin. Atle Folkestad, Marc Ryan, Sigurd D. Nerhus, and Knut Eitrheim all contributed to successful field work on Svalbard in the summer of 2006. Janok Bhattacharya's insightful review significantly improved the manuscript. Roger Slatt's suggestions for refinement of the final version of the manuscript are highly appreciated. Simon Leary made a valuable contribution to this chapter by correcting the English text. We also thank Lars Reistad at Statoil Research Centre for a tremendous effort in the drafting of figures.

REFERENCES CITED

Anderson, J. B., 2005, Diachronous development of Late Quaternary shelf-margin deltas in the northwestern Gulf of Mexico: Implications for sequence stratigraphy and deep water reservoir occurrence, *in* L. Giosan and J. Bhattacharya, eds., River deltas: Concepts, models, and examples: SEPM Special Publication 83, p. 257–278.

Bates, C. C., 1953, Rational theory of delta formation: AAPG Bulletin, v. 37, no. 9, p. 2119–2162.

Bhattacharya, J., and J. A. MacEachern, 2009, Hyperpycnal rivers and prodeltaic shelves in the Cretaceous seaway of North America: Journal of Sedimentary Research, v. 79, p. 184–209, doi:10.2110/jsr.2009.026.

Bouma, A. H., 1962, Sedimentology of some flysch deposits: A graphic approach to facies interpretation: Amsterdam, Netherlands, Elsevier, 168 p.

Bullimore, S., S. Henriksen, F. M. Liestøl, and W. Helland-Hansen, 2005, Clinoform stacking patterns, shelf-edge trajectories and facies associations in Tertiary coastal deltas, offshore Norway: Implications for the prediction of lithology in prograding systems: Norwegian Journal of Geology, v. 85, p. 169–187.

Carvajal, C. R., and R. J. Steel, 2006, Thick turbidite successions from supply-dominated shelves during sea level rise: Geology, v. 8, p. 665–668, doi:10.1130/G22505.1.

Crabaugh, J., and R. Steel, 2004, Basin-floor fans at the Central Tertiary Basin, Spitsbergen: Relationship of basin floor sand-floor sand bodies to prograding clinoforms in structurally active basins, *in* S. A. Lomas and P. Joseph, eds., Confined turbidite systems, Geological Society (London) Special Publication, v. 222, p. 187–208.

Dalland, A., 1977, Erratic clasts in the lower Tertiary deposits of Svalbard: Evidence for transport by winter ice: Norwegian Polar Institute Annual Report 1976, p. 151–166.

Davies, C., S. Poynter, D. MacDonald, R. Flecker, L. Voronova, V. Galverson, P. Kovotunovich, L. Fot'yanova, and E. Blanc, 2005, Facies analysis in the Neogene delta of Amur River, Sakhalin, Russian Far East: Controls on sand distribution, *in* L. Giosan and J. Bhattacharya, eds., River deltas: Concepts, models, and examples: SEPM Special Publication 83, p. 207–229.

DiCroce, J., A. W. Bally, and P. Vail, 1999, Sequence stratigraphy of Eastern Venezuela Basin, *in* P. Mann, ed., Caribbean basins, sedimentary basins of the world: Amsterdam, Netherlands, Elsevier Science B.V., p. 419–476.

Elliot, T., 2000, Depositional architecture of a sand-rich, channelized turbidite system: The upper carbinoferous Ross Sandstone Formation, Western Ireland, *in* P. Weimer, R. M. Slatt, A. H. Bouma, and D. T. Lawrence, eds., Deep-water reservoirs of the world: Gulf Coast SEPM Foundation 20th Annual Research Conference, p. 342–373.

Haq, B. U., J. Hardenbol, and P. R. Vail, 1987, Chronology of fluctuating sea level since Triassic (250 million years to present): Science, v. 235, p. 1156–1166.

Helland-Hansen, W., 1990, Sedimentation in Paleogene Foreland Basin, Spitsbergen: AAPG Bulletin, v. 74, no. 3, p. 260–272.

Helland-Hansen, W., 1992, Geometry and facies of Tertiary clinothems, Spitsbergen: Sedimentology, v. 39, p. 1013–1029, doi:10.1111/j.1365-3091.1992.tb01994.x.

Helle, K., 2008, The response of prograding wave-dominated delta systems to rising relative sea level: Ph.D. thesis, University of Bergen, Bergen, Norway, 99 p.

Henriksen, S., C. Fichler, A. Grønlie, T. Henningsen, I. Laursen, H. Løseth, D. Ottesen, and I. Prince, 2005, The Norwegian Sea during the Cenozoic, *in* B. Wandås, et al., eds., Onshore-offshore relationships on the North Atlantic Margin: Norwegian Petroleum Society Special Publication 12, p. 111–133.

Henriksen, S., W. Helland-Hansen, and S. A. Bullimore, 2011, Relationships between shelf edge and shoreline trajectories and sediment dispersal along dip and strike: A different approach to sequence stratigraphy: Basin Research, v. 23, p. 3–21, doi:10.1111/j.1365-2117.2010.00463.x.

Hiscott, R. N., K. T. Pickering, A. H. Bouma, B. M. Hand, B. C. Kneller, G. Postma, and W. Soh, 1997, Basin-floor fans in the North Sea: Sequence stratigraphic models vs. sedimentary facies (discussion): AAPG Bulletin, v. 81, p. 662–665.

Irwing, E., 1977, Drift of the major continental blocks since the Devonian: Nature, v. 270, p. 304–309, doi:10.1038/270304a0.

Johannessen, E. P., and R. J. Steel, 2005, Shelf-margin clinoforms and prediction of deep water sands: Basin Research, v. 17, p. 521–550, doi:10.1111/j.1365-2117.2005.00278.x.

Kane, I. A., W. D. McCaffrey, and J. Peakall, 2008, Controls on sinuosity evolution within submarine channels, Geology, v. 36, no. 4, p. 287–290.

Lindquist, S. J., 2000, The North Sakhalin Neogene Total Petroleum System of Eastern Russia: U.S. Department of Interior U.S. Geological Survey Open-File Report 99-50-O. Online Edition.

MacEachern, J., J. P. Bhattacharya, and C. D. Howell, 2005, Ichnology of deltas, *in* L. Giosan and J. Bhattacharya, eds., River deltas: Concepts, models, and examples: SEPM Special Publication 83, p. 49–85.

Massari, F., L. Capraro, and D. Rio, 2007, Climatic modulation of timing of systems-tract development with respect to sea level changes (middle Pleistocene of Crotone, Calabria, Southern Italy): Journal of Sedimentary Research, v. 77, p. 461–468, doi:10.2110/jsr.2007.047.

Mellere, D., P. Plink-Björklund, and R. Steel, 2002, Anatomy of shelf deltas at the edge of a prograding Eocene shelf margin, Spitsbergen: Sedimentology, v. 49, p. 1181–1206, doi:10.1046/j.1365-3091.2002.00484.x.

Miall, A. D., 1991, Stratigraphic sequences and their chronostratigraphic correlation: Journal of Sedimentary Petrology, v. 61, no. 4, p. 479–505.

Mitchum, R. M., P. R. Vail, and S. Thompson, 1977, Seismic stratigraphy and global changes of sea level, Part 2: The depositional sequences as a basic unit for stratigraphic analysis, *in* C. E. Payton, ed., Seismic stratigraphy: Applications to hydrocarbon exploration: AAPG Memoir 26, p. 56–62.

Mulder, T., and J. P. M. Syvitski, 1995, Turbidity currents generated at river mouths during exceptional discharges to the world's oceans: Journal of Geology, v. 103, p. 285–299, doi:10.1086/629747.

Mulder, T., O. Weber, P. Anschutz, F. J. Jorissen, and J.-M. Jouaneau, 2001, A few months-old storm-generated turbidite deposited in the Capbreton Canyon (Bay of Biscay, S-W France): Geo-Marine Letters, v. 21, no. 3, p. 149–156, doi:10.1007/s003670100077.

Mulder, T., J. P. M. Syvitski, S. Migeon, J. C. Faugeres, and B. Savoy, 2003, Marine hyperpycnal flows: Initiation, behaviour and related deposits: A review: Marine and Petroleum Geology, v. 20, p. 861–882, doi:10.1016/j.marpetgeo.2003.01.003.

Muto, T., and R. J. Steel, 2002, In defense of shelf-edge delta development during falling and lowstand of relative sea level: Journal of Geology, v. 110, p. 421–436, doi:10.1086/340631.

Mutti, E., G. Davoli, R. Tinterri, and C. Zavala, 1996, The importance of ancient fluvio-deltaic systems dominated by catastrophic flooding in tectonically active basins: Memori di Scienze Geologiche, v. 48, p. 233–291.

Mutti, E., R. Tinterri, G. Benevelli, D. di Biase, and G. Cavanna, 2003, Deltaic, mixed and turbidite sedimentation of ancient foreland basins: Marine and Petroleum Geology, v. 20, p. 733–755, doi:10.1016/j.marpetgeo.2003.09.001.

Mutti, E., D. Bernoulli, F. Ricci Lucchi, and R. Tinterri, 2009, Turbidites and turbidity currents from Alpine "flysch" to the continental margins: Sedimentology, v. 56, p. 267–318, doi:10.1111/j.1365-3091.2008.01019.x.

Olsen, A. H., 2008, Sedimentology and paleogeography of the Battfjellet Fm, Southern Van Mijenfjorden, Svalbard: Master's thesis, University of Bergen, Bergen, Norway, 91 p.

Peakall, J., K. J. Amos, G. M., Keevil, P. W. Bradbury, and S. Gupta, 2007, Flow processes and sedimentation in submarine channel bends: Marine and Petroleum Geology, v. 24, p. 470–486, doi:10.1016/j.marpetgeo.2007.01.008.

Petter, A. L., and R. J. Steel, 2006, Hyperpycnal flow variability and slope reorganization on an Eocene shelf margin, Central Basin, Spitsbergen: AAPG Bulletin, v. 90, no. 10, p. 1451–1472, doi:10.1306/04240605144.

Plink-Björklund, P., 2005, Stacked fluvial and tide dominated estuarine deposits in high frequency (fourth-order) sequences in the Eocene Central Basin, Spitsbergen: Sedimentology, v. 53, p. 391–428, doi:10.1111/j.1365-3091.2005.00703.x.

Plink-Björklund, P., and R. J. Steel, 2002, Sea level fall below the shelf edge, without basin-floor fan: Geology, v. 30, no. 2, p. 115–118, doi:10.1130/0091-7613(2002)030<0115:SLFBTS>2.0.CO;2.

Plink-Björklund, P., and R. J. Steel, 2004, Initiation of turbidity currents: Outcrop evidence for Eocene hyperpycnal flow turbidites: Sedimentary Geology, v. 165, p. 29–52, doi:10.1016/j.sedgeo.2003.10.013.

Plink-Björklund, P., and R. J. Steel, 2005, Deltas on falling stage and lowstand shelf margins, Eocene Central Basin of Spitsbergen: Importance of sediment supply, *in* L. Giosan and J. Bhattacharya, eds., River deltas: Concepts, models and examples: SEPM Special Publication 83, p. 179–206.

Plink-Björklund, P., D. Mellere, and R. J. Steel, 2001, Turbidite variability and architecture of sand-prone, deep-water slopes: Eocene clinoforms in Central Basin, Spitsbergen: Journal of Sedimentary Research, v. 71, p. 895–912, doi:10.1306/030501710895.

Pontén, A., and P. Plink-Björklund, 2009, Process regime changes across a regressive to transgressive turnaround in a shelf-slope basin, Eocene Central Basin of Spitsbergen: Journal of Sedimentary Research, v. 79, p. 2–23.

Porobski, S. J., and R. J. Steel, 2006, Deltas and sea level change: Journal of Sedimentary Research, v. 76, p. 390–403, doi:10.2110/jsr.2006.034.

Posamentier, H. W., and G. P. Allen, 1999, Siliciclastic sequence stratigraphy: Concepts and applications. SEPM Concepts in Sedimentlogy and Paleontology No. 7, ISBN 15676-070-0, 210 p.

Posamentier, H. W., R. D. Erskine, and R. M., Mitchum Jr., 1991, Submarine fan deposits in a sequence-stratigraphic framework, *in* P. Weimer and M. H. Link, eds., Seismic facies and sedimentary processes of submarine fans and turbidite systems: New York, New York, Springer-Verlag, p. 127–136.

Posamentier, H. W., G. P. Allen, D. P. James, and M. Tesson, 1992, Forced regression in a sequence-stratigraphic framework: Concepts, examples, and exploration significance: AAPG Bulletin, v. 76, no. 11, p. 1687–1709.

Schumm, S. A., 1993, River response to base level change: Implications for sequence stratigraphy: Journal of Geology, v. 101, p. 279–294, doi:10.1086/648221.

Schweitzer, H.-J., 1980, Environment and climate in the Early Tertiary of Spitsbergen: Paleogeography, Paleoclimatology, Paleoecology, v. 30, p. 297–311, doi:10.1016/0031-0182(80)90062-0.

Spielhagen, R. F., and A. Tripati, 2009, Evidence from Svalbard for near-freezing temperatures and climate oscillations in the Arctic during the Paleocene and Eocene: Paleogeography, Paleoclimatology, Paleoecology, v. 278, p. 48–56, doi:10.1016/j.palaeo.2009.04.012.

Steel, R., and T. Olsen, 2002, Clinoforms, clinoform trajectories and deepwater sands, *in* J. M. Armentrout and N. C. Rosen, eds., Sequence-stratigraphic models for exploration and production: Evolving methodology, emerging models and application histories: Gulf Coast Section SEPM Special Publication, 22d Annual Research Conference, CD, p. 367–381.

Steel, R., J. Crabaugh, M. Schellpeper, D. Mellere, P. Plink-Björklund, T. Deibert, and T. Løseth, 2000, Deltas versus rivers on the shelf edge: their relative contributions to the growth of shelf margins and basin-floor fans (Barremian and Eocene, Spitsbergen), *in* Deep-water reservoirs of the world: Gulf Coast Section SEPM Foundation 20th Annual Research Conference, p. 981–1009.

Steel, R. J., A. Dalland, K. Kalgraff, and V. Larsen, 1981, The Central Tertiary Basin on Spitsbergen. Sedimentary development of a sheared margin basin, *in* J. W. Kerr and A. J. Fergusson, eds., Geology of North Atlantic Borderland: Canadian Society of Petroleum Geologists Memoir 7, p. 647–664.

Steel, R. J., J. Gjelberg, W. Helland-Hansen, K. Kleinspehn, A. Nøttvedt, and M. R. Larsen, 1985, The Tertiary strike-slip basins and orogenic belt of Spitsbergen, *in* K. T. Biddle and N. Christie-Blick, eds., Strike-slip deformation, basin formation and sedimentation: SEPM Special Publication 37, p. 339–359.

Stene, S. A. K., 2009, Facies and architecture of the Battfjellet Formation Northern Nathorst Land, Spitsbergen: Master's thesis, University of Bergen, Bergen, Norway, 103 p.

Sydow, J., J. Finneran, and A. P. Bowman, 2003, Stacked shelf edge delta reservoirs of the Columbus Basin, Trinidad, West Indies, *in* H. Roberts, ed., Shelf-margin deltas and linked downslope petroleum systems: Global significance and future exploration potential: Gulf Coast Section SEPM Special Publication, 32d Annual Research Conference, CD, p. 441–465.

Uroza, C. A., and R. J. Steel, 2008, A highstand shelf-margin delta system from the Eocene of West Spitsbergen, Norway: Sedimentary Geology, v. 203, p. 229–245, doi:10.1016/j.sedgeo.2007.12.003.

Vail, P. T., R. M. Mitchum, R. G. Todd, J. M. Widmier, S. Thomson, J. B. Sangree, J. N. Bubb, and W. G. Hatlelid, 1977, Seismic stratigraphy and global changes in sea level: Parts I–II. Overview, *in* C. E. Payton, ed., Seismic stratigraphy: Application to hydrocarbon exploration: AAPG Memoir 26, p. 51–212.

Van Wagoner, J. C., R. M. Mitchum Jr., K. M. Campion, and V. D. Rahmanian, 1990, Siliciclastic sequence stratigraphy in well logs, cores, and outcrops: concepts for high resolution correlation of time and facies: AAPG Methods in Exploration Series 7, 55 p.

Wong, H. K., T. Lüdmann, B. Baranov, V. Ya, B. Karp, P. Konerding, and G. Ion, 2003, Bottom current-controlled sedimentation and mass wasting in the northwestern Sea of Okhotsk: Marine Geology, v. 201, no. 4, p. 287–305, doi:10.1016/S0025-3227(03)00221-4.

Wood, L. J., 2000, Chronostratigraphy and tectonostratigraphy of the Columbus Basin, eastern offshore Trinidad: AAPG Bulletin, v. 84, p. 1905–1921.

Wynn, R. B., B. T. Cronin, and J. Peakall, 2007, Sinuous deep-water channels: Genesis, geometry and architecture: Marine and Petroleum Geology, v. 24, p. 341-387.

Yoshida, S., R. J. Steel, and R. W. Dalrymple, 2007, Changes in depositional processes: An ingredient in a new generation of sequence-stratigraphic models: Journal of Sedimentary Research, v. 77, p. 447–460, doi:10.2110/jsr.2007.048.

8

Buatois, L. A., L. L. Saccavino, and C. Zavala, 2011, Ichnologic signatures of hyperpycnal flow deposits in Cretaceous river-dominated deltas, Austral Basin, southern Argentina, *in* R. M. Slatt and C. Zavala, eds., Sediment transfer from shelf to deep water—Revisiting the delivery system: AAPG Studies in Geology 61, p. 153–170.

Ichnologic Signatures of Hyperpycnal Flow Deposits in Cretaceous River-dominated Deltas, Austral Basin, Southern Argentina

Luis Alberto Buatois

Department of Geological Sciences, University of Saskatchewan, Saskatoon, Saskatchewan, Canada

Luis Lucas Saccavino

Petrobras Argentina SA, Neuquén, Argentina

Carlos Zavala

GCS Argentina, Universidad Nacional del Sur, Bahía Blanca, Argentina

ABSTRACT

Hyperpycnal flow deposits and associated facies in upper Cretaceous reservoirs of the Magallanes Formation in the Campo Boleadoras-Estancia Agua Fresca-Puesto Peter area of southern Patagonia, Argentina, are documented based on subsurface data. The study represents one of the first ichnologic characterizations of a deltaic system dominated by hyperpycnal processes. The fluvial-deltaic system was sourced from uplifted areas located in Central Patagonia and the Río Chico High and prograded toward the south and southeast. Integration of sedimentologic and ichnologic data allows establishing proximal-distal trends within a sediment transport system. The bulk of sandy hyperpycnal lobe deposits consists of coarse- to fine-grained sandstones that are either structureless or display a subtle parallel lamination that is commonly delineated by abundant plant remains. Vertical grain-size changes reflect flow fluctuations, with coarsening- and fining-upward intervals indicating waxing and waning flows, respectively. The concentration of plant remains indicates phytodetrital pulses in connection with direct fluvial discharges. These deposits are unburrowed to sparsely

Copyright ©2011 by The American Association of Petroleum Geologists.

DOI:10.1306/13271355St611948

bioturbated, containing the *Thalassinoides* ichnofabric, which records opportunistic colonization during times of decreased sedimentation rate. However, the pervasive laminated fill of the burrows may reflect relatively high sedimentation rates, most likely because of suspension fallout of fine-grained material. The extreme compaction suggests burrow emplacement in a water-saturated soft substrate that underwent compaction subsequent to the bioturbation event. Associated heterolithic intervals contain the *Planolites-Teichichnus* ichnofabric, which characterizes marginal areas with respect to the hyperpycnal sand-rich lobes and times of quiescence between flows. During times of reduced sediment supply, material was reworked by wave processes, and hypopycnal conditions were dominant. The *Thalassinoides-Teichichnus* ichnofabric records colonization of these wave-reworked sandstone units, whereas the *Terebellina-Phycosiphon* ichnofabric reflects stable conditions that allowed intense bioturbation and the establishment of a moderately diverse benthic fauna. Fully marine offshore deposits are characterized by the *Teichichnus-Phycosiphon* ichnofabric, which display total biogenic reworking, high ichnodiversity, and a complex tiering structure.

INTRODUCTION

Increased attention is being paid to river-flood-generated hyperpycnal flows as a mechanism of sediment transfer from river mouths to shelf and deep-sea areas (e.g., Mulder and Syvitski, 1995; Mulder et al., 2003; Plink-Björklund and Steel, 2004). Hyperpycnal flows create several stress factors, such as increased sedimentation rate, freshwater discharges, and high hydrodynamic energy that impact on the benthic faunas. However, the function played by these factors and the ichnology of hyperpycnal deposits have remained mostly underexplored. From an ichnologic standpoint, it is useful to differentiate two main depositional settings, where hyperpycnal flows are limiting factors on trace fossil distribution: deep marine and deltaic. The ichnology of deep-marine hyperpycnites has been analyzed by Ponce et al. (2007), Wetzel (2008), and Carmona and Ponce (2011). The ichnology of hyperpycnal flow deposits in deltaic systems is the focus of this study.

The Austral Basin (Figure 1) is one of the most important petroleum basins of Argentina (e.g., Biddle et al., 1986; Keeley and Light, 2008). The upper Cretaceous reservoirs of the Magallanes Formation in the Campo Boleadoras-Estancia Agua Fresca-Puesto Peter area have been analyzed based on an integration of ichnologic and sedimentologic data. As a result of this study, the function of hyperpycnal flows in sedimentary dynamics is underscored. The aim of this chapter is to document sedimentologic and ichnologic evidence that helps reinterpret the depositional mechanisms and sedimentary environments of the Magallanes Formation. This is one of the first studies that document in detail the ichnologic signatures of hyperpycnal flow deposits in a deltaic system. Accordingly, our observations in the Magallanes Formation may be of use to understand depositional dynamics in other successions formed in river-dominated deltas.

GEOLOGIC SETTING AND PETROLEUM SYSTEMS

The Austral Basin, also known as Magallanes Basin, is located in the southernmost region of South America (Figure 1) and hosts up to 8000 m (26,247 ft) of rift (Triassic to middle–upper Jurassic), postrift (upper Jurassic to lower Cretaceous), and foreland (upper Cretaceous to Cenozoic) deposits. During this latter stage, the establishment of a fold and thrust belt resulted in the formation of a foredeep in the foreland basin (Figure 2). Sediment was sourced from uplifted areas in Central Patagonia and the Río Chico High located toward the northwest. Clastic wedges prograded toward the basin center, leading to a southeastern migration of the depocenters and forming a series of stacked sedimentary cycles controlled by regional tectonics.

Clastic wedges prograded from the north and northwest, and shoreface and deltaic deposits (Anita Formation) accumulated during the Campanian–Early Maastrichtian. Offshore mudstone and thin-bedded sandstone (Palermo Aike Formation) interfinger with the sandstone tongues toward the southeast. This succession is capped by the so-called D3 unconformity.

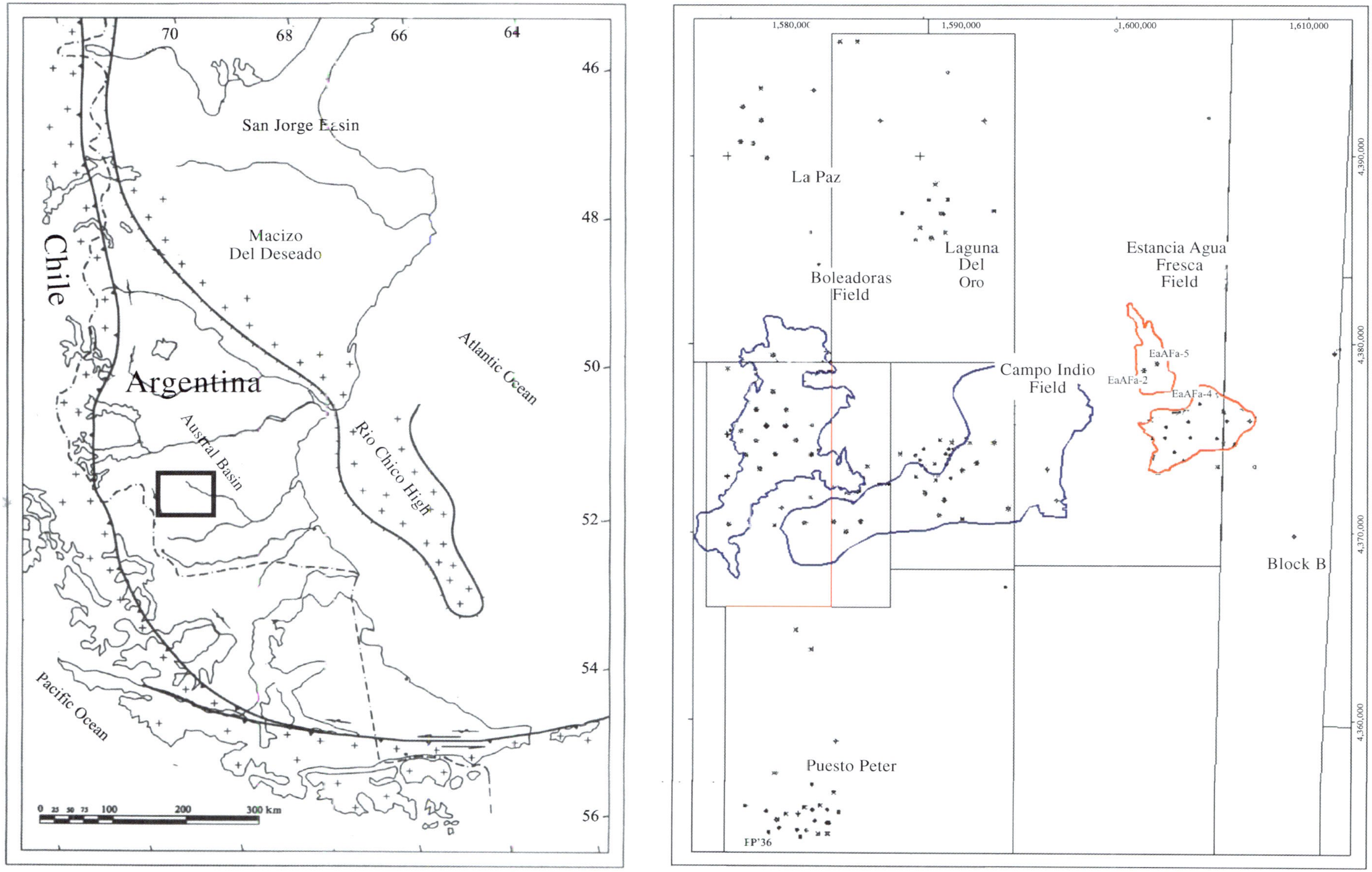

FIGURE 1. Location maps of the Austral Basin and the studied area in southern Argentina. EaAFa-2, EaAFa-4, EaAFa-5 = Estancia Agua Fresca well cores; PP'36 = Puesto Peter well core. Numbers at top and side of map at right indicate coordinates of the oil fields.

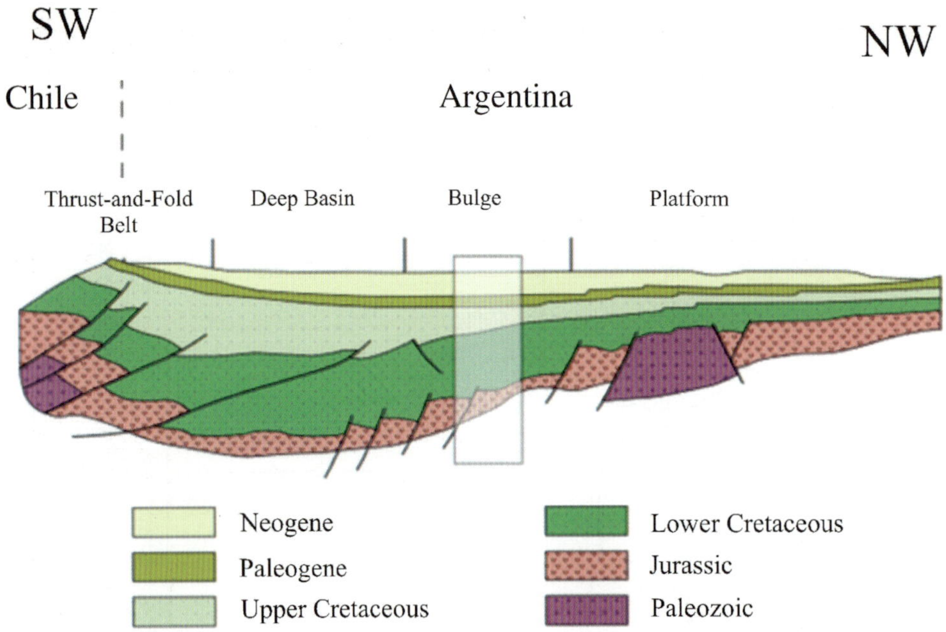

FIGURE 2. Tectonic setting of the Austral Basin. Box outlines position of the study area.

The Calafate sequence (Magallanes Formation, Upper Maastrichtian–Danian) occurs above this unconformity (Figure 3). During the rest of the Paleocene, the area was uplifted and the unconformity D4 was formed. Deep incisions were carved, leaving erosional remnants that favored formation of stratigraphic traps. Glauconite wackes accumulated during the Eocene transgression, forming the seal.

SEDIMENTARY FACIES AND TRACE FOSSIL DISTRIBUTION

Our sedimentologic and ichnologic study is based on the analysis of 156.4 m (513.1 ft) of conventional cores and well logs from eight wells in an area of 200 km^2 (77 mi^2). The analyzed deposits correspond to the Lower Maastrichtian interval of the Magallanes Formation. Lower Maastrichtian deposits in the study area are stacked, forming a coarsening-upward succession, reflecting a prograding deltaic system during highstand (Figure 2). These deposits are bounded at the top by an erosional discontinuity related to the incision of estuarine valleys during a relative sea level fall. Logs from all wells in the study area were analyzed and correlated. Core information was matched with their log signatures. Sedimentary facies were distinguished based on lithology, physical sedimentary structures, and bed contacts. Three main facies have been defined: sandy hyperpycnal lobes, storm-influenced delta front and prodelta, and lower offshore. Our detailed trace fossil study was based on ichnotaxa identification, ethological and trophic types, population strategies, ichnodiversity, degree of bioturbation, and tiering structure. Degree of bioturbation was assessed following Taylor and Goldring (1993). In this scheme, a bioturbation index (BI), ranging from 0 (no bioturbation) to 6 (complete bioturbation), is defined. Five ichnofabrics were defined, characterized, and interpreted. Ichnofabrics were named after their most visible components and/or their most characteristic components (McIlroy, 2004). Ichnofabrics were placed in an ichnofacies context that allows their location into a paleoenvironmental framework. Environmental zonation and ichnofacies definitions of open-marine wave-dominated deposits follow the terminology of MacEachern et al. (1999). Standard terminology is used for environmental zonation of deltaic deposits (e.g., Bhattacharya, 2006). A comparison of ichnofabrics and ichnofacies from sandy hyperpycnal lobes, storm-influenced delta front and prodelta, and lower offshore allows the evaluation of benthic fauna response to different stress levels along a depositional gradient.

Sandy Hyperpycnal Lobes

Description

The bulk of these deposits consists of coarse- to fine-grained sandstone that is either structureless or display a subtle parallel lamination that is commonly delineated by abundant plant remains. Hummocky cross-stratification and combined flow ripples are rare. Gradual changes in grain size are common, showing alternating coarsening- and fining-upward

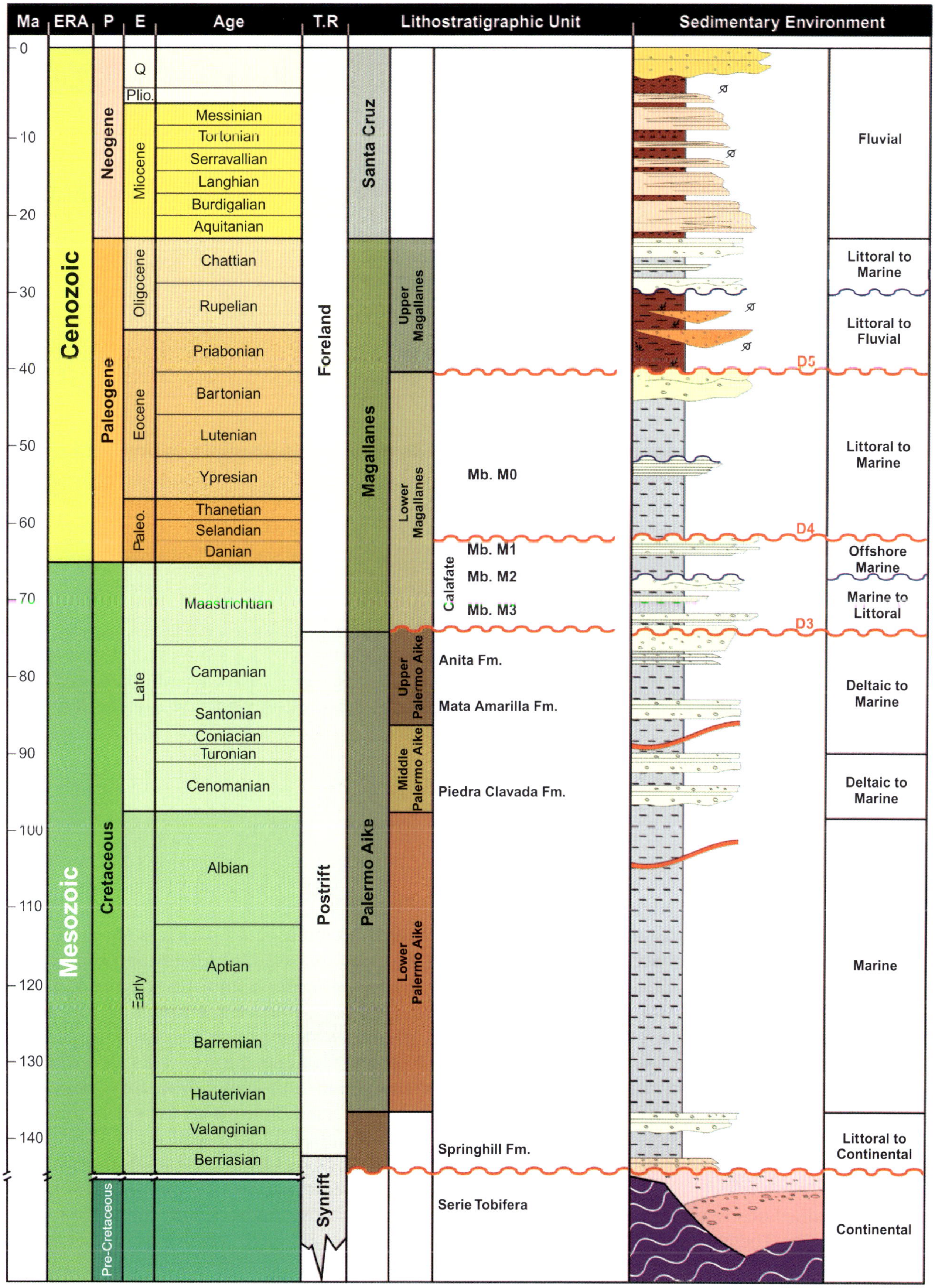

FIGURE 3. Stratigraphy of the Austral Basin in the studied region. P = period; E = Epoch; T.R = Tectonic Regime; Fm. = Formation; Q = Quaternary; Plio. = Pliocene; Paleo. = Paleocene; Mb. = Member; M = Magallanes; D3–D5 = unconformities.

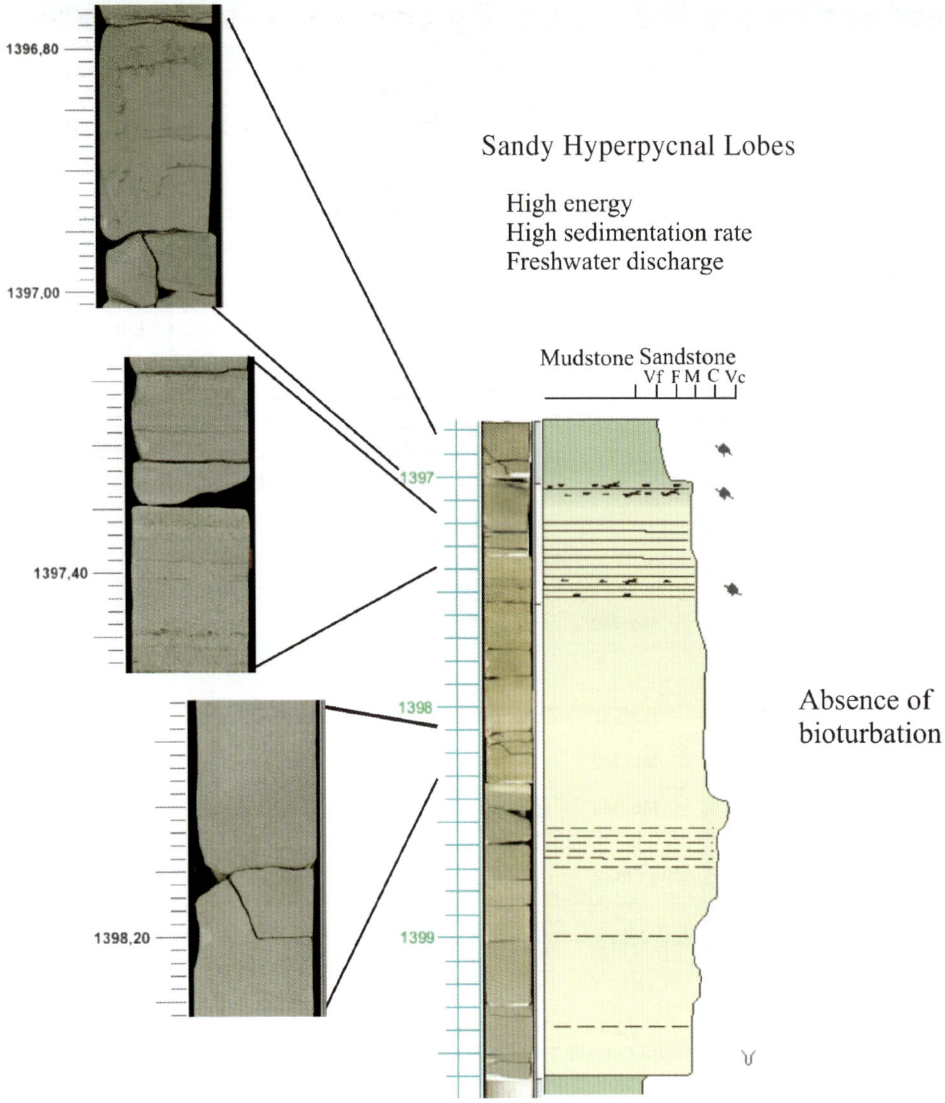

FIGURE 4. Sedimentologic attributes of internally complex sandstone deposits, interpreted as accumulated by long-lived and fluctuating hyperpycnal flows. Vf = very fine; F = fine; M = medium; C = coarse; Vc = very coarse. Numbers at left indicate core depths.

intervals (Figure 4). Erosive-based pebble conglomerate layers that contain bioclasts and flattened and deformed mudstone intraclasts occur at the base of very coarse- to coarse-grained sandstone units. Convolute lamination and fluid-escape structures are common throughout the interval. Discrete sandstone beds are difficult to identify, but amalgamated sandstone bedsets are up to 9.2 m (30.2 ft) thick. Sandstone bedsets are separated by intervals of dark gray, massive, or parallel-laminated mudstone-dominated units (up to 1.75 m [5.7 ft] thick) that may contain thinner sandstone layers (0.3–2 cm [0.12–0.79 in.] thick). These finer grained deposits contain syneresis cracks and plant remains. These deposits are present in well cores EaAFa-2, EaAFa-3, EaAFa-4, and EaAFa-5 of the Agua Fresca field in the central-eastern zone of the study area (Figure 1).

Ichnology

Sandstone-dominated packages are unburrowed (Figure 4) to sparsely bioturbated and characterized by the *Thalassinoides* ichnofabric (Figure 5; Table 1). *Thalassinoides* is the dominant ichnogenus, consisting of burrow systems that contain both horizontal and vertical segments. Burrow walls are well defined and mud lined, although burrow fill is sandy, laminated, and passive. Burrows are extremely compacted, displaying flattened cross sections instead of circular or subcircular. The BI is typically 0–1 and, only locally, 2. Bioturbation is commonly restricted to the top of sandstone beds, and suites are typically monospecific, although *Ophiomorpha*, *Skolithos*, *Diplocraterion*, and *Palaeophycus* are present in places. The ichnofabric is dominated by dwelling structures of suspension feeders. These bioturbated deposits are

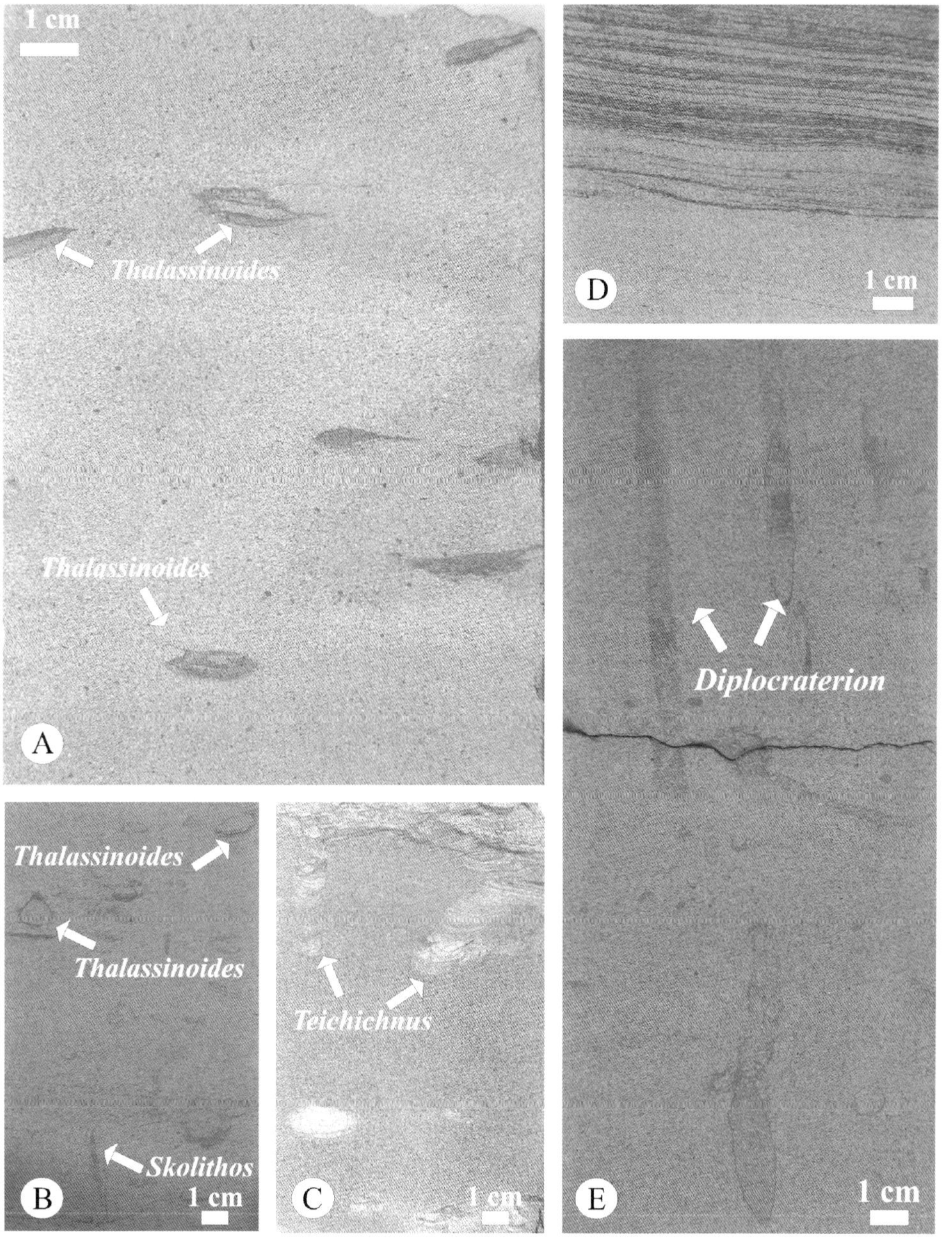

FIGURE 5. *Thalassinoides* ichnofabric in sandy hyperpycnal lobe deposits. (A) Highly compressed mud-lined *Thalassinoides*, showing infill with passive parallel lamination. EaAFa-4. (B) *Thalassinoides* and *Skolithos*. Note local subtle parallel lamination delineated by particulate carbonaceous material. EaAFa-2. (C) Retrusive *Teichichnus*. EaAFa-4. (D) Layers of particulate carbonaceous material, illustrating phytodetrital pulses. EaAFa-4. (E) Long, spreiten, U-shaped *Diplocraterion* burrows. EaAFa-4. EaAFa-2, EaAFa-4 = Estancia Agua Fresca well cores.

commonly interbedded with nonbioturbated structureless sandstone units. Tiering structure is very simple, displaying a dominance of deep-tier structures. *Palaeophycus* occupies shallow tiers, whereas *Thalassinoides*, *Ophiomorpha*, *Skolithos*, and *Diplocraterion* are deep-tier structures. In terms of archetypal ichnofacies, the *Thalassinoides* ichnofabric is representative of the *Skolithos* ichnofacies.

The associated thinner bedded sandstone and mudstone intervals contain the *Planolites-Teichichnus* ichnofabric (Figure 6; Table 1). *Planolites* and *Teichichnus* are commonly the only ichnogenera identified. Locally, *Thalassinoides* may occur. In contrast to the *Thalassinoides* ichnofabric, the *Planolites-Teichichnus* ichnofabric is dominated by feeding traces of deposit feeders. The BI of these finer grained deposits is typically 1–2. However, some intervals contain unbioturbated dark gray mudstone units. Tiering structure is very simple. *Planolites* and *Teichichnus* are shallow-tier structures, whereas *Thalassinoides* occupies deep tiers. The *Planolites-Teichichnus* ichnofabric illustrates a highly stressed expression of the *Cruziana* ichnofacies.

Interpretation

Vertical grain-size changes reflect flow fluctuations, with coarsening- and fining-upward intervals indicating waxing and waning flows, respectively. This

Table 1. Summary of ichnologic and sedimentologic characteristics of the Magallanes Formation ichnofabrics.

Ichnofabric	*Taxonomic Composition*	*Dominant Ethology and Trophic Type*	*Tiering*	*Bioturbation Index*	*Colonization Style*	*Depositional Conditions*	*Facies and Paleoenvironment*
Thalassinoides	Dominant: *Thalassinoides*; accessory: *Ophiomorpha*, *Skolithos*, *Diplocraterion*, *Palaeophycus*.	Dwelling structures of suspension feeders	Very simple	0–1 (locally, 2)	Opportunistic colonization during times of decreased sedimentation rate	Waxing and waning flows (hyperpycnal discharges). Phytodetrital pulses. High suspended-sediment concentrations. Water-saturated soft grounds. High to relatively high energy. Sporadic reworking by oscillatory flows. Common salinity fluctuations.	Sand-rich hyperpycnal flow deposits. Deltaic lobes emplaced in distal delta front to proximal prodelta.
Planolites-Teichichnus	Dominant: *Planolites*, *Teichichnus*; accessory: *Thalassinoides*.	Feeding structures of deposit feeders	Very simple	1–2	Opportunistic colonization during interflow times	Low energy. Sporadic to common reworking by oscillatory flows. Common salinity fluctuations.	Sand-rich hyperpycnal flow deposits. Deltaic lobes emplaced in distal delta front to proximal prodelta.
Thalassinoides-Teichichnus	Dominant: *Thalassinoides*, *Teichichnus*; subordinate: *Rhizocorallium*, *Skolithos*, *Planolites*, *Asterosoma*.	Feeding structures of deposit feeders	Simple	0–2	Opportunistic colonization followed by climax populations	High suspended-sediment concentrations (hypopycnal discharges). Moderate energy. Common reworking by oscillatory flows. Sporadic salinity fluctuations.	Storm-dominated distal delta front to proximal prodelta.
Terebellina-Phycosiphon	Dominant: *Terebellina-Phycosiphon*; subordinate: *Planolites*, *Palaeophycus*, *Teichichnus*, *Thalassinoides*, *Zoophycos*, *Schaubcylindrichnus*.	Feeding structures of deposit feeders	Relatively well developed	1–3 (locally, 4–5)	Opportunistic colonization followed by climax populations	High suspended-sediment concentrations (hypopycnal discharges). Moderate to low energy. Repeated reworking by oscillatory flows. Sporadic salinity fluctuations.	Storm-dominated distal delta front to proximal prodelta.
Teichichnus-Phycosiphon	Dominant: *Teichichnus*, *Phycosiphon*; subordinate: *Chondrites*, *Asterosoma*, *Planolites*; accessory: *Terebellina*, *Thalassinoides*, *Zoophycos*.	Feeding structures of deposit feeders	Complex	6	Climax populations	Low energy. Good oxygenation. Soft grounds. Negligible wave reworking. No salinity fluctuations.	Lower offshore

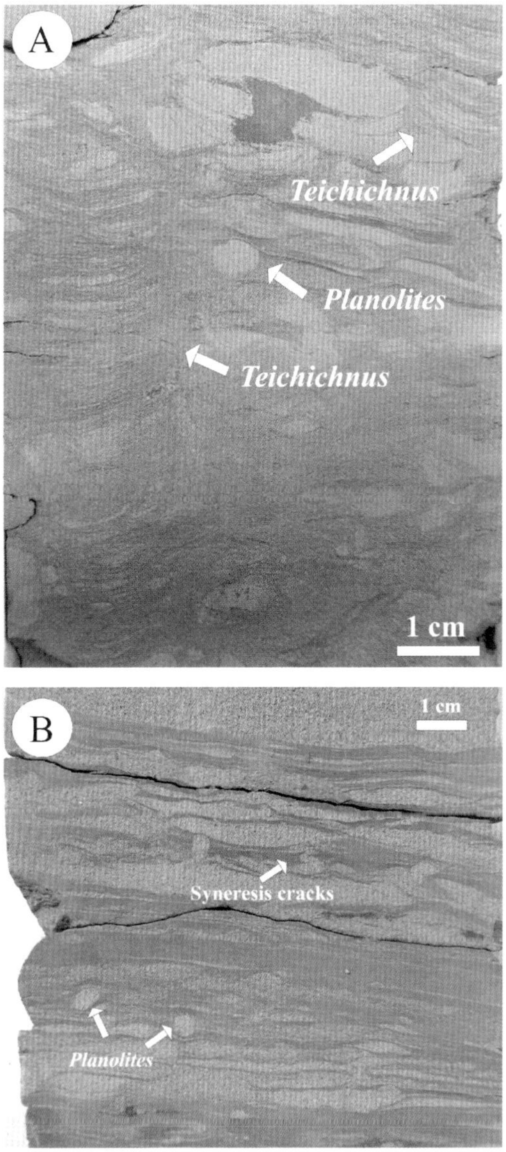

FIGURE 6. *Planolites-Teichichnus* ichnofabric in hyperpycnal lobe deposits accumulated between flow events. (A) Moderately bioturbated deposits with deep and wide retrusive *Teichichnus* and isolated *Planolites*. Note discontinuous mudstone interbeds as a result of partial homogenization by bioturbation. EaAFa-4. (B) Sparsely bioturbated heterolithic deposits. Mudstone interbeds are continuous and well preserved. A low-density monospecific suite of *Planolites* is present in some of the mudstone interbeds. Note associated syneresis cracks. EaAFa-5. EaAFa-4, EaAFa-5 = Estancia Agua Fresca well cores.

pattern results from a progressive increase in river discharge during waxing and peak flood stages followed by a waning flow stage (Mulder et al., 2003; Zavala et al., 2006; Soyinka and Slatt, 2008). The concentration of plant remains indicates phytodetrital pulses in connection with fluvial discharges (MacEachern et al., 2005). These deposits are interpreted as the result of turbulent flows of fluctuating energy and high suspended-sediment concentrations. Associated pebble conglomerate layers may indicate sediment bypass. This is also suggested by sandstone beds with asymptotic cross-stratification that may reflect fill of localized topographic lows. The local presence of hummocky cross-stratification and combined flow ripples indicates sand reworking by oscillatory flows and points to deposition above storm-wave base (Cheel and Leckie, 1993; Lamb et al., 2008). The syneresis cracks may reflect salinity fluctuations (Foster et al., 1955; Donovan and Foster, 1972). Interbedded dark gray massive or parallel-laminated mudstone and thin to very thin and very fine- to fine-grained sandstone accumulated during times of quiescence between flow events.

The *Thalassinoides* ichnofabric records opportunistic colonization in sand-rich hyperpycnal deposits. The characteristics of the ichnofauna (low ichnodiversity, dominance of dwelling burrows) and its presence at the top of massive beds suggest rapid colonization during times of decreased sedimentation rate. However, the pervasive laminated fill may reflect relatively high sedimentation rates, most likely caused by suspension fallout of fine-grained material. The extreme compaction suggests burrow emplacement in a water-saturated soft substrate that underwent compaction subsequent to the bioturbation event. The sparse bioturbation and simple tiering structure indicate short-term colonization windows. Low ichnodiversity and the presence of syneresis cracks are consistent with stress caused by dilution of marine salinity, specifically freshwater input during hyperpycnal discharges.

The dominance of feeding structures of deposit feeders in the *Planolites-Teichichnus* ichnofabric indicates exploitation of organic detritus in sediment accumulated during low-energy conditions. A low degree of bioturbation, preservation of thin-bedded sandstone layers, simple tiering structure, and low ichnodiversity suggest stressed conditions, most likely because of dilution of normal marine salinity caused by freshwater input. This is consistent with the presence of syneresis cracks and plant fragments.

In terms of bathymetry, ichnologic data indicate deposition in a shallow-marine environment. Elements typical of the distal *Cruziana* or the *Zoophycos* ichnofacies, which occur immediately above and below the storm-wave base, respectively (MacEachern et al., 1999), are absent. This is consistent with the localized presence of structures that indicate reworking of the sands by strong oscillatory flows, which point to flow emplacement well above the storm-wave base. However, intercalation of mudstone intervals indicates low-energy conditions in the absence of waves between flow events, suggesting deposition below fair-weather wave base most of the time. Therefore, integration of ichnologic and sedimentologic data indicates deposition in deltaic lobes emplaced in distal delta-front to proximal prodelta environments.

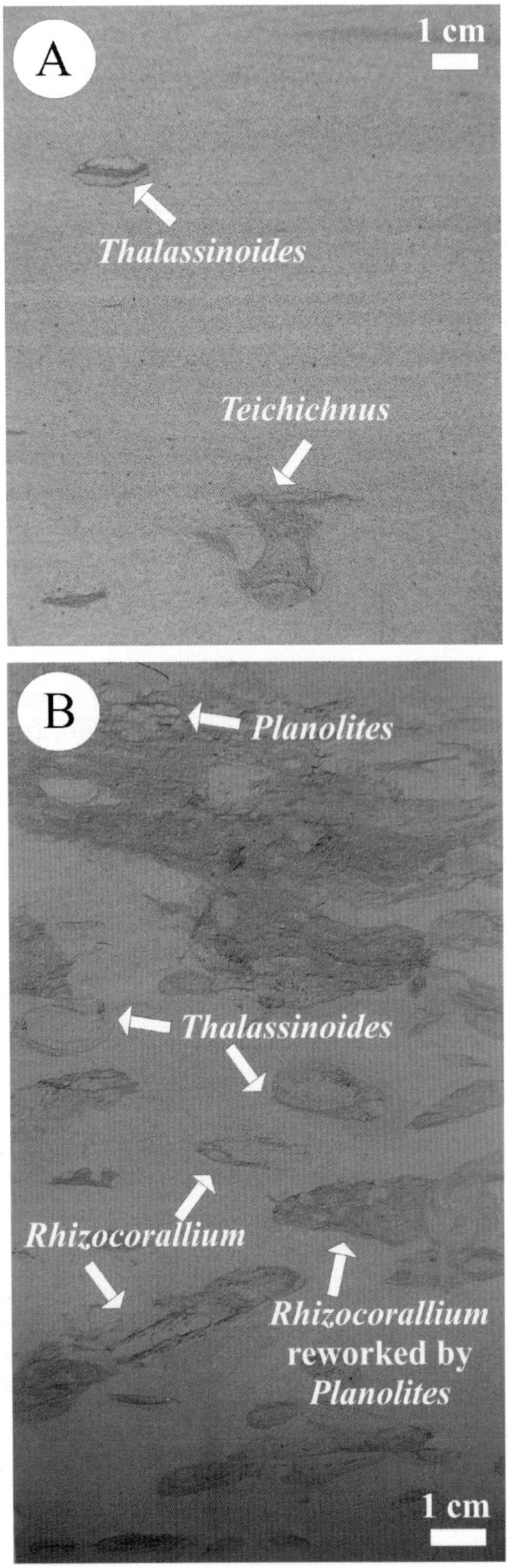

FIGURE 7. *Thalassinoides-Teichichnus* **ichnofabric in storm-influenced delta-front and prodelta deposits. (A) Sparsely bioturbated sandstone showing subtle hummocky cross-stratification. Protrusive** *Teichichnus* **and mud-lined** *Thalassinoides* **displaying infill with passive parallel lamination. (B) Moderately bioturbated sandstone showing an increase in bioturbation intensity toward the top. The fabric is dominated by** *Thalassinoides* **and** *Rhizocorallium*. *Planolites* **is concentrated either toward the top of the bed or inside** *Rhizocorallium* **burrows. Both photographs are from Estancia Agua Fresca well core EaAFa-3.**

Storm-influenced Delta Front and Prodelta

Description

Sandy hyperpycnal lobe packages are locally interbedded and interfingered with silty very fine-grained sandstone with hummocky cross-stratification and combined flow-ripple cross-lamination and light gray mudstone. Sandstone beds are sharp and erosive based but have gradational tops. Hummocky sandstone may occur as 25- to 35-cm (9.8- to 13.8-in.)-thick discrete beds separated by 3- to 30-cm (1.9- to 11.8-in.)-thick mudstone-dominated units or amalgamated bedsets forming up to 1-m (0.39-in.)-thick packages. Thinner sandstone beds tend to form lenses. Syneresis cracks occur in the heterolithic intervals. These deposits occur in well cores EaAFa-3, EaAFa-4, and EaAFa-5 of the Agua Fresca field in the central-eastern zone of the study area.

Ichnology

Medium-bedded hummocky sandstone beds are either unburrowed or only bioturbated at the top, containing the *Thalassinoides-Teichichnus* ichnofabric (Figure 7; Table 1). *Thalassinoides*, *Teichichnus*, *Rhizocorallium*, *Skolithos*, *Planolites*, and *Asterosoma* are the most common ichnotaxa. *Thalassinoides* burrow segments are mostly horizontal and, less commonly, inclined. As in the case of the *Thalassinoides* ichnofabric of the sandy hyperpycnal lobe deposits, *Thalassinoides* has a mud lining and shows a laminated passive sandy fill. Locally, *Rhizocorallium* burrows are reworked by *Planolites*. The BI is 0–2 and, only locally, 3. The tiering structure is simple, showing the presence of both shallow- and deep-tier structures. *Asterosoma* and *Teichichnus* occupy shallow tiers, whereas *Ophiomorpha*, *Thalassinoides*, *Rhizocorallium*, *Planolites*, and

Skolithos are deep-tier structures. In terms of the ichnofacies model, the *Thalassinoides-Teichichnus* ichnofabric is included within a mixed *Skolithos-Cruziana* ichnofacies.

Mudstone intervals with thin layers of sandstone contain the *Terebellina-Phycosiphon* ichnofabric (Figure 8; Table 1). *Terebellina and Phycosiphon* are the dominant ichnotaxa. Subordinate components are *Planolites, Palaeophycus, Teichichnus, Thalassinoides, Zoophycos*, and *Schaubcylindrichnus*. Because it has been shown that the type specimen of *Terebellina* is a body fossil (Miller, 1995), the name is informally used to designate a trace fossil. *Phycosiphon* forms the background ichnofabric and is crosscut by the deeper tier structures, such as *Thalassinoides* and *Zoophycos*. *Terebellina* is easily visible because of its thick white wall. Degree of bioturbation in these deposits is highly variable, commonly 1–3, although some intervals may reach BIs of 4–5. Tiering structure is relatively well developed. Shallow-tier structures are *Terebellina, Planolites, Palaeophycus, Teichichnus*; the middle tier is occupied by *Schaubcylindrichnus* and *Phycosiphon*; and *Zoophycos* and *Thalassinoides* occupy the deep tier. The *Terebellina-Phycosiphon* ichnofabric illustrates the archetypal *Cruziana* ichnofacies.

Interpretation

In contrast to the sandy hyperpycnal lobe intervals, these deposits display more abundant wave-generated structures, indicating that oscillatory flows were capable of reworking a substantial part of the material supplied by the distributary system. The lateral relationships between these deposits and the sand-rich hyperpycnal lobes suggest that oscillatory flows reworked the outer regions of the lobes. In addition, the vertical passage of hyperpycnal lobe deposits into intervals dominated by wave-reworked sandstone and suspension fallout mudstone indicates that basin processes were particularly effective during times of reduced sediment supply, when hypopycnal conditions were dominant. Although syneresis cracks are less common in this facies than in the hyperpycnal lobe deposits, their localized presence suggests that salinity fluctuations were important at times.

The *Thalassinoides-Teichichnus* ichnofabric records colonization of the wave-reworked sandstone units. As in the case of the *Thalassinoides* ichnofabric of the sandy hyperpycnal flow deposits, the laminated fill of the *Thalassinoides* burrows points to relatively high sedimentation rates. The low diversity of biogenic

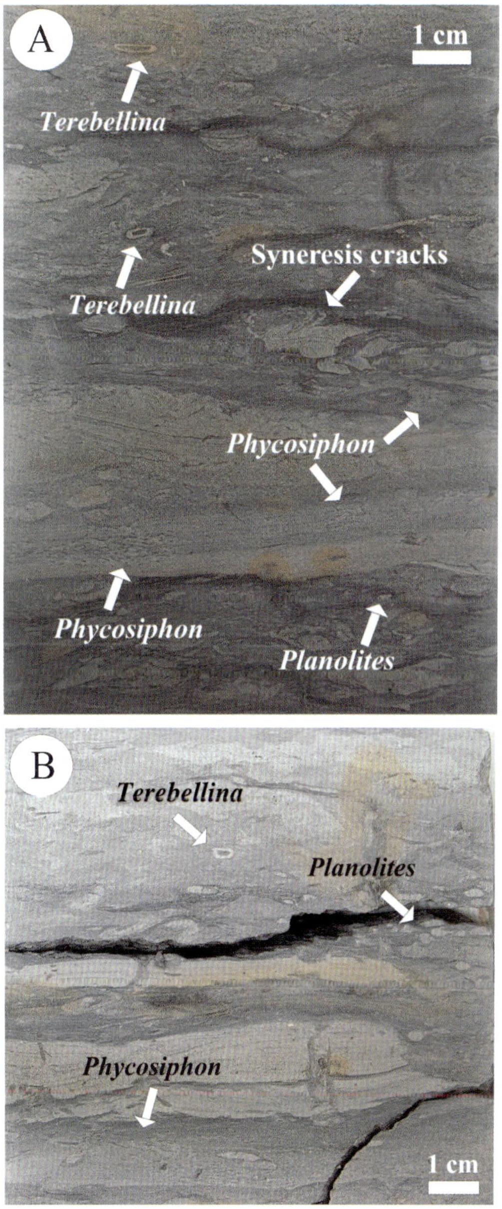

FIGURE 8. *Terebellina-Phycosiphon* ichnofabric in storm-influenced delta-front and prodelta deposits. (A) Interbedded mudstone and combined flow-ripple cross-laminated sandstone. Fair-weather mudstone is intensely bioturbated, showing dominance of *Planolites* and *Terebellina*. Interbedded sandstone shows locally preserved primary fabric and clusters of *Phycosiphon*. Isolated and contorted sandstone lenses encased in mudstone may record biogenically reworked thin tempestites. Note associated syneresis cracks. (B) Interbedded mudstone and laterally discontinuous combined flow-ripple cross-laminated sandstone. Fair-weather mudstone is intensely bioturbated, showing dominance of *Planolites, Terebellina*, and clusters of *Phycosiphon*. Both photos are from Estancia Agua Fresca well core EaAFa-3.

structures and simple tiering structures are also indicative of stress conditions, most likely associated with freshets. However, the increase in degree of bioturbation indicates alternation of times of environmental stability, particularly during times of reduced sediment discharge.

The *Terebellina-Phycosiphon* ichnofabric indicates bioturbation during times of moderate- to low-energy conditions. The dominance of deposit-feeding structures, the locally high degree of bioturbation, the more complex tiering structure, and the increase in ichnodiversity levels point to a relatively stable setting and long-term colonization windows. In any case, trace fossil diversity is lower than in coeval open-marine deposits, total homogenization of the background fines is rarely attained, and syneresis cracks are locally present, suggesting that freshwater discharge was a controlling factor during deposition. An additional stress factor was represented by storms that caused repeated wave erosion of the substrate affecting the benthic fauna (e.g., Pemberton and Frey, 1984; Pemberton and MacEachern, 1997; Buatois et al., 2008).

With respect to water depth, these deposits were most likely emplaced similar to slightly deeper bathymetric zones than the sand-rich hyperpycnal lobes. Deposition well above storm-wave base is suggested by the widespread presence of wave-generated structures. As in the case of the hyperpycnal lobes, intercalation of mudstone intervals indicates alternation of low-energy conditions, suggesting that for the most part, deposition occurred below fair-weather wave base. In short, deposition in a storm-dominated distal delta front to proximal prodelta is the most likely environmental scenario.

Lower Offshore

Description

Lower offshore deposits are represented by intensely bioturbated mudstone forming a 6-m (19.7-ft)-thick interval. No physical sedimentary structures are preserved, and syneresis cracks were not observed. This facies has been observed in the PP'36 well core in the Puesto Peter zone, southwest corner of the study area.

Ichnology

Primary sedimentary fabric has been totally obliterated by bioturbation. These deposits are characterized by the *Teichichnus-Phycosiphon* ichnofabric (Figure 9; Table 1), with these two ichnotaxa being the dominant components. The BI is 6. In contrast to ichnofabrics of the sandy hyperpycnal lobe deposits and the storm-influenced delta-front and prodelta deposits, this ichnofabric exhibits a complex tiering structure. The shallowest tier consists of an indistinct mottling that represents the background ichnofabric, reflecting emplacement in the water-saturated soupy substrate of the mixed zone. Shallow-tier structures (but deeper into the substrate than the mottling) are *Thalassinoides*, *Phycosiphon*, and *Zoophycos*. *Phycosiphon* tends to evenly cover the sediment, whereas *Thalassinoides* and *Zoophycos* show a patchy distribution. Midtier levels are represented by *Asterosoma*, *Planolites*, and *Terebellina*. The deep tier is occupied by *Chondrites*, *Teichichnus*, and another generation of *Thalassinoides*. In particular, the well-defined morphology and sharp burrow margins of *Teichichnus* and the absence of other trace fossils crosscutting this spreiten burrow indicate emplacement in deep and cohesive sediment zones. This ichnofabric is placed within a distal *Cruziana* ichnofacies.

Interpretation

The intense bioturbation and the fine grain size indicate slow suspension fallout sedimentation under low-energy conditions well below the fair-weather wave base. If storm deposits were emplaced, these may have been totally reworked by biogenic activity. The dominance of deposit-feeding structures, the intense bioturbation, the highly complex tiering structure, the dominance of climax population strategies, and the high trace fossil diversity indicate a stable setting and long-term colonization windows. Ichnologic data suggest deposition under fully marine conditions; this is also supported by the absence of syneresis cracks. The high ichnodiversity and intense bioturbation indicate well-oxygenated bottom and interstitial waters.

Sedimentologic and ichnologic features indicate deposition in a lower offshore environment. The lack of stress factors and overall environmental stability indicate that sedimentation occurred far beyond the area of influence of hyperpycnal flows.

DISCUSSION

The studied succession is interpreted as formed in a shallow-marine setting affected by sustained turbulent flows linked to a coeval fluvial distributary system and reworked by wave action. The fluvial-deltaic system was sourced from uplifted areas in Central

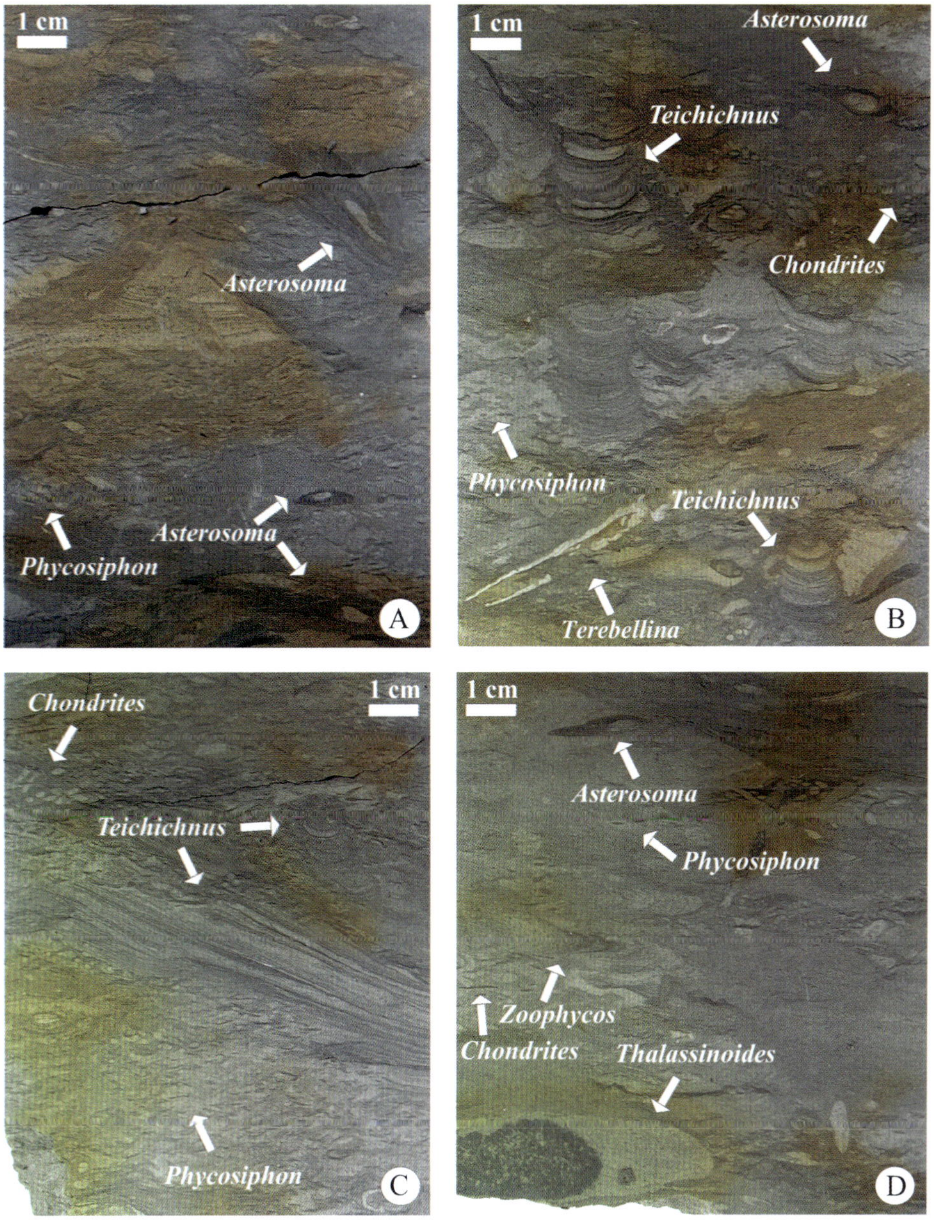

FIGURE 9. *Teichichnus-Phycosiphon* ichnofabric in intensely bioturbated lower offshore mudstone. (A) Evenly distributed *Phycosiphon* crosscut by *Asterosoma*. (B) Evenly distributed *Phycosiphon* crosscut by *Asterosoma*, *Chondrites*, and deep *Teichichnus*. Note thick wall in *Terebellina*. (C) Evenly distributed *Phycosiphon* crosscut by *Chondrites* and *Teichichnus*. Note longitudinal view of *Teichichnus* spreiten that may be confused with primary sedimentary lamination. (D) Poorly preserved *Phycosiphon* intensely reworked by *Zoophycos*, *Asterosoma*, *Chondrites*, and *Thalassinoides*. All photographs are from Puesto Peter well core PP'36.

Patagonia and the Río Chico High, was located northwest of the study area, and prograded toward the south and southeast (Figure 10). Extensive sand-rich lobes were formed in the Agua Fresca, La Paz, and Laguna del Oro fields. Deposits accumulated forming lobes that filled topographic lows, occasionally allowing the bypass of sand-rich flows into more distal areas of the Boleadoras field. Material supplied by the distributary system was reworked by wave action along the fringes of the deltaic lobe and during times of quiescence between hyperpycnal flows. Wave reworking of hyperpycnal flow deposits has been documented in both modern (e.g., Wright et al., 1988) and ancient (e.g., Pattison, 2005a, b) deposits. Toward the south in the Puesto Peter area, offshore environments were formed under fully marine conditions (Figure 10).

Integration of sedimentologic and ichnologic data allows establishing proximal-distal trends within a sediment transport system (Figure 11). High rates of sedimentation and freshwater discharge from a prograding river-dominated delta were among the most important stress factors that affected the benthic fauna. The *Thalassinoides* ichnofabric records opportunistic colonization of sand-rich hyperpycnal flow deposits in lobes emplaced in the distal delta front and the proximal prodelta. The *Planolites-Teichichnus* ichnofabric occurs in associated finer grained distal delta-front and proximal prodelta deposits that were formed lateral to the hyperpycnal sand-rich lobes

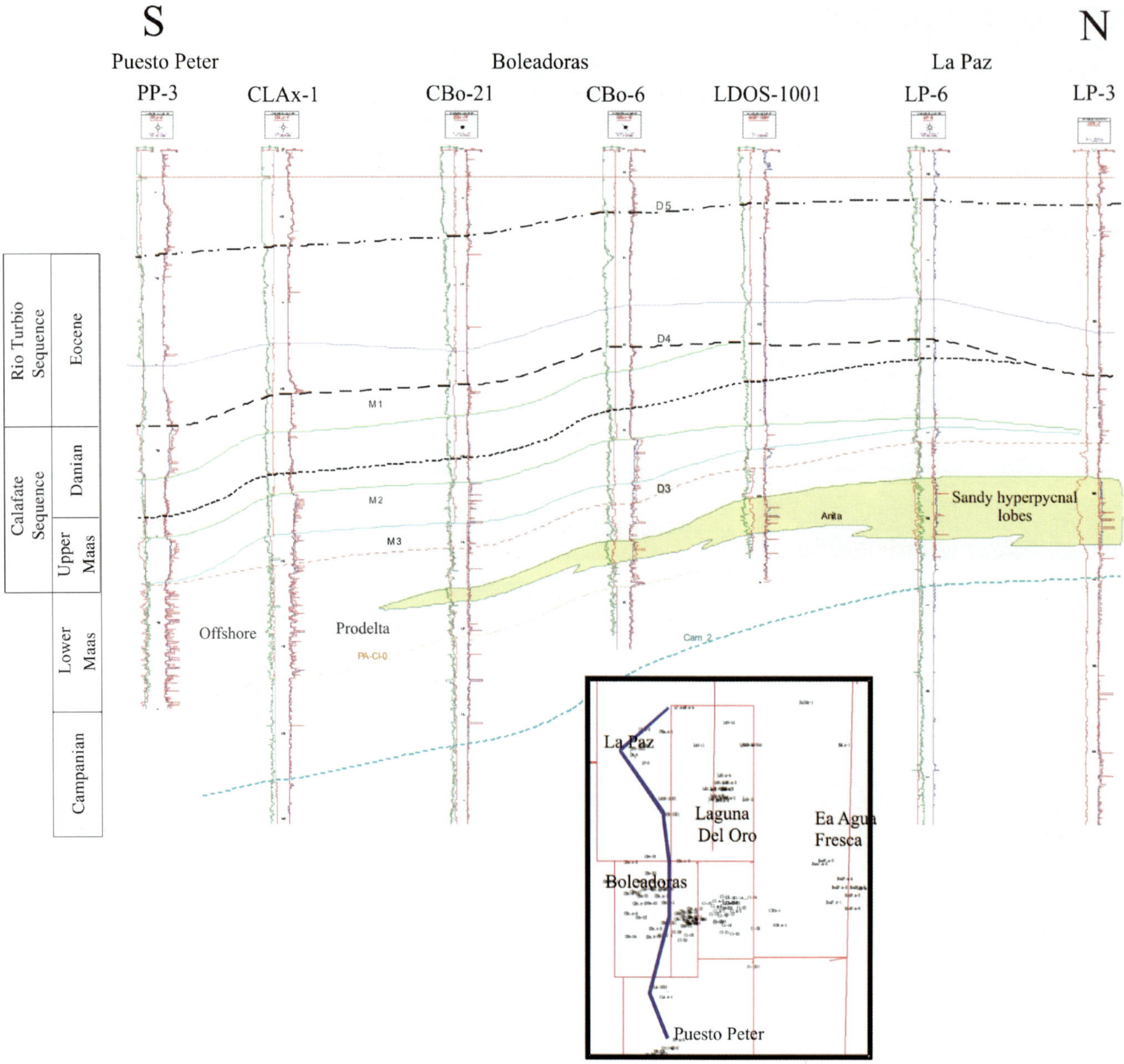

FIGURE 10. South-north cross section showing progradation toward the south. Extensive sand-rich lobes were formed in the La Paz field, whereas bypass sand-rich channels occurred toward the Boleadoras field. Toward the south in the Puesto Peter area, offshore deposits are predominant. Maas = Maastrichtian; Ea = Estancia; M = Magallanes; D3–D5 = unconformities; PP-3, CLAx-1, Cbo-21, CB0-6, LDOS-1001, LP-6, LP-3 = wells.

and during times of quiescence between flows by suspension fallout. The *Thalassinoides-Teichichnus* ichnofabric represents colonization of wave-reworked sandstone lobes. The *Terebellina-Phycosiphon* assemblage reflects stable conditions that allowed intense bioturbation and the establishment of a moderately diverse benthic fauna. During times of reduced sediment supply, material was reworked by wave processes and hypopycnal conditions were dominant. Finally, the *Teichichnus-Phycosiphon* ichnofabric represents intense bioturbation of lower offshore deposits.

Ichnofabrics vary along the sediment transport pathway and in times of dominance of hyperpycnal or hypopycnal conditions in response to a gradient in stress factors. Proximal positions and times of hyperpycnal conditions are characterized by rapid sedimentation and freshwater discharge. Distal positions and times of hypopycnal conditions are characterized by reduced sedimentation rates and normal salinity. Colonization windows were short-term in proximal positions and during times of hyperpycnal flows and long-term in distal positions and during

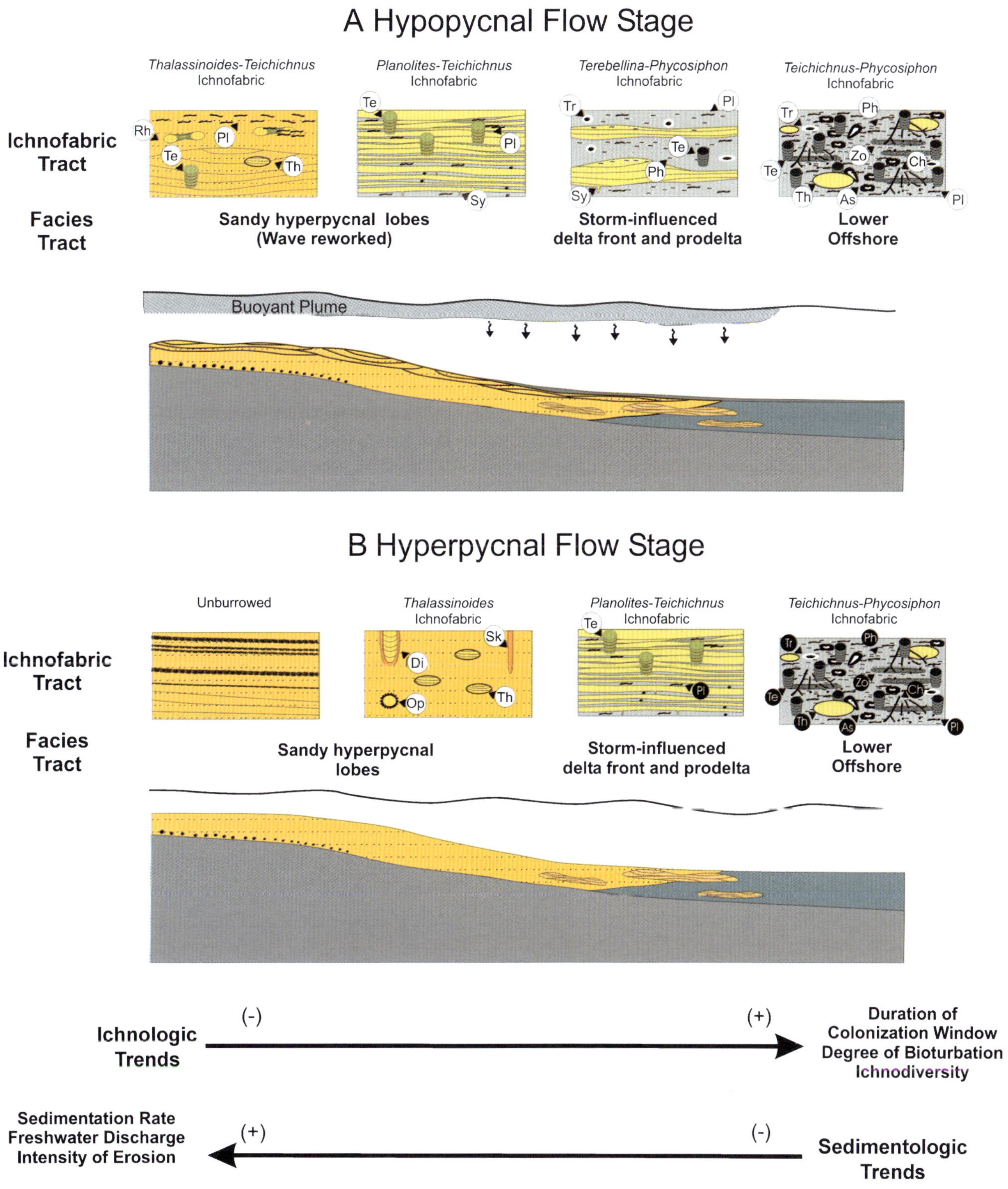

FIGURE 11. Paleoenvironmental distribution of the ichnofabrics and associated ichnologic and sedimentologic trends. Although the diagram illustrates proximal-distal relationships, these may also be understood not as simple bathymetric trends, but as radial trends from the inner zone to the margins of the lobe. Representative ichnogenera in the studied ichnofabrics include *Asterosoma* (As), *Chondrites* (Ch), *Diplocraterion* (Di), *Ophiomorpha* (Op), *Phycosiphon* (Ph), *Planolites* (Pl), *Rhizocorallium* (Rh), *Skolithos* (Sk), *Teichichnus* (Te), *Terebellina* (Tr), *Thalassinoides* (Th), and *Zoophycos* (Zo). Syneresis cracks (Sy) are commonly associated with the *Planolites-Teichichnus* and *Terebellina-Phycosiphon* ichnofabrics.

times of hypopycnal conditions. An increase in trace fossil diversity, intensity of bioturbation, tiering structure complexity, dominance of climax strategies, and dominance of deposit-feeding strategies is displayed along the transport pathway and as a response to the transition from hyperpycnal to hypopycnal conditions.

Freshwater discharge was conducive to limited biogenic activity, resulting in reduced trace fossil diversity and sparse bioturbation (e.g., Bhattacharya and MacEachern, 2009). High sedimentation rates also decrease animal activity. The laminated fill of *Thalassinoides* burrows is suggestive of rapid suspension fallout of fine-grained material. Phytodetrital pulses are evident in the sandy hyperpycnal lobes, as revealed by the high concentration of particulate plant fragments along bedding planes. In turn, phytodetrital pulses are linked to oxidation of organic carbon and oxygen depletion (MacEachern et al., 2005). However, in the Magallanes Formation, phytodetrital pulses did not seem to have been associated with oxygen depletion. This may have resulted from wave agitation in sand-rich lobes, which allowed the persistence of oxygenated bottom and interstitial waters. In addition to freshwater discharge, water turbidity represents another stress factor in river-dominated deltas. Deltas developed under hypopycnal conditions are commonly characterized by a buoyant plume leading to rapid flocculation of clays (Bates, 1953; Wright, 1977; Kineke et al., 1996). High suspended loads of fine-grained material related to river influx clog the filter-feeding apparatus of suspension feeders, therefore resulting in an impoverishment or direct suppression of the *Skolithos* ichnofacies (Gingras et al., 1998; MacEachern et al., 2005). In addition, accumulation of fluid muds in the distal delta-front and prodelta environment reduces boundary shear stress, preventing benthic organisms from constructing permanent structures or actively backfilling tunnels (Bhattacharya and MacEachern, 2009). Both buoyant plumes and fluid muds apparently were not important limiting factors in the present case. However, the sandy substrate of the hyperpycnal lobes seems to have been rich in water and very soft, as revealed by the extreme compaction of *Thalassinoides* burrows.

CONCLUSIONS

1) Integration of sedimentologic and ichnologic data allows a more detailed characterization of the sedimentary facies and depositional dynamics of a river-dominated delta characterized by hyperpycnal flows. Five ichnofabrics have been defined in subsurface deposits of the upper Cretaceous Magallanes Formation.
2) The *Thalassinoides* ichnofabric records opportunistic colonization of sand-rich hyperpycnal lobes emplaced in the distal delta front and the proximal prodelta. The *Planolites-Teichichnus* ichnofabric occurs in associated finer grained deposits that were formed lateral to the hyperpycnal sand-rich lobes and during times of quiescence between flows by suspension fallout.
3) The *Thalassinoides-Teichichnus* ichnofabric represents colonization of wave-reworked sandstone lobes. The *Terebellina-Phycosiphon* assemblage reflects stable conditions that allowed intense bioturbation and the establishment of a moderately diverse benthic fauna. During times of reduced sediment supply, material was reworked by wave processes and hypopycnal conditions were dominant.
4) The *Teichichnus-Phycosiphon* ichnofabric represents intense bioturbation of lower offshore deposits. High ichnodiversity and intense bioturbation indicate normal marine salinity and well-oxygenated bottom and interstitial waters.
5) Ichnofabrics vary along the sediment transport pathway and in times of dominance of hyperpycnal or hypopycnal conditions in response to a gradient in stress factors. Colonization windows were short-term in proximal positions and during times of hyperpycnal flows and long-term in distal positions and during times of hypopycnal conditions. An increase in ichnodiversity, intensity of bioturbation, tiering structure complexity, dominance of climax strategies, and dominance of deposit-feeding strategies is displayed along the transport pathway and as a response to the transition from hyperpycnal to hypopycnal conditions.

ACKNOWLEDGMENTS

Financial support was provided by Petrobras and Natural Sciences and Engineering Research Council Discovery Grants 311726-05 and 311726-08 awarded to Buatois. Gabriela Mángano provided useful feedback on the ichnology of these deposits. Roger Slatt is thanked for his review of the manuscript. Patricio Desjardins prepared Figure 11. We thank Petrobras for allowing publication of this study.

REFERENCES CITED

Bates, C. C., 1953, Rational theory of delta formation: AAPG Bulletin, v. 37, p. 2119–2162.

Bhattacharya, J. P., 2006, Deltas, *in* H. Posamentier and R. G. Walker, eds., Facies models revisited: SEPM Special Publication 84, p. 237–292.

Bhattacharya, J. P., and J. A. MacEachern, 2009, Hyperpycnal rivers and prodeltaic shelves in the Cretaceous seaway of North America: Journal of Sedimentary Research, v. 79, p. 184–209, doi:10.2110/jsr.2009.026.

Biddle, K. T., M. A. Uliana, R. M. Mitchum, M. G. Fitzgerald, and R. C. Wright, 1986, The stratigraphic and structural evolution of the central and eastern Magallanes Basin, southern South America, *in* P. A. Allen and P. Homewood, eds., Foreland basins: International Association of Sedimentologists Special Publication 8, p. 41–61.

Buatois, L. A., N. Santiago, K. Parra, and R. Steel, 2008, Animal-substrate interactions in an Early Miocene wave-dominated tropical delta: Delineating environmental stresses and depositional dynamics (Tácata field, Eastern Venezuela): Journal of Sedimentary Research, v. 78, p. 458–479, doi:10.2110/jsr.2008.053.

Carmona, N. B., and J. J. Ponce, 2011, Ichnology and sedimentology of Miocene hyperpycnites of the Austral foreland basin (Tierra del Fuego, Argentina): Trace fossil distribution and paleoecological implications: *in* R. M. Slatt and C. Zavalla, eds., Sediment transfer from shelf to deep water—Revisiting the delivery system: AAPG Studies in Geology 61, p. 171–192.

Cheel, R. J., and D. A. Leckie, 1993, Hummocky cross-stratification: Sedimentology Review, v. 1, p. 103–122.

Donovan, R. N., and R. J. Foster, 1972, Subaqueous shrinkage cracks from the Caithness Flagstone Series (Middle Devonian) of Northeast Scotland: Journal of Sedimentary Petrology, v. 42, p. 309–317.

Foster, W. R., J. G. Savings, and J. M. Waite, 1955, Lattice expansion and rheological behavior relationships in water-montmorillonite systems: Proceedings of the Third National Conference on Clays and Clay Minerals: National Academy of Sciences, National Research Council Publication, v. 395, p. 296–316.

Gingras, M. K., J. A. MacEachern, and S. G. Pemberton, 1998, A comparative analysis of the ichnology of wave and river-dominated allomembers of the upper Cretaceous Dunvengan Formation: Bulletin of Canadian Petroleum Geology, v. 46, p. 51–73.

Keeley, M. L., and M. P. R. Light, 2008, Basin evolution and prospectivity of the Argentine continental margin: Journal of Petroleum Geology, v. 16, p. 451–464, doi:10.1111/j.1747-5457.1993.tb00352.x.

Kineke, G. C., R. W. Sternberg, J. H. Trowbridge, and W. R. Geyer, 1996, Fluid mud processes on the Amazon continental shelf: Continental Shelf Research, v. 16, p. 667–697, doi:10.1016/0278-4343(95)00050-X.

Lamb, M. P., P. M. Myrow, C. Lukens, K. Houck, and J. Strauss, 2008, Deposits from wave-influenced turbidity currents: Pennsylvanian Minturn Formation, Colorado: Journal of Sedimentary Research, v. 78, p. 480–498, doi:10.2110/jsr.2008.052.

MacEachern, J. A., B. A. Zaitlin, and S. G. Pemberton, 1999, A sharp-based sandstone of the Viking Formation, Joffre field, Alberta, Canada, criteria for recognition of transgressively incised shoreface complexes: Journal of Sedimentary Research, Section B, Stratigraphy and Global Studies, v. 69, p. 876–892.

MacEachern, J. A., K. L. Bann, J. P. Bhattacharya, and C. D. Howell Jr., 2005, Ichnology of deltas: Organism responses to the dynamic interplay of rivers, waves, storms, and tides, *in* L. Giosan and J. P. Bhattacharya, eds., River deltas: Concepts, models and examples: SEPM Special Publication 83, p. 49–85.

McIlroy, D., 2004, Some ichnological concepts, methodologies, applications and frontiers, *in* D. McIlroy, ed., The application of ichnology to paleoenvironmental and stratigraphic analysis: Geological Society (London) Special Publication 228, p. 3–27.

Miller, W. III., 1995, *Terebellina*; (= *Schaubcylindrichnus freyi* ichnosp. nov.) in Pleistocene outer-shelf mudrocks of northern California: Ichnos, v. 4, p. 141–149.

Mulder, T., and J. P. M. Syvitski, 1995, Turbidity currents generated at river mouths during exceptional discharges to the world oceans: Journal of Geology, v. 103, p. 285–299, doi:10.1086/629747.

Mulder, T., J. P. M. Syvitski, S. Migeon, J.-C. Faugères, and B. Savoye, 2003, Marine hyperpycnal flows: Initiation, behavior and related deposits—A review: Marine and Petroleum Geology, v. 20, p. 861–882, doi:10.1016/j.marpetgeo.2003.01.003.

Pattison, S. A. J., 2005a, Storm-influenced prodelta turbidite complex in the lower Kenilworth Member at Hatch Mesa, Book Cliffs, Utah: Implications for shallow marine facies models: Journal of Sedimentary Research, v. 75, p. 420–439, doi:10.2110/jsr.2005.033.

Pattison, S. A. J., 2005b, Isolated highstand shelf sandstone body of turbiditic origin, lower Kenilworth member, Cretaceous Western Interior, Book Cliffs, Utah: Sedimentary Geology, v. 177, p. 131–144, doi:10.1016/j.sedgeo.2005.02.005.

Pemberton, S. G., and R. W. Frey, 1984, Ichnology of storm-influenced shallow marine sequence, Cardium Formation (upper Cretaceous) at Seebe, Alberta, *in* D. F. Stott and D. J. Glass, eds., The Mesozoic of middle North America: Canadian Society of Petroleum Geologists Memoir 9, p. 281–304.

Pemberton, S. G., and J. A. MacEachern, 1997, The ichnological signature of storm deposits, the use of trace fossils in event stratigraphy, *in* C. E. Brett and G. C. Baird, eds., Paleontological events, stratigraphic, ecological and evolutionary implications: New York, New York, Columbia University Press, p. 73–109.

Plink-Björklund, P., and R. Steel, 2004, Initiation of turbidity currents: Outcrop evidence for Eocene hyperpycnal flow turbidites: Sedimentary Geology, v. 165, p. 29–52, doi:10.1016/j.sedgeo.2003.10.013.

Ponce, J. J., E. B. Olivero, D. R. Martinioni, and M. I. López Cabrera, 2007, Sustained and episodic gravity flow deposits and related bioturbation patterns in Paleogene turbidites (Tierra del Fuego, Argentina), *in* R. G. Bromley, L. A. Buatois, M. G. Mángano, J. F. Genise, and R. N. Melchor, eds., Sediment-organism interactions: A multifaceted ichnology: SEPM Special Publication 88, p. 253–266.

Soyinka, O., and R. M. Slatt, 2008, Identification and micro-stratigraphy of hyperpycnites and turbidites in Cretaceous Lewis Shale, Wyoming: Sedimentology, v. 55, p. 1117–1133, doi:10.1111/j.1365-3091.2007.00938.x.

Taylor, A., and R. Goldring, 1993, Description and analysis of bioturbation and ichnofabric: Journal of the Geological Society, v. 150, p. 141–148, doi:10.1144/gsjgs.150.1.0141.

Wetzel, A., 2008, Recent bioturbation in the deep South China Sea: A uniformitarian ichnologic approach: Palaios, v. 23, p. 601–615, doi:10.2110/palo.2007.p07-096r.

Wright, L. D., 1977, Sediment transport and deposition at river mouths: A synthesis: Bulletin of the Geological Society of America, v. 88, p. 857–868, doi:10.1130/0016-7606(1977)88<857:STADAR>2.0.CO;2.

Wright, L. D., W. J. Wiseman, B. D. Bornhold, D. B. Prior, J. N. Suhayda, G. H. Keller, Z.-S. Yang, and Y. B. Fan, 1988, Marine dispersal and deposition of Yellow River silts by gravity-driven underflows: Nature, v. 332, p. 629–632, doi:10.1038/332629a0.

Zavala, C., J. Ponce, M. Arcuri, D. Drittanti, H. Freije, and M. Asensio, 2006, Ancient lacustrine hyperpycnites: A depositional model from a case study in the Rayoso Formation (Cretaceous) of west-central Argentina: Journal of Sedimentary Research, v. 76, p. 40–58.

9

Carmona, N. B., and J. J. Ponce, 2011, Ichnology and sedimentology of Miocene hyperpycnites of the Austral foreland basin (Tierra del Fuego, Argentina): Trace fossil distribution and paleoecological implications, *in* R. M. Slatt and C. Zavala, eds., Sediment transfer from shelf to deep water—Revisiting the delivery system: AAPG Studies in Geology 61, p. 171–192.

Ichnology and Sedimentology of Miocene Hyperpycnites of the Austral Foreland Basin (Tierra del Fuego, Argentina): Trace Fossil Distribution and Paleoecological Implications

Noelia B. Carmona and Juan José Ponce

Instituto Argentino de Oceanografía, Consejo Nacional de Investigaciones Científicas y Técnicas (IADO-CONICET), Bahía Blanca, Argentina

ABSTRACT

Combined ichnologic and sedimentologic analysis of Miocene deep-marine hyperpycnites in the Austral foreland basin, Tierra del Fuego, Argentina, allows recognition and characterization of different trace fossil assemblages from proximal to distal within this depositional system. In proximal and middle settings, hyperpycnal channel and inner levee deposits contain trace fossils in climbing-ripple and parallel-laminated sandstone facies; dwelling structures of suspension-feeding organisms (e.g., *Diplocraterion*) dominate. In outer levee deposits, fine-grained heterolithic intervals display a suite of trace fossils related to opportunistic producers composed of concentrically laminated burrows of detritus or deposit feeders and also locomotion and grazing traces (e.g., *Protovirgularia*, *Gordia*). The general low intensity of bioturbation and the dominance of structures produced by opportunistic organisms suggest a stressful environment probably because of episodes with high sedimentation rates and fluctuating salinity. In addition, in the middle slope setting, escape structures of imported bivalves occur in isolated sandstone bodies that accumulated at the base of the depositional slope.

Copyright ©2011 by The American Association of Petroleum Geologists.

DOI:10.1306/13271356St613442

In the distal setting, the ichnofossils occur mainly in heterolithic intervals developed at the base and top of the hyperpycnal lobe deposits. Bioturbation intensity is commonly higher than in more proximal deposits, although the ichnocoenoses display low to moderate diversity. The fairly sandy heteroliths show discrete intervals completely reworked by deposit feeders forming, for instance, *Scolicia* and *Nereites*. This assemblage probably reflects periods of normal marine salinity and high food content. Intervals with abundant *Phycosiphon*, *Nereites*, and subordinate *Tasselia* commonly occur in fine-grained heteroliths and massive mudstones that accumulated at the top of the hyperpycnal lobe deposits. The presence of these ichnogenera also indicates full-marine conditions, although patchy distribution of these trace fossil suites may suggest fluctuations in salinity and in sediment and organic input. Uncommon intervals with predepositional graphoglyptids (e.g., *Paleodictyon*, *Helicolithus*) are preserved on the soles of sandstone-heterolithic beds of the hyperpycnal lobes, thus reflecting periods of fairly stable ecological conditions.

The trace fossil suites of the studied hyperpycnite successions are commonly impoverished, showing monogeneric assemblages and dominance of opportunistic strategies. Sedimentologic and ichnologic evidences indicate that fluctuations in hydraulic energy, sedimentation rates that are moderate on average but nonetheless strongly fluctuating, high food supply, and changes in salinity generated by the hyperpycnal flows were probably the main paleoecological factors that heavily affected the development of these infaunal communities.

INTRODUCTION

In foreland basins, the absence of extended shelf areas, synsedimentary tectonic activity, and steep slopes resulting from rapidly uplifted areas within the orogen favor the formation of hyperpycnal flows (Mutti et al., 1996). These flows develop during extraordinary fluvial discharges, constitute an important mechanism of basinward sediment transfer (Mulder and Syvitski, 1995; Mulder et al., 2003), and generate very complex sedimentary deposits (e.g., Ponce et al., 2004, 2008a, b).

In the last years, a growing recognition and an understanding of the hyperpycnal depositional systems are seen. Many studies have mainly focused on the sedimentary processes and on the characterization of the internal facies arrangements and geometries of these deposits (Mulder et al., 2003; Mutti et al., 2003; Plink-Björklund and Steel, 2004; Ponce et al., 2005, 2007a, 2008a, b, c; Zavala et al., 2006). However, specific details of the trace fossils found in hyperpycnites, such as the distribution of the biogenic structures, ichnodiversity, and bioturbation intensities, have not yet been evaluated. Only a few studies have analyzed the ichnologic information in the context of hyperpycnal systems (e.g., Carmona et al., 2006, 2008; Ponce et al., 2007a; Uchman and Steel, 2007).

The detailed study of the hyperpycnite successions exposed along the Atlantic coast of Tierra del Fuego, Argentina, enabled the recognition of the main architectural elements in the deep-marine depositional system of the Miocene Austral Basin (Ponce et al., 2008b). These outcrops are excellent because of their great lateral continuity, and thus provide a good opportunity to investigate the composition and spatial distribution of the ichnofauna. The ichnotaxonomy of these deposits has been recently published (López-Cabrera et al., 2008). In this study, the main sedimentary facies and the architecture of the deposits are described, and then the principal characteristics of the ichnoassemblages are outlined. Integration of both ichnologic and sedimentologic information provides a better understanding of the dominant ecological parameters that controlled the development and the distribution of the trace fossils in these hyperpycnal systems. Ichnologic analysis may help to better understand hyperpycnal depositional systems.

GEOLOGIC FRAMEWORK AND STRATIGRAPHIC SETTING OF THE STUDY AREA

The Austral foreland basin fill (Figure 1) comprises a series of late Cretaceous to early Miocene asymmetric

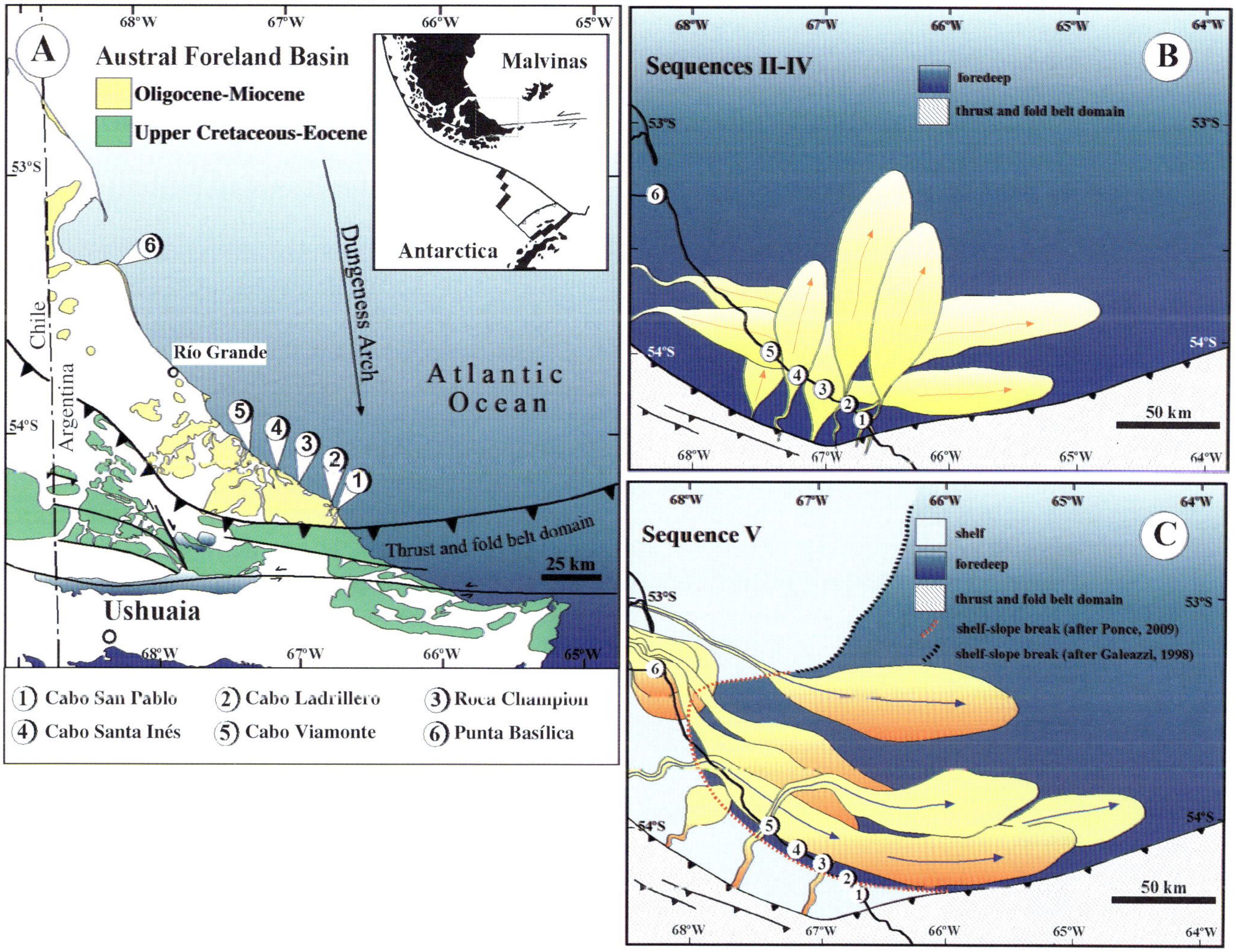

FIGURE 1. (A) Simplified geologic map of Tierra del Fuego, Argentina, showing the studied localities exposed along the Atlantic coast (modified from Ponce et al., 2008b). (B, C) Paleogeographic reconstruction of the Miocene deposits of the Austral Basin showing the distribution of hyperpycnites (J. J. Ponce, 2009, unpublished data).

clastic wedges, which are thickest and relatively complete toward the orogen, and which thin greatly toward the foreland (Olivero and Malumián, 2002, 2008; see also Malumián and Olivero, 2006). Since the early Miocene, marine sediments constitute relatively undeformed successions that formed after cessation of compressive deformation. These marine sediments are included within the Cabo Domingo Group (Malumián, 1999). This group reaches a minimum thickness of 1000 m (3281 ft); these marine sediments grade northwestward into marginal-marine and finally into continental deposits (Malumián and Olivero, 2006). Five depositional sequences (SI–SV) have been distinguished in the upper Oligocene–Miocene deposits of the Austral foreland basin (Figure 2) (Ponce et al., 2005, 2008b). These sequences show two main types of fill geometries, a wedge-shaped fill geometry for SI–SIV, and a sigmoidal fill geometry for SV (Ponce et al., 2008b). The oldest sequence, SI (Desdémona Formation), includes two members: A, a mudstone-dominated member deposited by cohesive flows; and B, a glauconite-rich, sandstone-dominated member deposited by episodic turbidity currents. The remaining sequences, SII to SV (Lower and Upper Cabo Ladrillero, Cabo San Pablo, and Cabo Viamonte beds), are composed of complex arrangements of pebble conglomerates, coarse- to fine-grained sandstones, and mudstones that were deposited mainly by hyperpycnal flows (Figure 2). The SI is situated closer to the orogen and was deposited on the frontal limb of a growing anticline, which led to the formation of a growth structure that involved slumping. The SII to SV are situated increasingly toward the foreland and are characterized by subhorizontal conformable beds of a large lateral extent. This study focuses on the Miocene hyperpycnal deposits of the Austral foreland basin,

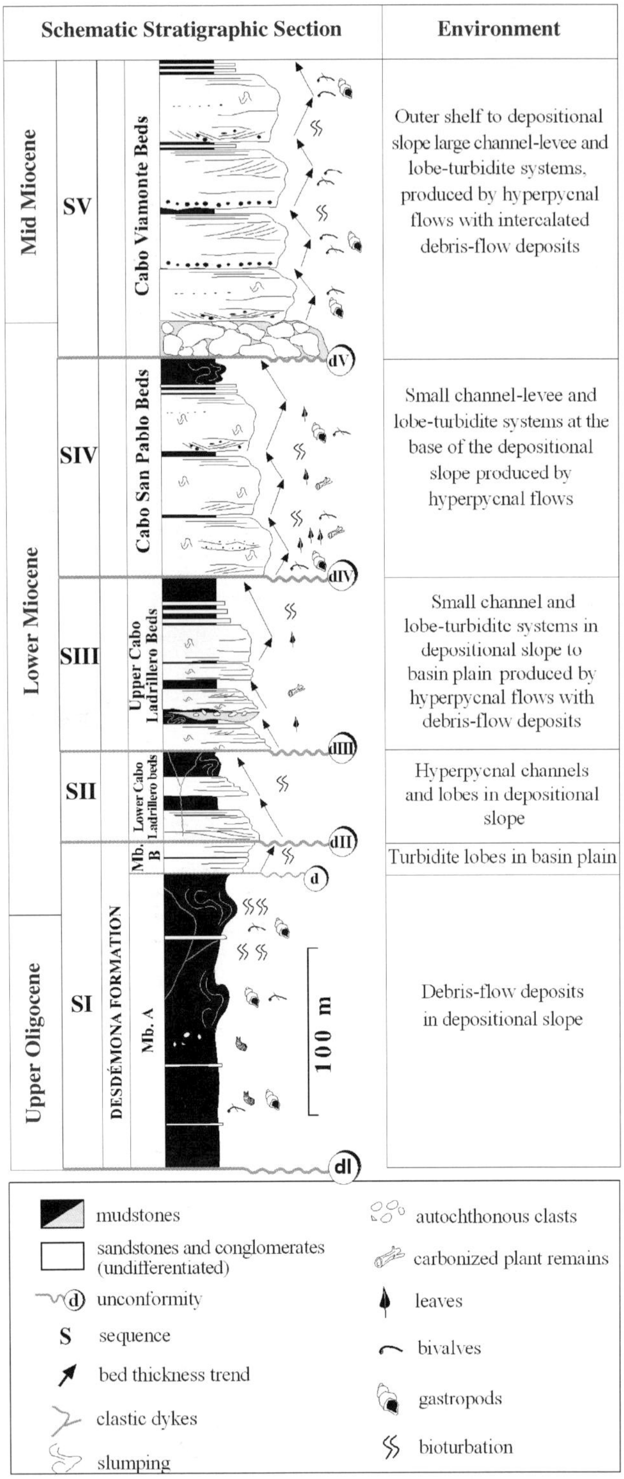

FIGURE 2. Schematic section illustrating the five identified sequences and their interpreted depositional environment (modified from Ponce et al., 2008b). This study focuses on sequences SII–SV. Mb. = Member.

and therefore, only the upper four sequences (SII–SV) are analyzed (Figures 1B, C; 2). The studied area covers a southeast-northwest transect along the Atlantic coast of Tierra del Fuego between the localities of Cabo San Pablo and Punta Basílica (Figure 1).

DEPOSITIONAL SETTINGS AND ICHNOLOGIC CONTENT

In the current study, three main depositional settings are recognized within the Miocene foredeep deposits of the Austral Basin (Figure 3); each one with a distinctive trace fossil association: (1) proximal setting, characterized by small meandering channel deposits and associated levee facies; (2) intermediate setting, displaying major aggradational channel and levee deposits at the base of the depositional slope; and (3) distal setting, exhibiting lobe deposits. With respect to the levee deposits, inner and outer zones can be differentiated. The inner levee deposits, situated closer to the channel axis, consist mainly of sandy heteroliths. In contrast, the outer levee deposits, developed farther from the channel axis, comprise muddy heteroliths. Table 1 summarizes additional information about the recognized facies associations. Most of the trace fossils mentioned in this chapter were recently described by López-Cabrera et al. (2008). The bioturbation index (BI) was assessed following the method of Taylor and Goldring (1993). This index ranges from 0 (no bioturbation) to 6 (completely bioturbated).

Proximal Setting

Two main facies associations are recognized in proximal settings: small hyperpycnal channels with lateral accretion and levee facies (Figures 3, 4; Table 1). The channel deposits have sharp erosive lower contacts and show a general fining-thinning-upward trend. These deposits are up to 100 m (328 ft) wide and commonly less than 10 m (32.8 ft) thick (Figure 4A, B). They consist of pebble conglomerates and transitions between massive, parallel-laminated, and climbing-ripple sandstones (Figure 4C). Internally, the individual sigmoidal beds are generally less than 50 cm (20 in.) thick, and the inclined surfaces are perpendicularly oriented to the flow direction (Ponce et al., 2008a, b).

The levee-facies association consists of rhythmical successions of heterolithic, thinly bedded, fine-grained sandstones and mudstones. Carbonaceous plant fragments are locally present. The inner levee consists of medium- to fine-grained sandstones, which are massive or display parallel lamination and climbing ripples (Figure 4D). Isolated horizons of deformed beds are common (Ponce et al., 2008a, b). Intervals with syneresis cracks occur locally (Figure 4E). The outer levee

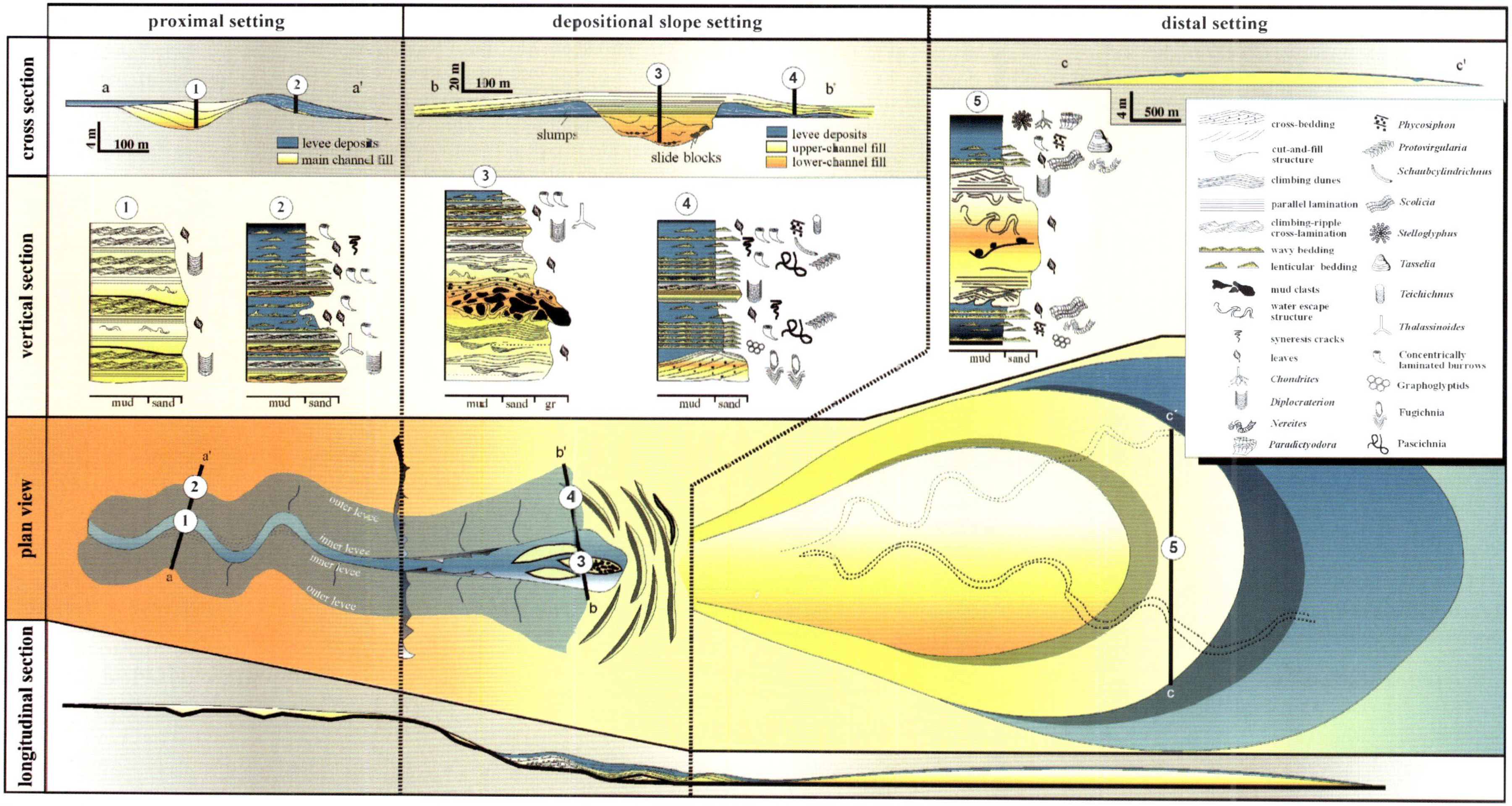

FIGURE 3. Paleoenviromental scheme for the hyperpycnal channel-lobe complexes of the Miocene of the Austral Basin, Tierra del Fuego, including ichnologic and sedimentologic information of the main subenvironments: in a proximal setting, hyperpycnal channels with lateral accretion (section 1) and levee (section 2); in the depositional slope setting, aggradational channels (section 3) and levee (section 4); and in a distal setting, hyperpycnal lobes (section 5). gr = gravel.

Table 1. Principal sedimentologic and ichnologic characteristics of the facies associations recognized in the Miocene hyperpycnal deposits of the Austral foreland basin.

Facies Association	*Sedimentologic Description*	*Geometry and Dimensions*	*Ichnologic Content*
Hyperpycnal channels with lateral accretion fill	These deposits have sharp erosive lower contacts and consist of pebble conglomerates and massive, parallel-laminated, and climbing-ripple sandstones, with *Nothofagus* leaves and phytodetritus. Internally, the deposits show a gradual passage and repetition of sedimentary structures. The sigmoidal bedsets are generally less than 50 cm (20 in.) thick, and the inclined surfaces are oriented perpendicular to flow direction.	These are isolated lenticular bodies with general fining-thinning-upward arrangements that are up to 100 m (328 ft) wide and rarely exceeding 10 m (32.8 ft) in thickness.	*Diplocraterion*
Hyperpycnal channels with aggradational fill	The lower channel fill has sharp erosive basal contacts and thick packages (>20 m [66 ft]) with a coarsening-thickening-upward trend followed by a fining-thinning-upward trend. These packages consist of sandy conglomerates and gravelly sandstones, with subordinate intraformational boulders; massive sandstones (with *Nothofagus* leaves); and sandstones with climbing dunes, parallel lamination, climbing ripples, and cut-and-fill structures; with a gradual passage and repetition of sedimentary structures.	These are lenticular bodies with general fining-thinning-upward arrangements, up to 1000 m (3280 ft) wide and 40 m (131 ft) thick.	*Diplocraterion* Escape structures
	The upper channel fill consists of tabular graded sandstone and mudstone beds (<50 cm [<20 in.] thick), with onlap relationships toward the channel margins. The sandstones are massive or normally graded with parallel lamination, climbing ripples, and ripple lamination, with abundant fossil *Nothofagus* leaves and phytodetritus.		Concentrically laminated burrows

Levees	The inner levee deposits have sharp, slight erosive lower contacts and consist of rhythmical successions of fine-grained heterolithic sandstone beds, which can be massive or display climbing dunes, parallel lamination, and climbing ripples. Isolated, contorted (slumped) beds, and horizons with syneresis cracks are locally intercalated within these heterolithic packages. *Nothofagus* leaves and phytodetritus are abundant.	These bodies have a wedge-shaped geometry, with a thickness up to 20 m (66 ft) closer to the channel axis and thinner toward distal positions. The minimum lateral extension, perpendicular to the channel, reaches 400 m (1312 ft) at the scale of the outcrops.	*Diplocraterion*
	The outer levee deposits have sharp lower contacts and consist of rhythmical successions of fine-grained heterolithic mudstone beds, with wavy and lenticular bedding. *Nothofagus* leaves and phytodetritus are abundant.		*Scolicia* *Nereites* *Phycosiphon* *Gordia* *Helminthoidichnites* *Protovirgularia* Concentrically laminated burrows Graphoglyptids
Hyperpycnal lobes	These bodies have sharp lower contacts, are up to 3 m (9.8 ft) thick, and exhibit a coarsening-thickening-upward trend followed by a fining-thinning-upward trend. The lower coarsening-thickening-upward deposits consist of fine-grained heteroliths, with ripple cross-lamination, wavy and lenticular bedding, and high phytodetrital concentration. The overlying fine-grained sandstones record transitional passages between climbing ripples, parallel lamination, and climbing dunes. The thick massive sandstone beds contain well-preserved *Nothofagus* leaves. The upper fining-thinning-upward portion is characterized by a transitional and recurrent passage between fine-grained massive sandstone and sandstones with climbing dunes, parallel lamination, and climbing ripples.	These bodies exhibit tabular bed geometries (at the scale of the outcrops); locally they show subtle interfingering and pinch-out relationships. The lateral extension is usually wider than 400 m (1312 ft) with 3 m (9.8 ft) as an average maximum thickness.	*Diplocraterion* *Scolicia* *Nereites* *Phycosiphon* *Tasselia* *Gyrophyllites* *Thalassinoides* *Chondrites* *Paradictyodora* *Phycodes* Graphoglyptids

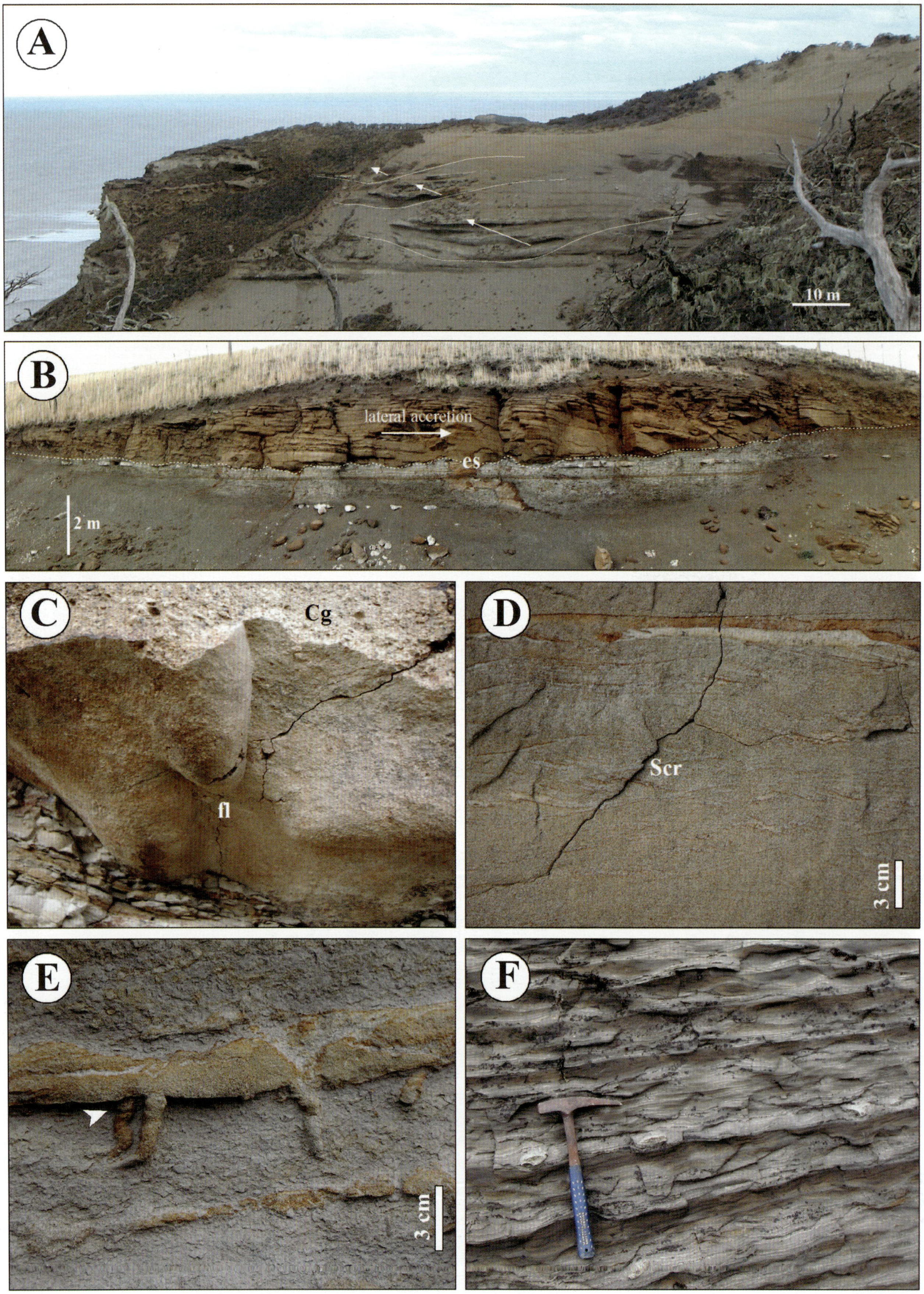
A
10 m
B
lateral accretion
es
2 m
C
Cg
fl
D
Scr
3 cm
E
3 cm
F

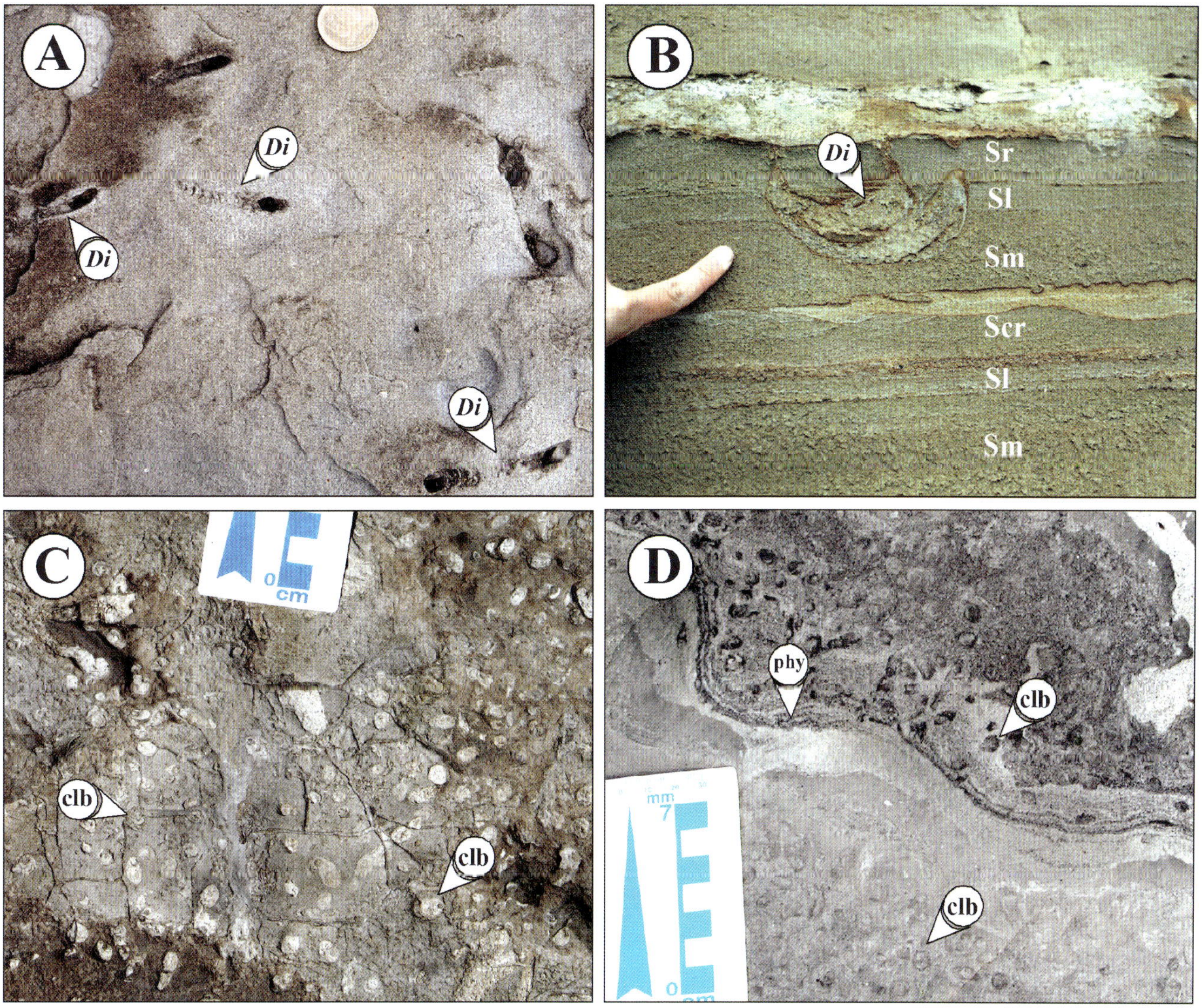

FIGURE 5. Trace fossils in the proximal setting of the hyperpycnal depositional system. (A) Bedding-plane view of *Diplocraterion* (*Di*) in parallel-laminated sandstones of channel-margin deposits (north of Cabo Ladrillero). Coin is 25.4 mm (0.99 in.). (B) Cross section view of *Di* preserved in parallel-laminated (Sl) and cross-ripple (Sr) sandstones of inner levee deposits (Cabo San Pablo). Note the transitional and recurrent passage of sedimentary structures, indicating fluctuations in flow velocity. Sm = massive sandstones; Scr = climbing-ripple sandstones. (C, D) Bedding plane view of the heterolithic levee facies displaying high abundance of concentrically laminated burrows (clb). Note in (D) the abundant phytodetrital material (phy) (Cabo Santa Inés).

consists mainly of fine-grained sandstones and mudstones with wavy bedding and lenticular bedding (Figure 4F). Both in the channel and the levee facies, *Nothofagus* leaves and phytodetritus are abundant.

In marginal parts of the channel deposits, the degree of bioturbation is low (BI 0–2). *Diplocraterion*, a dwelling structure of a suspension feeder, is the only ichnogenus found within these beds; it preferentially occurs in the parallel-laminated and climbing-ripple sandstones (Figure 5A).

A monogeneric suite of *Diplocraterion* also occurs in the most energetic intervals of the inner levee

FIGURE 4. Architectural features and principal sedimentary facies in the proximal setting. (A, B) Small-scale hyperpycnal channels with lateral accretion (arrows indicate the main direction of migration). (A) North of Cabo Ladrillero, (B) Cabo Viamonte. (C) Detail of (B) showing the base of the channel with flute marks (fl) and graded conglomerates (Cg). (D) Climbing-ripple sandstones in inner levee deposits (Cabo San Pablo). (E) Close-up of sandier heteroliths showing syneresis cracks (arrow) (Roca Champlon). (F) Fine-grained heteroliths with wavy and lenticular bedding in outer levee deposits (Cabo Santa Inés). Scr = climbing-ripple sandstones; es = erosion surface.

A
upper-channel fill
ies
lower-channel fill
es
levee
6 m
B
cf
1 m
C
Scd
cf
1 m
D
flow direction
1 m
E
F

facies—the massive, parallel-laminated, and climbing-ripple sandstones (Figure 5B). In contrast, the fine-grained heteroliths of the outer levee, which are mostly wavy bedded intervals, display a high abundance of concentrically laminated burrows (in the sense of Goldring, 1996), constituting a monospecific suite (Figure 5C, D). Although variable, the intensity of bioturbation is locally high (BI 0–4).

Depositional Slope Setting

In the depositional slope setting, large aggradational hyperpycnal channels and levee deposits are recognized (Figures 3, 6; Table 1). The aggradational channel deposits are up to 1000 m (3280 ft) wide and up to 40 m (130 ft) thick (Figure 6A). In general, they show a fining-thinning-upward arrangement. In steep slope settings, the lower channel fill, deposited during the waxing stage of the hyperpycnal flow, consists of thick sandy conglomerates and gravelly sandstones; massive sandstones; and sandstones with climbing dunes, parallel lamination, climbing ripples, and cut-and-fill structures (Figure 6B, C) (Ponce et al., 2008c). The upper channel fill consists of tabular graded sandstone and mudstone beds, less than 50 cm (20 in.) thick, with sharp basal contacts (Figure 6A). The sandstones are structureless or normally graded with parallel lamination, climbing ripples, and ripple lamination. In less steep depositional slope settings, the lower channel fill commonly comprises sandstone deposits with sedimentary structures resembling antidune structures (Figure 6D). These deposits are blanketed by thick heterolithic successions, with onlap relationships (Figure 6D). The levee facies displays mainly heterolithic stratification, with dominance of wavy and lenticular bedding (Figure 6E). Syneresis cracks and deformed massive mudstones occur in isolated intervals. Plant remains, especially *Nothofagus* leaves and wood fragments, are common in these heteroliths (Figure 6F).

Different suites of trace fossils have been observed in these channel and levee complexes (Figure 7). In the major channels, the biogenic structures are restricted to the marginal areas and to the heterolithic intervals at the top of the upper channel fill. In the parallel-laminated sandstones of the marginal channels, protrusive *Diplocraterion* is found (Figure 7A). This trace fossil commonly occurs in low abundance, forming a monogeneric assemblage (BI 1–2). In contrast, the heterolithic upper-channel-fill intervals are commonly more bioturbated (BI 3–4) and display a monogeneric suite of concentrically laminated burrows (Figure 7B).

The levee deposits laterally show a recurrent distribution of trace fossils (Figures 7C–G; 8). In the inner levee deposits, a moderate density (BI 2–3) of protrusive *Diplocraterion* occurs in the parallel-laminated and climbing-ripple sandstones (Figure 8A). Between the inner and outer levee deposits, the wavy bedded heteroliths display an ichnoassemblage dominated mainly by *Scolicia*, *Nereites*, and *Phycosiphon* (Figures 7C, D; 8B). Isolated specimens of *Teichichnus* and *Schaubcylindrichnus* are also identified in these deposits. These heterolithic intervals range from weakly to almost totally bioturbated (BI 1–5). In the fine-grained heteroliths of the outer levee deposits, small grazing traces (e.g., *Gordia* and *Helminthoidichnites*) and locomotion traces (e.g., *Protovirgularia*) occur (Figures 7E; 8C). The abundance of these biogenic structures is commonly high, and they are commonly preserved as casts at the base of the fine-grained sandstones. In addition, certain levels of the outer levee deposits can also contain concentrically laminated burrows (Figures 7F; 8C). In some isolated sandstone beds of the outer levee deposits, a low-diversity graphoglyptid suite (e.g., *Paleodictyon* and *Helicolithus*) has been identified (Figures 7G; 8C). Finally, a different assemblage, restricted to the isolated sandstone bodies resembling antidune deposits, consists of abundant fugichnia structures of bivalves (Figure 7H).

Distal Setting

The lobe deposits recognized in distal settings of the hyperpycnal system are characterized by tabular bodies, up to 3 m (9.8 ft) thick, with sharp lower contacts and coarsening-thickening followed by fining-thinning-upward trends (Figures 3; 9A, B; Table 1). The

FIGURE 6. Main architectural characteristics and sedimentary facies associated with the depositional slope setting. (A) Aggradational hyperpycnal channel and levee complex. The dashed white lines indicate internal erosion surfaces (ies), and the complete line corresponds to a major erosion surface (es) (Cabo Viamonte). (B, C) Lower channel fill facies, displaying cut-and-fill structures (cf), climbing dunes (Scd), and intraclasts oriented following the stratification (Cabo Viamonte). (D) Isolated sandstone bodies resembling antidune deposits occur at the base of the depositional slope. These deposits are blanketed by thick heterolithic intervals (Punta Basílica). (E) Fine-grained heteroliths with wavy and lenticular bedding accumulated as outer levee deposits (Cabo Viamonte). (F) Bedding-plane view of heterolithic levee intervals displaying abundant and well-preserved *Nothofagus* leaves and wood fragments (Punta Basílica).

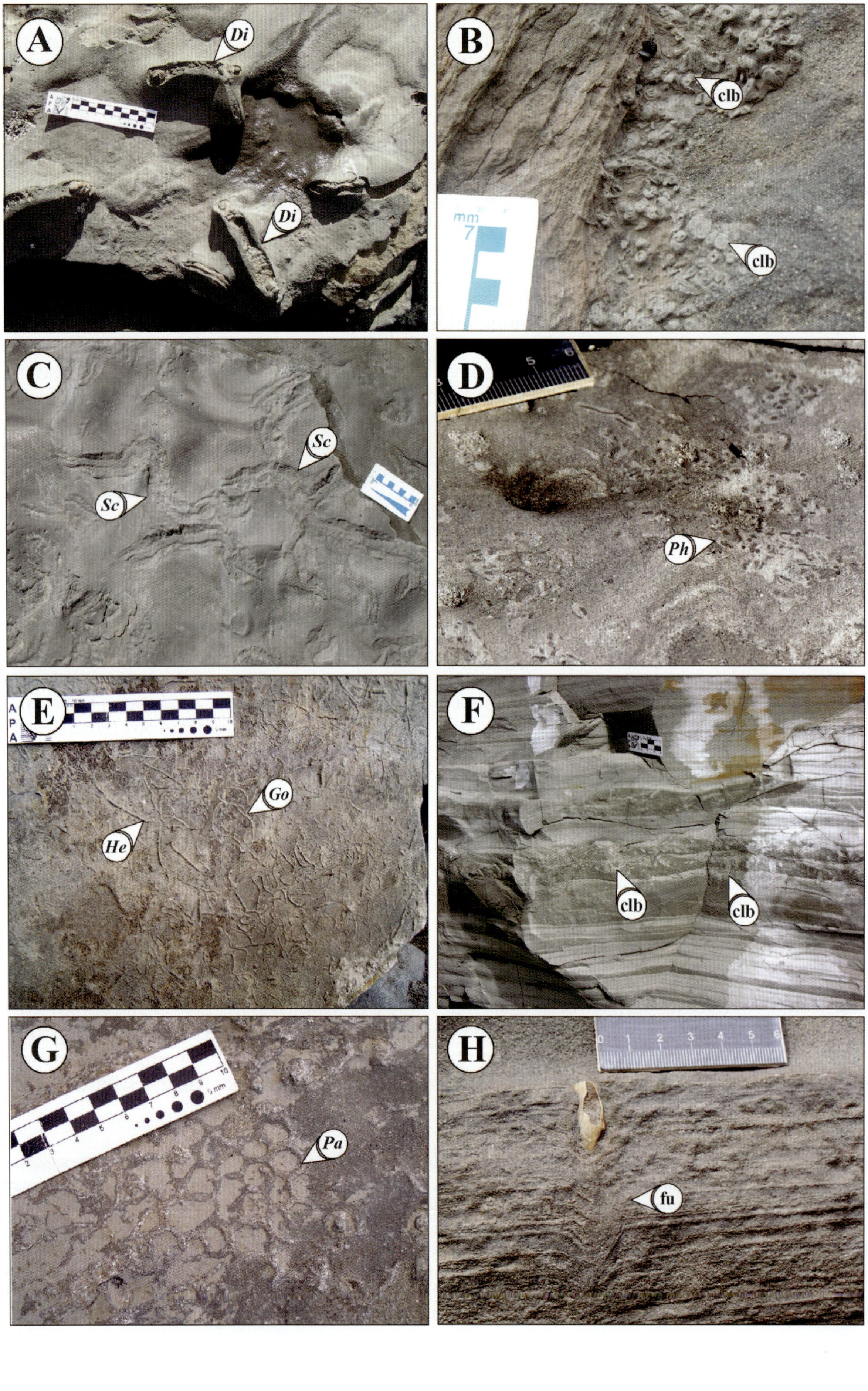
A
Di
Di
B
clb
clb
C
Sc
Sc
D
Ph
E
Go
He
F
clb
clb
G
Pa
H
fu

basal coarsening-thickening-upward beds, accumulated during the waxing stage of the hyperpycnal flow, consist of rhythmical successions of heterolithic, thinly bedded, fine-grained sandstones with transitional and recurrent passage between ripple cross-lamination, wavy and lenticular bedding, and massive mudstones. The overlying fine-grained sandstones consist of horizons with climbing ripples, parallel lamination, climbing dunes, and transitions between them (Figure 9C–E). Deposition of thick massive sandstone beds, with well-preserved *Nothofagus* leaves, could have originated through constant aggradation or load collapse of a high-concentration hyperpycnal flow. At maximum flow velocity, numerous erosive surfaces are produced. The overlying fining-thinning succession, originated during the waning stage of the hyperpycnal flow, is characterized by a transitional and recurrent passage between fine-grained massive sandstones and sandstones with climbing dunes, parallel lamination, and climbing ripples. Heterolithic beds and massive mudstones commonly occur at the uppermost part of these packages. Dewatering structures and high phytodetrital concentration are commonly observed (Figure 9D, E).

Distinctive ichnologic suites characterize the lobe facies. In the sandy facies of the lobe successions, bioturbation intensity is commonly low (BI 0–2). Only protrusive *Diplocraterion* occurs in the parallel-laminated sandstones deposited during the waning stage of the flow (Figure 10A), whereas in the coarser grained intervals, no bioturbation was observed. In the heterolithic facies that occur at the base and at the top of the hyperpycnal lobes, the ichnocoenosis is dominated by fodinichnia, mainly *Scolicia*, *Nereites*, and *Phycosiphon* (Figure 10B–D). The producers of these burrows commonly completely reworked the heterolithic intervals, although some layers remain unbioturbated (BI 0–6). The massive mudstone intervals are commonly thoroughly bioturbated (BI 6), with *Phycosiphon* and *Nereites* as dominant elements (Figure 10D). Large *Tasselia* also occurs in distinctive levels of the fine-grained heteroliths and massive mudstones, in the uppermost part of the lobe successions (Figure 10E). Sporadically, the top of these muddy intervals can be completely reworked by producers of *Gyrophyllites*, *Thalassinoides*, *Chondrites*, *Paradictyodora*, *Stelloglyphus*, and *Phycodes*, among others (Figure 10F). In addition, a low-diversity suite of graphoglyptids (e.g., *Paleodictyon*, *Helicolithus*) occurs sporadically, casted at the base of heterolithic sandstone beds, accumulated during the initial phase of the hyperpycnal discharge (Figure 10G, H).

DISCUSSION

Trace Fossil Distribution and Paleoenvironmental Conditions

In marine environments, the distribution of trace fossils depends on certain major ecological factors, such as bathymetry, oxygen content, salinity, energy, depositional rates, benthic food content, and type of substrate, among other parameters (Ekdale, 1985). In the studied hyperpycnal system, each subenvironment is characterized by a distinctive trace fossil association, with different ichnodiversity levels and bioturbation intensities (Figure 3). The maximum diversity and abundance of trace fossils are observed in the distal lobe deposits, whereas in settings proximal to that, the trace fossil assemblages are less diverse and the bioturbation intensities are low. Particularly, in the analyzed successions, the distribution of trace fossils seems to be heavily affected by the fluctuating hydraulic energy and, hence, fluctuating sedimentation rates, high input of organic material, substrate consistency, and probably fluctuating salinity.

Proximal settings were most likely affected by high-energy currents and rapid deposition, which are evidenced by erosive structures and the absence of trace fossils (e.g., in the coarser grained channel facies; see Figure 3, sections 1 and 3); or by the development of low-diversity ichnoassemblages, with biogenic structures sporadically occurring, and dominance of dwelling or equilibrium burrows such as *Diplocraterion*, and escape structures produced by tellinid bivalves. Furthermore, observations in recent environments indicate that some deposit-feeding organisms, such as the

FIGURE 7. Trace fossil assemblages in depositional slope settings. (A) Bedding-plane view of the marginal channel facies, with abundant protrusive *Diplocraterion* (*Di*) (Roca Champion). (B) Concentrically laminated burrows (clb) in the upper channel fill deposits (Cabo Santa Inés). (C) Plane view of *Scolicia* (*Sc*) within the reworked wavy bedded heteroliths in levee deposits (Roca Champion). (D) Levee deposits completely reworked by *Phycosiphon* (*Ph*) (Punta Basílica). (E) Abundant grazing traces (e.g., *Gordia* [*Go*] and *Helminthoidichnites* [*He*]) in outer levee deposits (Cabo Viamonte). (F) Concentrically laminated burrows (clb) in outer levee deposits (Cabo Viamonte). (G) Isolated levels with graphoglyptids can be observed in outer levee deposits. In the photograph, one specimen of *Paleodictyon* (*Pa*) is casted at the base of a sandstone layer (Cabo Viamonte). (H) Sandstone deposit displaying escape structures (fu) of tellinid bivalves (Punta Basílica).

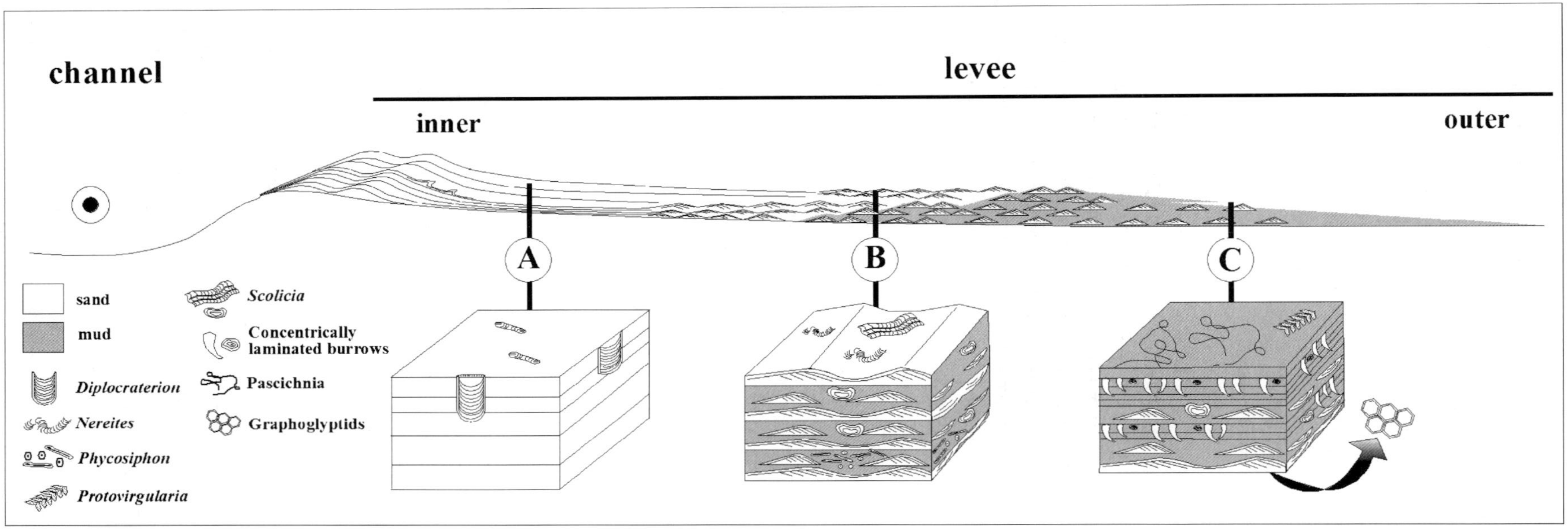

FIGURE 8. Typical distribution of the trace fossils in the levee deposits associated with major channels. In the inner levees (A), more proximal to the channels, a monogeneric suite of protrusive *Diplocraterion* is developed. In an intermediate setting (B), the heterolithic facies of the levee deposits are reworked by *Scolicia, Nereites,* and *Phycosiphon.* More distally, the fine-grained heteroliths of the outer levee (C) are reworked mainly by grazing and small locomotion traces. In some intervals, concentrically laminated burrows are also present. In addition, a low diverse suite of graphoglyptids is preserved sporadically at the base of the very fine-grained sandstone beds.

Scolicia trace makers, seem to prefer settings where sedimentation rates are relatively high (Fu and Werner, 2000). Consequently, the great abundance of *Scolicia* in these successions may reveal not only a high benthic food content within the sediment (see later), but also relatively high sedimentation rates.

Evidence of salinity changes is suggested by the presence of syneresis cracks, which are mostly found in heteroliths from proximal and depositional slope settings. This indicates that salinity reduction caused by the freshwater input during the hyperpycnal discharge probably had a major effect in the proximal than in the distal settings of the studied system. The development of a low-diversity trace fossil suite in these heteroliths (e.g., the suites composed of concentrically laminated burrows), with structures reflecting opportunistic strategies, also supports this conclusion. In the distal setting, episodes with freshwater influx may have existed; however, normal marine salinity must have been rapidly reestablished as is indicated by intervals completely reworked by producers of trace fossils such as *Scolicia*, *Nereites*, and *Phycosiphon*, which are known to indicate fully marine conditions.

Accumulation of flocculated mud seems to have been common in these hyperpycnal settings, mainly in the heterolithic levee facies and in the lower and upper parts of the distal lobe deposits. Clay flocculation is a frequent mechanism in environments with fluctuating salinity (e.g., in deltaic environments; Pryor, 1975), and it commonly tends to produce turbid water and soupground (sediment with a high water content [soupy substrate]) conditions (MacEachern et al., 2005). Soupy substrates and watery soft grounds are generally colonized by mobile deposit feeders and some small penetrative structures (generating mottled bioturbated fabrics) and excluding large endobenthic deposit-feeding and suspension-feeding organisms (Bromley, 1996; MacEachern et al., 2005). In addition, the organisms that inhabit the soupy substrates commonly use mucous to construct their structures (MacEachern et al., 2005). In the heterolithic facies of the levee deposits, levels with biodeformational structures are frequent. These trace fossils resemble the "mantle and swirl" traces of Lobza and Schieber (1999) generated in soupy muds. In the upper part of the distal lobe deposits, *Tasselia* trace makers most likely colonized and inhabited the substrate during soupground conditions. This matches the observations in modern environments where *Tasselia* is commonly found in soupy substrates (Wetzel and Bromley, 1996). These authors emphasized that the sediment lining typical of this ichnotaxon provides a stable vertical orientation of the tube within the soupground. Considering the producers of *Tasselia* as capable of tolerating soupground conditions also helps to understand previous observations of the tiering structure of the upper part of the lobe deposits, in which the *Tasselia* producer was interpreted as the first organism to colonize the muddy layers, being crosscut by the other trace fossils (Ponce et al., 2007a). Consequently, the presence of *Tasselia* would reflect colonization during the soupground conditions, whereas the rest of the trace fossils that crosscut this ichnotaxon represent postcompactional structures that have been formed during soft ground conditions.

Observations from these outcrops indicate that food supply was not a limiting factor in these environments. The large amount of organic material in this hyperpycnite succession, preferentially in the fine-grained heteroliths, may have favored the presence of deposit-feeding organisms such as *Scolicia*, *Nereites*, and *Phycosiphon* producers. Particularly, Wetzel and Uchman (1997) considered that *Scolicia* and *Nereites* display an "opportunistic population strategy in response to high trophic levels" (Wetzel and Uchman, 1997, p. 155). Moreover, the rare occurrence of graphoglyptids points to a high benthic food content (see below). Our conclusion that the benthic food was high in these hyperpycnal successions is also supported by observations of modern hyperpycnal deposits in the South China Sea (Wetzel, 2008). This author observed that these hyperpycnal layers are rapidly colonized by the fauna, and he interpreted that most likely, this occurs because the organic-matter content in the hyperpycnal flows increases during the traveling of the flows along the shelf and the continental slope. Wetzel (2007, 2008) also associated the presence of *Scolicia* to environments affected by a high delivery of organic matter because of fertilization from land and the presence of the *Nereites missouriensis* ichnofabrics to environments with high seasonal organic-matter input. Interestingly, most of the *Nereites* specimens studied in the Miocene successions correspond to the ichnospecies *Nereites missouriensis*. The occurrence of *Tasselia* and *Phycosiphon* also suggest well-oxygenated sediments rich in benthic food (Wetzel and Bromley, 1996; Wetzel and Uchman, 2001).

Oxygen concentration seems to have been mostly normal in most of the analyzed successions. The occurrence of trace fossils commonly associated with environments with lowered oxygenation (e.g., *Chondrites*) has been observed in isolated levels in the distal

A
1 m
B
1 m
C
he
Scd
Sl
Sm
Scd
Sl
Sm
he
1 m
D
Sm
he
Sl
he
40 cm
E
M
Sl
Scr
Sl
Sr
Sm
5 cm

setting of the studied hyperpycnal system. Furthermore, most of the trace fossils do not have an open connection to the sea floor (e.g., *Nereites*, *Phycosiphon*) and the burrows do not show size reductions, therefore indicating that the sediment was most likely fully oxygenated at depth. The sustained character of the hyperpycnal flow probably improved the oxygen content of these sediments. However, deposition of organic-rich flocculated muds also may lead to temporally anoxic or dysoxic conditions (MacEachern et al., 2005). In addition, recent studies in the South China Sea indicate that the seasonal organic input produces short periods of anoxia in the deep-marine sediments (Wetzel, 2008) because of the increment in oxygen consumption. Although these anoxic periods are well recorded in modern deep-sea environments, and most likely also occurred in the studied deposits, their recognition in the fossil successions is extremely difficult because of the short period involved. Furthermore, the presence of relatively deep bioturbators, such as the *Scolicia* trace makers, produced a complete obliteration of the sediment, therefore erasing any anoxic signals.

Paleoecology of the Ichnoassemblages

In environments with long-term stable conditions, such as the deep sea, the communities are generally highly diverse, with organisms showing well-specialized behaviors, and having typically equilibrium (K-selected) strategies. However, in environments where physical parameters are highly variable and conditions are less predictable, the communities tend to be less diverse and the organisms are commonly adapted to a wide range of physical stresses, displaying mainly opportunistic strategies (r-selected) (Sanders, 1968; Wetzel, 1991). Based on these observations, Sanders and Hessler (1969) proposed the "stability-time hypothesis," which considered that in evolutionary time, the physical stability of the deep sea led to the development of highly diverse communities. Recent observations show that some statements of this hypothesis may not be totally accurate (e.g., the environmental conditions fluctuate more than previously supposed; Gage and Tyler, 1991). However, this hypothesis is useful to understand the evolutionary trends of the deep-sea communities on a general scale (Seilacher, 1974; Wetzel, 1991).

Analyzing the paleoecology of the recognized trace fossil suites has revealed that most of these ichnoassemblages reflect colonization by opportunistic organisms (r-selected), forming communities poorly to moderately diverse, patchy distribution and commonly poorly developed tiering structure. This is particularly evident not only for the suites developed in proximal settings (such as the *Diplocraterion* association or the concentrically laminated burrow suite), but also for some of the ichnoassemblages found in the distal heteroliths, certainly the *Scolicia-Nereites* association (found in sandy heteroliths), and probably the *Phycosiphon-Nereites* association (present in muddy heteroliths and in massive mudstones).

In this context, one point that deserves further analysis is the presence of imported fauna in these hyperpycnal systems. Recently, Wetzel (2008) analyzed hyperpycnal deposits from the South China Sea, indicating that areas with hyperpycnal discharges are influenced by downwelling currents that transport organisms (larvae and adults) and benthic food to the deep settings. In these hyperpycnal deposits, he noted the presence of ichnotaxa commonly found in shallower settings (such as *Diplocraterion* and structures resembling *Lapispira*), and he concluded that most likely, the trace makers of these structures were transported. The great abundance of short protrusive *Diplocraterion* specimens in the studied Miocene successions can also be interpreted as produced by transported fauna. When analyzing a particular bed with *Diplocraterion*, almost all the specimens show similar dimensions (probably revealing size selection during transport of the organisms). Moreover, none of the studied specimens show downward widening, representing growth (Bromley, 1996); this fact may indicate that the survival period of these transported organisms was probably short. Similarly, Ponce et al. (2007b) interpreted the escape structures of bivalves found in

FIGURE 9. Main sedimentary facies and architectural elements in distal setting. (A, B) View of the hyperpycnal lobe complexes, showing the typical coarsening-thickening and fining-thinning-upward arrangements (indicated by white arrows in B) (Cabo Ladrillero). (C, D) Detailed view of the lobe deposits. These successions display a vertical transition of sedimentary structures that mainly reflects fluctuations in flow velocity (he = heteroliths; Scd = climbing-dune sandstone; Sm = massive sandstone; Sl = laminated sandstone) in hyperpycnal lobes (south of Cabo Santa Inés). (E) Detail showing the vertical transition between fine-grained massive sandstones (Sm), rippled sandstones (Sr), parallel-laminated sandstones (Sl), climbing-ripple sandstones (Scr), and massive mudstones (M). Note the carbonaceous detritus in the rippled sandstone facies (north of Roca Champion).

A
Di
Di
B
Ph
Sc
Sc
C
Ph
Sc
D
Ne
Ph
APA
E
Ne
Ta
F
Par
Gy
G
Pa
H
he
Pa
2 cm

the sandstone levels at the base of the depositional slopes (Figure 7H) to have been produced by tellinids imported from shallower settings.

In the studied succession, the equilibrium (K-selected) community is typified by the graphoglyptid suite found in outer levee deposits of the depositional slope setting and in distal lobe deposits. These shallow, highly patterned trace fossils represent equilibrium strategies, developed under stable ecological parameters, generally in a food-restricted deep-marine environment (Seilacher, 1977; Ekdale et al., 1984; Ekdale, 1985; Miller, 1991; but see also Fürsich et al., 2007; Olivero et al., 2008). In the studied deposits, intervals with graphoglyptids are uncommon, and the observed suites show a lower diversity when compared with other Miocene deep-marine ichnofaunas (e.g., Uchman, 1995). This could be a consequence of an area-size effect because well-exposed sandstone soles are not abundant in the analyzed outcrops. In addition, as these predepositional trace fossils are characterized by a very patchy distribution, we should expect an increase in ichnodiversity with a larger sampled area (Gage and Tyler, 1991). However, although the previous aspects should be considered when analyzing these ichnoassemblages, the low abundance and diversity of graphoglyptids may suggest that the appropriate conditions for the development and preservation of these equilibrium communities (e.g., oligotrophic environments, absence or low sediment deposition, slightly erosive currents, etc.) were achieved only occasionally in distal lobe settings or in outer levee settings of these hyperpycnal systems.

CONCLUSIONS

1) On the basis of the sedimentologic analysis of the lower Miocene succession of the Austral Basin, three main depositional settings have been recognized: a proximal setting consisting of hyperpycnal channels with lateral accretion and levees, an intermediate setting in the depositional slope area consisting of aggradational hyperpycnal channels and levees, and a distal setting with development of lobe deposits.
2) The ichnologic analysis revealed that each of these depositional settings is characterized by distinctive trace fossil associations.
   a) A monogeneric suite of *Diplocraterion* occurs in massive and parallel-laminated sandstones at channel margins and in the inner levee deposits in proximal and intermediate settings.
   b) Concentrically laminated burrows of detritus- or deposit-feeding organisms are invariably present in fine-grained heterolithic facies with lenticular and wavy bedding, particularly in outer levee deposits of proximal and slope settings. In addition, some intervals of the outer levee also present small locomotion and grazing traces (e.g., *Protovirgularia*, *Gordia*).
   c) Intervals completely reworked by producers of *Scolicia* and *Nereites* are recognized in the fairly sandy heteroliths of the distal lobe setting. This suite is also recognized in isolated fine-grained heterolithic levee intervals of the slope setting.
   d) *Phycosiphon* and *Nereites* producers commonly reworked the muddier heteroliths and massive mudstones deposited in outer levee and distal lobe settings. In addition, large specimens of *Tasselia* occur in distinctive levels of the fine-grained heteroliths and massive mudstones in the uppermost part of the lobe successions.
   e) Low-diversity graphoglyptid associations are found only locally at the base of heterolithic sandstone beds, in outer levee deposits in depositional slope setting and distal lobe deposits.
3) Environmental stresses that affected these settings are evidenced by ichnologic and sedimentologic observations.
   a) A high organic content is indicated by the high degree of bioturbation, large trace fossils, dominance of deposit-feeding traces, and rare graphoglyptids.

FIGURE 10. Trace fossils in hyperpycnal lobe deposits. (A) *Diplocraterion* (*Di*) specimens occurring in the parallel-laminated sandstones of the hyperpycnal lobe deposits (Cabo Ladrillero). (B) Cross section of the sandier heterolithic intervals showing *Scolicia* (*Sc*), reworking preferentially the sand-mud interfaces, and *Phycosiphon* (*Ph*) (north of Roca Champion). (C) Bedding-plane view of a heterolithic interval showing *Phycosiphon* (*Ph*) and large *Scolicia* (*Sc*) specimens (Cabo Ladrillero). (D) Muddier intervals completely reworked, with abundant *Nereites* (*Ne*) and *Phycosiphon* (*Ph*) (Cabo Ladrillero). (E) Oblique section of a massive mudstone interval with *Tasselia* (*Ta*) and *Nereites* (*Ne*) (Cabo Ladrillero). (F) Muddy intervals completely reworked. *Paradictyodora* (*Par*), *Gyrophyllites* (*Gy*) (Cabo Ladrillero). (G, H) Specimens of *Paleodictyon* (*Pa*) preserved at the base of heterolithic sandstone intervals, which were deposited during the acceleration stage of the hyperpycnal flow (north of Roca Champion). he = heteroliths.

b) Episodically high sedimentation rates are revealed by the presence of escape trace fossils in the more energetic, fairly sandy intervals. In addition, trace fossils, such as *Scolicia* and *Tasselia*, in finer grained layers, suggest also relatively high sedimentation rates.
c) Syneresis cracks and low-diversity trace fossil assemblages in proximal and intermediate settings suggest salinity fluctuations. In distal settings, sporadic distribution of trace fossils may also reflect periods with lowered salinity. However, normal marine salinity conditions may have been rapidly reestablished, as is evidenced by the occurrence of ichnogenera commonly associated with fully marine environments, such as *Scolicia*.

4) The dominance of poorly to moderately diverse trace fossil associations related to opportunistic producers, in combination with the scarcity of graphoglyptids, indicates that in these hyperpycnal flow-affected environments, physical parameters have been highly variable. Moreover, the presence of ichnotaxa commonly occurring in shallow-water settings most likely implies colonization by imported organisms (e.g., *Diplocraterion*, bivalve escape structures). Equilibrium communities with graphoglyptid ichnotaxa only persisted in more distal settings, far from the direct influence of these fluvially derived flows.

ACKNOWLEDGMENTS

We thank Eduardo Olivero for the critical review of a previous version of the manuscript and for the opportunity to study the Miocene deposits of Tierra del Fuego. We thank Luis Buatois, Gabriela Mángano, Renata Netto, and Carlos Zavala for providing useful suggestions. Daniel R. Martinioni, María I. López-Cabrera, Daniel Aureliano, José Contreras, and Patricio Desjardins helped during field work. We thank Alvar Sobral for preparation of polished samples. Andreas Wetzel is particularly thanked for his thorough review. Luis Gliubich (Estancia La Fueguina), Juan Carlos Apolinare (Estancia San Pablo), Renán Bilbao (Estancia Inés), and Simon Goodall (Estancia Viamonte) granted access through their properties to the outcrops. This study was part of a CONICET postdoctoral scholarship to Carmona and a CONICET doctoral scholarship to Ponce. Financial support for this study was provided by CONICET (PIP-5100), Agencia Nacional de Promoción Científica y Tecnológica (ANPCyT) (PICT-840), and Total Austral.

REFERENCES CITED

Bromley, R. G., 1996, Trace fossils: Biology, taphonomy and applications: London, UK, Chapman and Hall, 361 p.

Carmona, N. B., J. J. Ponce, M. I. López-Cabrera, and E. B. Olivero, 2006, Distribución y diversidad de trazas fósiles en hiperpicnitas: Implicancias etológicas y comparación con patrones de trazas fósiles en turbiditas clásicas, 9° Congreso Argentino de Paleontología y Bioestratigrafía, Córdoba, Argentina, Abstract Book, v. 1, p. 279.

Carmona, N. B., J. J. Ponce, E. B. Olivero, M. I. López-Cabrera, and D. R. Martinioni, 2008, Ichnology of Miocene hyperpycnites in the foreland Austral Basin, Tierra del Fuego, Argentina, Ichnia 2008, Second International Congress on Ichnology, Krakow, Poland, Abstract Book, v. 1, p. 26.

Ekdale, A. A., 1985, Paleoecology of the marine endobenthos: Palaeogeography, Palaeoclimatology, Palaeoecology, v. 50, p. 63–81, doi:10.1016/S0031-0182(85)80006-7.

Ekdale, A. A., R. G. Bromley, and S. G. Pemberton, 1984, Ichnology: Trace fossils in sedimentology and stratigraphy: SEPM Short Course 15, 317 p.

Fu, S., and F. Werner, 2000, Distribution, ecology and taphonomy of the organism trace, *Scolicia*, in northeast Atlantic deep-sea sediments: Palaeogeography, Palaeoclimatology, Palaeoecology, v. 156, p. 289–300, doi:10.1016/S0031-0182(99)00146-7.

Fürsich, F. T., J. Taheri, and M. Wilmsen, 2007, New occurrences of the trace fossil *Paleodictyon* in shallow marine environments: Examples from the Triassic-Jurassic of Iran: Palaios, v. 22, p. 408–416, doi:10.2110/palo.2006.p06-041r.

Gage, J. D., and P. A. Tyler, 1991, Deep-sea biology: A natural history of organisms at the deep-sea floor: Cambridge, UK, Cambridge University Press, 504 p.

Galeazzi, J. S., 1998, Structural and stratigraphic evolution of the western Malvinas Basin, Argentina: AAPG Bulletin, v. 82, p. 596–636.

Goldring, R., 1996, The sedimentological significance of concentrically laminated burrows from lower Cretaceous Ca-bentonites, Oxfordshire: Geological Society (London), v. 153, p. 255–263.

Lobza, V., and J. Schieber, 1999, Biogenic structures produced by worms in soupy, soft muds: Observations from the Chattanooga shale (Upper Devonian) and experiments: Journal of Sedimentary Research, v. 69, p. 1041–1049.

López-Cabrera, M. I., E. B. Olivero, N. B. Carmona, and J. J. Ponce, 2008, Cenozoic trace fossils of the *Cruziana*, *Zoophycos* and *Nereites* ichnofacies from the Fuegian Andes, Argentina: Ameghiniana, v. 45, p. 377–392.

MacEachern, J. A., K. L. Bann, S. G. Pemberton, and M. K. Gingras, 2005, The ichnofacies paradigm: High-resolution paleoenvironmental interpretation of the rock record, *in* J. A. MacEachern, K. L. Bann, M. K. Gingras, and S. G. Pemberton, eds., Applied ichnology: SEPM Short Course Notes, v. 52, p. 27–64.

Malumián, N., 1999, La sedimentación en la Patagonia extrandina, *in* R. Caminos, ed., Geología Argentina,

Servicio Geológico Minero Argentino, Anales, v. 29, p. 557–612.

Malumián, N., and E. B. Olivero, 2006, El Grupo Cabo Domingo, Tierra del Fuego: Bioestratigrafía, paleoambientes y acontecimientos del Eoceno-Mioceno marino: Revista de la Asociación Geológica Argentina, v. 61, p. 139–160.

Miller III, W., 1991, Paleoecology of graphoglyptids: Ichnos, v. 1, p. 305–312.

Mulder, T., and J. P. M. Syvitski, 1995, Turbidity currents generated at river mouths during exceptional discharges to the world oceans: Journal of Geology, v. 103, p. 285–299, doi:10.1086/629747.

Mulder, T., J. P. M. Syvitski, S. Migeon, J.-C. Faugères, and B. Savoye, 2003, Marine hyperpycnal flows: Initiation, behavior and related deposits: A review: Marine and Petroleum Geology, v. 20, p. 861–882, doi:10.1016/j.marpetgeo.2003.01.003.

Mutti, E., G. Davoli, R. Tinterri, and C. Zavala, 1996, The importance of ancient fluviodeltaic systems dominated by catastrophic flooding in tectonically active basins: Memorie di Scienze Geologiche, Universita du Padova, v. 48, p. 233–291.

Mutti, E., R. Tinterri, G. Benevelli, D. di Biase, and G. Cavanna, 2003, Deltaic, mixed and turbidite sedimentation of ancient foreland basins: Marine and Petroleum Geology, v. 20, p. 733–755, doi:10.1016/j.marpetgeo.2003.09.001.

Olivero, E. B., and N. Malumián, 2002, Upper Cretaceous-Cenozoic clastic wedges from the Austral-Malvinas foreland basins, Tierra del Fuego, Argentina: Eustatic and tectonic controls (Addendum), European Meeting on the Palaeontology and Stratigraphy of Latin America (EMPSLA), N°3, Septiembre 19–20, 2002, Universidad Paul Sabatier, Toulouse, France, p. 6–9.

Olivero, E. B., and N. Malumián, 2008, Mesozoic Cenozoic stratigraphy of the Fuegian Andes, Argentina: Geologica Acta, v. 6, p. 5–18.

Olivero, E. B., M. I. López-Cabrera, and N. Malumián, 2008, Eocene graphoglyptids from high-energy, organic-rich, and bioturbated turbidites, Fuegian Andes, Argentina, Ichnia 2008, Poland, Abstract Book, v. 1, p. 95.

Plink-Björklund, P., and R. Steel, 2004, Initiation of turbidity currents: Outcrop evidence for Eocene hyperpycnal flow turbidites: Sedimentary Geology, v. 165, p. 29–52, doi:10.1016/j.sedgeo.2003.10.013.

Ponce, J. J., 2009, Análisis estratigráfico secuencial del Cenozoico de la Cordillera Fueguina, Tierra del Fuego, Argentina, Ph.D. dissertation, Universidad Nacional del Sur, Bahía Blanca, Argentina, 245 p.

Ponce, J. J., E. B. Olivero, and D. R. Martinioni, 2004, *Phymatoderma*-bearing turbidites (Oligocene, Tierra del Fuego): Ichnologic implications for discrimination of sustained and episodic gravity flow deposits, Ichnia (abs.): First International Congress on Ichnology, Trelew, Argentina, p. 67–68.

Ponce, J. J., E. B. Olivero, and D. R. Martinioni, 2005, Estratigrafía y facies sedimentarias del Oligoceno–Mioceno medio? De la Cuenca Austral de Tierra del Fuego, XVI Congreso Geológico Argentino, La Plata, CD-ROM, Ponencia 468, 2 p.

Ponce, J. J., E. B. Olivero, D. R. Martinioni, and M. I. López-Cabrera, 2007a, Sustained and episodic gravity flow deposits and related bioturbation patterns in Paleogene turbidites (Tierra del Fuego, Argentina), *in* R. G. Bromley, L. A. Buatois, M. G. Mángano, J. F. Genise, and R. N. Melchor, eds., Organism-sediment interactions: A multifaceted ichnology: SEPM Special Publication 88, p. 253–266.

Ponce, J. J., N. B. Carmona, and D. R. Martinioni, 2007b, Trazas de escape generadas por bivalvos retransportados en hiperpicnitas del Mioceno de Tierra del Fuego, V Reunión Argentina de Icnología y III Reunión de Icnología del MERCOSUR, Ushuaia, Argentina, Abstract book, v. 1, p. 30.

Ponce, J. J., E. B. Olivero, and D. R. Martinioni, 2008a, Hyperpycnal-flow deposits in Oligocene–Miocene clinothems of the Austral Basin, Tierra del Fuego, Argentina, *in* J. J. Ponce and E. B. Olivero, conveners, Sediment transfer from shelf to deepwater: Revisiting the delivery mechanisms: Conference Proceedings, AAPG Hedberg Research Conference, Ushuaia, Patagonia, Argentina, Fieldguide Book, p. 25–48.

Ponce, J. J., E. B. Olivero, and D. R. Martinioni, 2008b, Upper Oligocene–Miocene clinoforms of the foreland Austral Basin of Tierra del Fuego, Argentina: Stratigraphy, depositional sequences and architecture of the foredeep deposits: Journal of South American Earth Sciences, v. 26, p. 36–54, doi:10.1016/j.jsames.2007.12.001.

Ponce, J. J., E. B. Olivero, and D. R. Martinioni, 2008c, Deep-marine hyperpycnal channel-levee complexes in the Miocene of Tierra del Fuego, Argentina: Architectural elements and facies associations, *in* J. J. Ponce and E. B. Olivero, conveners, Sediment transfer from shelf to deepwater: Revisiting the delivery mechanisms: Conference Proceedings, AAPG Hedberg Conference, Ushuaia, Patagonia, Argentina: http://searchanddiscovery.com/abstracts/html/2008/hedberg_argentina/extended/ponce/ponce.htm (accessed July 18, 2011).

Pryor, W. A., 1975, Biogenic sedimentation and alteration of argillaceous sediments in shallow marine environments: Geological Society of America Bulletin, v. 86, p. 1244–1254, doi:10.1130/0016-7606(1975)86<1244:BSAAOA>2.0.CO;2.

Sanders, H. L., 1968, Marine benthic diversity: A compar ative study: American Naturalist, v. 102, p. 243–282, doi:10.1086/282541.

Sanders, H. L., and R. R. Hessler, 1969, Ecology of the deep-sea benthos: Science, v. 163, p. 1419–1424, doi:10.1126/science.163.3874.1419.

Seilacher, A., 1974, Flysch trace fossils: Evolution of behavioural diversity in the deep-sea: Neues Jahrbuch fur Geologie und Paleontologie, Monatshefte, v. 1974 p. 233–245.

Seilacher, A., 1977, Pattern analysis of *Paleodictyon* and related trace fossils, *in* T. P. Crimes and J. C. Harper, eds., Trace fossils 2: Geological Journal Special Issue 9, p. 289–334.

Taylor, A. M., and R. Goldring, 1993, Description and analysis of bioturbation and ichnofabric: Geological Society (London), v. 150, p. 141–148, doi:10.1144/gsjgs.150.1.0141.

Uchman, A., 1995, Taxonomy and palaeoecology of flysch trace fossils: The Marnoso-Arenacea Formation and associated facies (Miocene, Northern Apennines, Italy): Beringeria, v. 15, p. 1–115.

Uchman, A., and R. Steel, 2007, Trace fossils in shelf-margin clinoforms: An example from the central Basin of Spitsbergen, V Reunión Argentina de Icnología y III Reunión de Icnología del MERCOSUR, Ushuaia, Argentina, Abstract Book, v. 1, p. 34.

Wetzel, A., 1991, Ecologic interpretation of deep-sea trace fossil communities: Palaeogeography, Palaeoclimatology, Palaeoecology, v. 85, p. 47–69, doi:10.1016/0031-0182(91)90025-M.

Wetzel, A., 2007, Ichnology of slope deposits along an active continental margin: An example from the South China Sea (abs.): International Ichnofabric Workshop, Calgary, Canada, p. 71.

Wetzel, A., 2008, Recent bioturbation in the deep South China Sea: A uniformitarian ichnologic approach: Palaios, v. 23, p. 601–615, doi:10.2110/palo.2007.p07-096r.

Wetzel, A., and R. G. Bromley, 1996, The ichnotaxon *Tasselia ordamensis* and its junior synonym *Caudichnus annulatus*: Journal of Paleontology, v. 70, p. 523–526.

Wetzel, A., and A. Uchman, 1997, Ichnology of deep-sea fan overbank deposits of the Ganei Slates (Eocene, Switzerland): A classical flysch trace fossil locality studied first by Oswald Heer: Ichnos, v. 5, p. 139–162, doi:10.1080/10420949709386413.

Wetzel, A., and A. Uchman, 2001, Sequential colonization of muddy turbidites in the Eocene Belovea Formation, Carpathians, Poland: Palaeogeography, Palaeoclimatology, Palaeoecology, v. 168, p. 171–186, doi:10.1016/S0031-0182(00)00254-6.

Zavala, C., J. J. Ponce, M. Arcuri, D. Drittanti, H. Freije, and M. Asensio, 2006, Ancient lacustrine hyperpycnites: A depositional model from a case study in the Rayoso Formation (Cretaceous) of west-central Argentina: Journal of Sedimentary Research, v. 76, p. 41–59, doi:10.2110/jsr.2006.12.

10

Gamero Diaz, H., C. Contreras, N. Lewis, R. Welsh, and C. Zavala, 2011, Evidence of shelfal hyperpycnal deposition of Pliocene sandstones in the Oilbird field, southeast coast, Trinidad: Impact on reservoir distribution, *in* R. M. Slatt and C. Zavala, eds., Sediment transfer from shelf to deep water—Revisiting the delivery system: AAPG Studies in Geology 61, p. 193–214.

Evidence of Shelfal Hyperpycnal Deposition of Pliocene Sandstones in the Oilbird Field, Southeast Coast, Trinidad: Impact on Reservoir Distribution

Helena Gamero Diaz and Carmen Contreras
Schlumberger Technology Corp., Addison, Texas, U.S.A.

Neil Lewis and Robert Welsh
EOG Resources Trinidad Limited, Port of Spain, Trinidad and Tobago

Carlos Zavala
GCS Argentina, Universidad Nacional del Sur, Bahía Blanca, Argentina

ABSTRACT

The Pliocene B4 sandstone is an important gas-bearing reservoir in the Oilbird field, Columbus Basin, southeast Trinidad. The B4 sandstone is dominated by fine-grained, massive, and parallel-laminated sandstones interbedded with thinly laminated siltstone and very fine-grained sandstones deposited in a shelfal setting. A detailed sedimentologic study of the B4 reservoir was performed by integrating 60 ft (18.3 m) of core, 3770 ft (1149 m) of borehole image data, open-hole logs, mud logs, and biostratigraphic data. A total of 12 sedimentary facies were described and interpreted on the basis of sedimentation processes using a genetically oriented facies analysis approach. Six facies associations were identified based on the gamma ray (GR) pattern and the vertical facies association. Core data show mostly massive sandstones with very little to no bioturbation. These massive sandstones commonly alternate with intervals having diffuse lamination, which appear as a recurrent feature within the massive bodies. The transitional recurrence of massive and parallel-laminated sandstones indicates velocity fluctuating and sustained turbulent flows. Sedimentologic evidences suggest that the origin of the B4 sandstone could be related to the paleo-Orinoco River-related

Copyright ©2011 by The American Association of Petroleum Geologists.
DOI:10.1306/13271357St613443

product of turbidity (hyperpycnal) outflows that extended into the Columbus Basin. The associated occurrence of plant debris favors the interpretation of a direct fluvial supply by rivers in flood stage (hyperpycnal systems).

The lateral correlation of facies associations throughout seven wells allowed the identification of six depositional units, named from base to top, A, B, C, D, E, and F. Facies maps for the B4 reservoir have been developed and were constrained by paleocurrent data extracted from image data. The B4 reservoir sand shows a progradational pattern reflecting the infill of a fault-controlled depocenter. Paleoflow data indicate axial transport roughly parallel to the main fault system, indicating that faults were active and controlled the accommodation space. The proposed hyperpycnal depositional model will introduce substantial changes for the prediction of the geometry and position of sandstone accumulations. The hyperpycnal model predicts the occurrence of sand accumulations in the lower parts of the paleolandscape, whereas the higher parts of the basin (margins) are characterized by fine-grained sediments. Thus, this new depositional model could represent a drastic shift in the prospectivity guide for new exploration plays.

INTRODUCTION

Located southeast of Trinidad, the Columbus Basin is one of the most proliferous petroleum basins of the Caribbean region. Main hydrocarbon reservoirs are highly porous and very thick fine-grained sandstones that accumulated during the Late Miocene and Pliocene. These deposits have been related to wave-dominated littoral deltas associated with an ancient big river located to the west, known as the paleo-Orinoco. The interpretation of sandstone deposits as wave-dominated delta-front accumulations led to several problems in explaining the occurrence of these sandstones:

1) If these sandstones belong to littoral deltas, relative sea level changes should be an important factor allowing the high-frequency progradation-retrogradation of delta-front and associated delta-plain deposits for more than 200 km (656,167 ft). However, delta-plain facies have never been recognized in cores nor in outcrops.
2) Although wave-dominated deltas are characterized by shoreface deposits with abundant wave ripples, these sandstones commonly lack any evidence of wave-reworking processes.
3) Facies analysis of core and borehole image data from Pliocene deposits of the Oilbird field allowed revision of the origin and significance of sandstone reservoirs. Main sandstone facies show depositional features that suggest an accumulation from long-lived sediment gravity flows. In this chapter, an alternative interpretation for the origin of sandstone deposits is proposed, related to an accumulation by hyperpycnal flow deposits in a marine shelfal setting.

GEOLOGIC FRAMEWORK

The Columbus Basin is located offshore southeast Trinidad in a zone of intense strike-slip deformation as a result of the oblique collision between the Caribbean and South American plates (Leonard, 1983; Robertson and Burke, 1989; Babb and Mann, 1999; Wood, 2000; Gibson et al., in press). This basin has an overall triangular shape and is limited by the Darien Ridge to the north, the Atlantic continental shelf to the east, the onshore southern Trinidad to the west, and the Amacuro shelf of Venezuela to the south (Figure 1). From a regional scale, the Columbus Basin is situated on the South American plate just south of its boundary with the Caribbean plate (Gibson et al., in press), which is moving eastward relative to South America and the Atlantic basin. The east–west-trending Caribbean–South American margin is characterized by a late Oligocene to middle Miocene transpressional thrust belt (Serrania del Interior Oriental of Venezuela and Central and Northern Ranges of Trinidad). This thrust belt continues offshore into the eastern shelf of Trinidad north of Point Radix into Darien Ridge and joining the Barbados accretionary prism beyond the modern shelf edge. The Miocene thrust front effectively defines the northern boundary of the Columbus Basin. The foreland basin associated with this deformed belt consists of the Eastern Venezuela Basin, which evolves eastward into the Columbus Basin. Foreland basin subsidence in eastern Venezuela

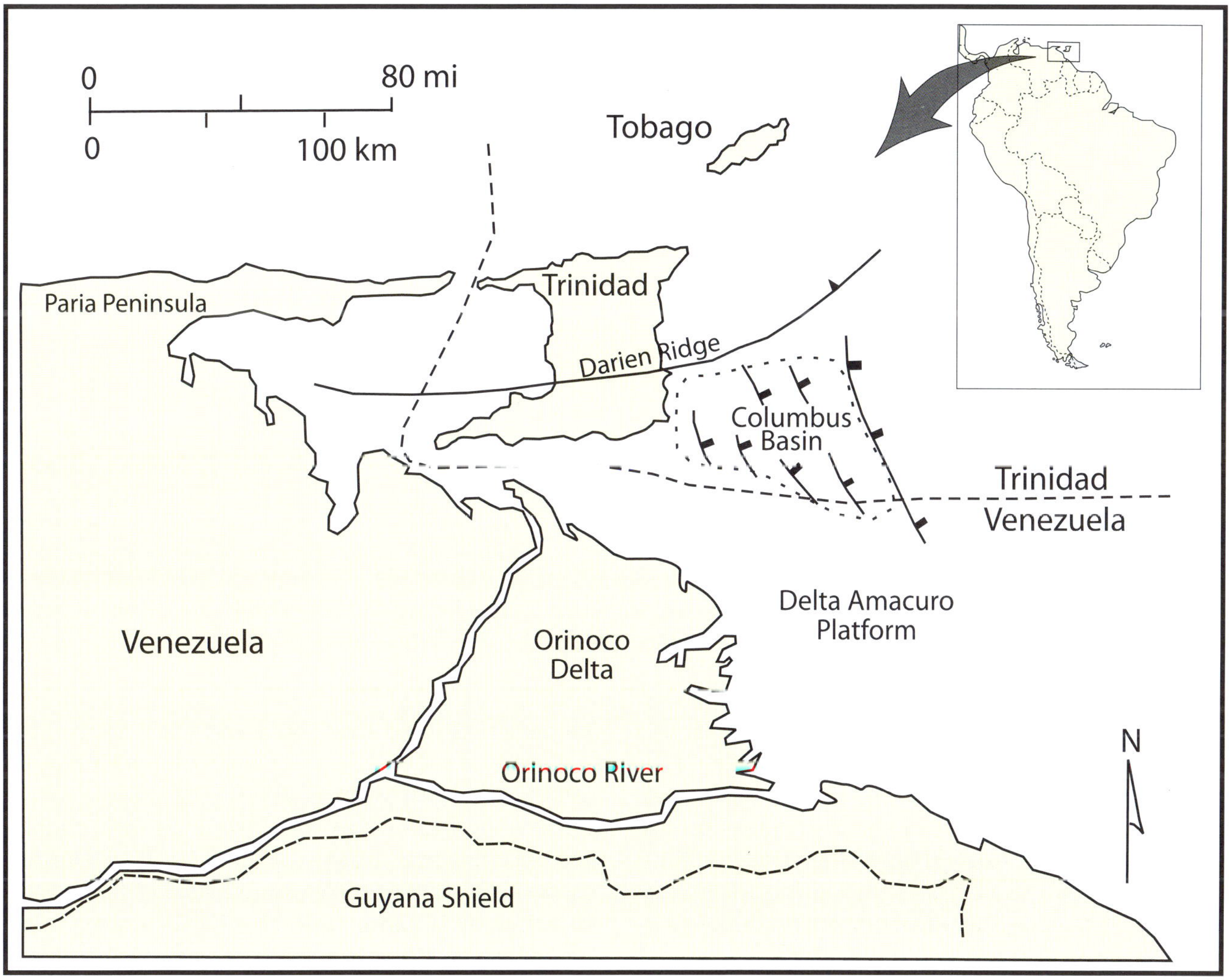

FIGURE 1. Location map of the Columbus Basin (modified after Wood, 2000).

and Trinidad began in the late Oligocene to early Miocene (Pindell et al., 1998), culminating with the emplacement of the Serrania del Interior Oriental–Central Range thrust belt in early Miocene. This thrusting came to an end by the late middle Miocene when the Caribbean–South American relative plate motion changed from transpression to transtension. Throughout the Late Cretaceous, widespread deposition of the petroleum-rich source rock occurred, and it is this source rock that is responsible for the generation of almost all the petroleum in Venezuela and Trinidad.

During the Pliocene–Pleistocene, the basin evolved into a thin-skinned, pull-apart basin that captured shallow-water sedimentary deposits of the prograding paleo-Orinoco River delta (Gibson et al., in press). Pliocene–Pleistocene structures within the basin include large-magnitude northwest–southeast-oriented down-to-the-northeast extensional faults, northeast–southwest-trending contractional folds, and strike-slip fault systems, all of which are regionally detached near the top of the underlying north-dipping passive margin succession. At the end of the middle Miocene, the north-draining paleo-Orinoco River was diverted eastward along the axis of the foreland basin by the growing Serrania–Central Range thrust belt (Diaz de Gamero, 1996). Since the late Miocene, the Columbus Basin area has been the depocenter for Orinoco-derived sediments exceeding 9150 m (30,000 ft) in thickness (reported sedimentation rates are between 5 and 10 m [16 and 33 ft]/1000 yr).

The Oilbird Field

The Oilbird field is located within the South East Coast Consortium Block some 40 km (25 mi) southeast of Trinidad's mainland (Figure 2). It sits within the Columbus Basin, which has produced significant

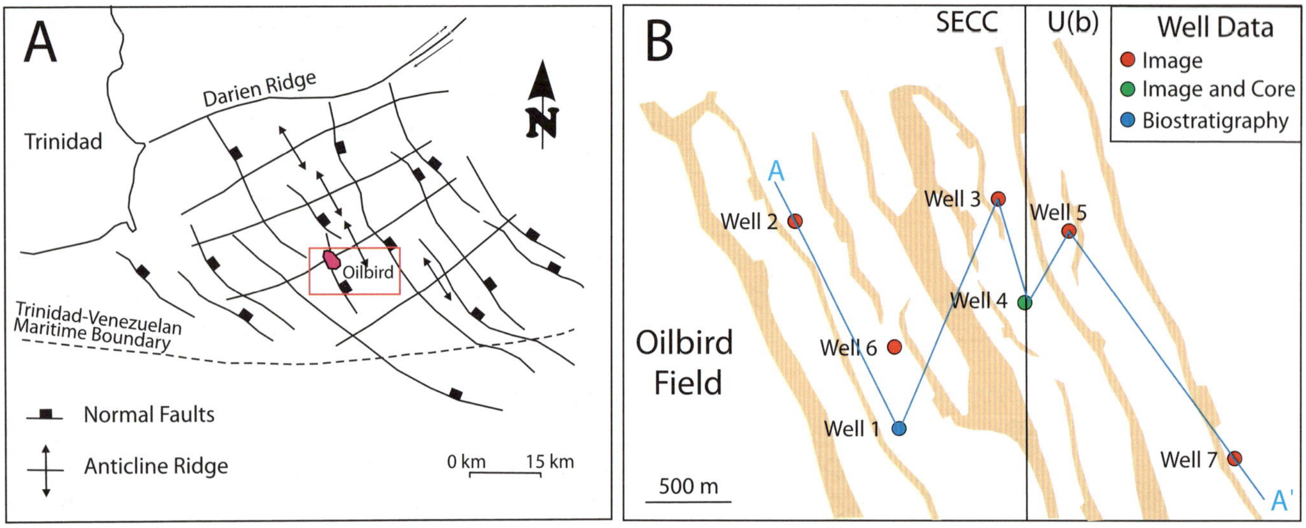

FIGURE 2. **(A) Main structural elements of the Columbus Basin (modified after Wood, 2000). The Columbus Basin is characterized by two sets of structural elements: (1) northwest–southeast-oriented down-to-the-northeast extensional faults and (2) northeast–southwest-trending contractional folds (Leonard 1983; Wood, 2000). (B) Map of the Oilbird field showing the location of the wells used in this study, and regional cross section AA′, shown in Figure 11. SECC = South East Coast Consortium; U(b) = concession block held and operated by EOG Resources Trinidad Limited.**

quantities of oil and gas since the first offshore oil discoveries during the early 1970s. The Columbus Basin is characterized by a prolific Cretaceous source rock with main associated reservoirs located in the Tertiary sand-rich shelfal stratigraphic section. The structural evolution of the basin has created abundant hydrocarbon traps, along with an efficient petroleum migration system. Proven and probable gas reserves at the end of 2008 stood at approximately 23.8 tcf.

The Oilbird field was discovered in 1977 by Texaco with the drilling of the Oilbird-1 well (well 1, Figure 2B). At that time, natural gas had no economic value, so appraisal drilling only occurred in 2001–2002. Platform installation occurred in late 2006, followed by the drilling of five additional development wells.

The B4 Pliocene sandstone is an important gas-condensate-bearing reservoir in the Oilbird field (Figure 3). Figure 3 shows the main reservoir sandstones in the Oilbird field. The B4 sandstone is dominated by fine-grained, massive, and parallel-laminated sandstones interbedded with thinly laminated siltstone and very fine-grained sandstones, which have been interpreted as deposited by a shelf-edge delta system (Bowman, 2003; Bowman and Johnson, 2006).

The analysis of core data shows a dominance of very well-sorted fine-grained sandstones, which are massive or show planar lamination with abundant plant debris. These sandstones are composed of very thick packages internally showing gradual recurrences of facies, thus indicating an origin related to sustained, long-lived, and fluctuating turbulent flows (Zavala et al., 2006c; Lamb et al., 2008; Jackson et al., 2009; Lamb and Mohrig, 2009). The association of massive and laminated sandstones is now considered diagnostic features originated by the collapse of sand-rich turbulent suspensions under different rates of sediment fallout (Arnott and Hand, 1989; Kneller and Branney, 1995; Sumner et al., 2008).

In this chapter, we propose an alternative interpretation for the origin of B4 sandstone deposits as related to an accumulation by hyperpycnal flows in marine shelfal areas. This interpretation is based on the analysis of facies and sedimentary features in the B4 Oilbird sandstones supported by core, log, and borehole electrical image data. High-resolution borehole images provide paleocurrent orientations that complemented the depositional model built from the facies analysis of cores.

Previous Work

Pliocene–Pleistocene sandstones of the Columbus Basin have been previously interpreted as accumulated by wave-dominated littoral deltas, characterized by extensive prograding shoreface/shelfal deposits (Wood, 2000; Wood and Roberts, 2001) and by shelf-edge deltas (Bowman, 2003; Bowman and Johnson, 2006). Wood (2000) established that the Columbus Basin was filled throughout the Pliocene and Pleistocene by more than 12 km (7.4 mi) of clastic sediments supplied by the paleo-Orinoco delta system.

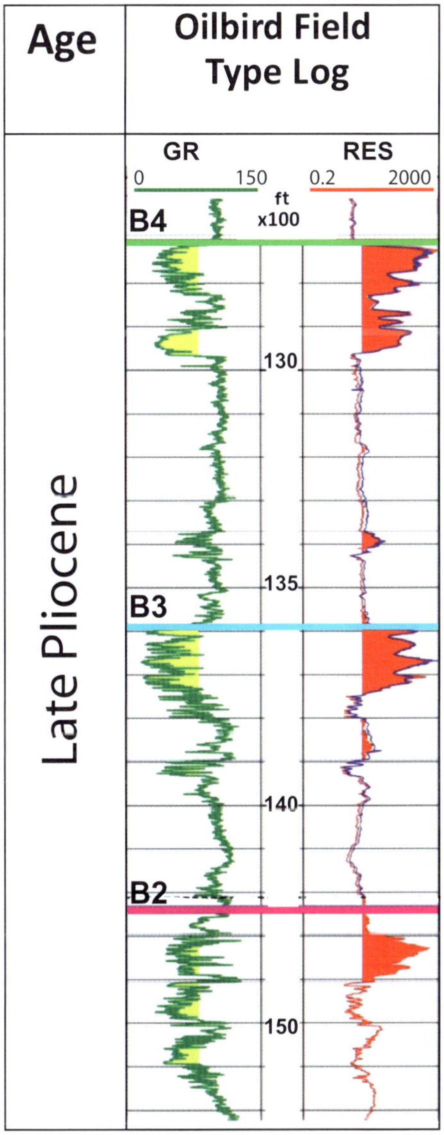

FIGURE 3. Type log of the Late Pliocene Oilbird sandstone reservoirs. The B4 sandstone is a gas-condensate reservoir. GR = gamma ray in API units; RES = resistivity in ohm m units.

The model proposed by Wood (2000) shows fluvial-dominated deltas feeding, wave-dominated strand-plain shoreline systems. Ultimately, strand-plain/shoreface deposits form elongated sand bodies oriented parallel to the coastline.

From a sedimentologic standpoint, the shoreface is a nearshore area located between the lowest tide and the wave base. Typical shoreface deposits are characterized by sandstone and mudstone layers composed of rhythmic heterolithic successions showing abundant wave-related sedimentary structures. The sand/shale ratio allowed the distinction of upper, middle, and lower shoreface deposits. The passage between the lower shoreface and the offshore transition is marked by the fair-weather wave base, which separates the high-energy coastal zone from the more internal areas characterized by mudstone accumulations commonly punctuated by storm deposits. Consequently, classic shoreface deposits are composed of sandy layers with flaser and truncated wave ripples reflecting wave action, a dominant process in a shoreface (Figure 4). Sedimentary structures, such as micro-hummocky cross-stratification, are also common and are considered diagnostic features of storm deposition in the shoreface (Walker, 1996).

The interpretation of a storm- or wave-dominated shoreface environment in the Columbus Basin was supported by faunal assemblages that reflect shallowing-upward conditions, as well as the existence of clean, massive, and parallel-laminated sandstones interbedded with levels showing hummocky cross-stratification and *Ophiomorpha* burrows. Nevertheless, no diagnostic structures of wave activity as wave-truncated ripples were described.

Bowman and Johnson (2006) proposed that the Pliocene-Pleistocene sandstones in the Columbus Basin were deposited by shelf-edge delta systems. They concluded that under conditions of high sediment supply and accommodation space, deltas in the shelf edge may generate thick delta slope and delta-front successions, more than 340 m (1115 ft) thick. The shelf-edge deltas were deposited within the structurally generated accommodation space between syndepositional northeast- and southwest-dipping normal faults (Gibson et al., in press). Wood (2000) and Bowman and Johnson (2006) established the Sao Francisco delta, offshore Brazil, and the Orinoco delta, Venezuela, respectively, as modern analogs for the Pliocene-Pleistocene deposits in the Columbus Basin. Although littoral delta systems are commonly preserved in the geologic record, the associated sandstone deposits in the delta front are commonly relatively thin (<5 m [16 ft]) because of the strong energetic control and the limited available accommodation space in coastal areas. Nevertheless, recent studies support the common occurrence of very thick clastic successions associated with submarine deltas (in the sense of Bates, 1953). These submarine deltas compose a new type and poorly known fluvial-related depositional system termed hyperpycnal systems (Zavala et al., 2006c; Zavala et al., 2011) capable of transporting huge volumes of sediment into the basin and accumulating very thick deposits having diagnostic structures commonly confused with shoreface deposits.

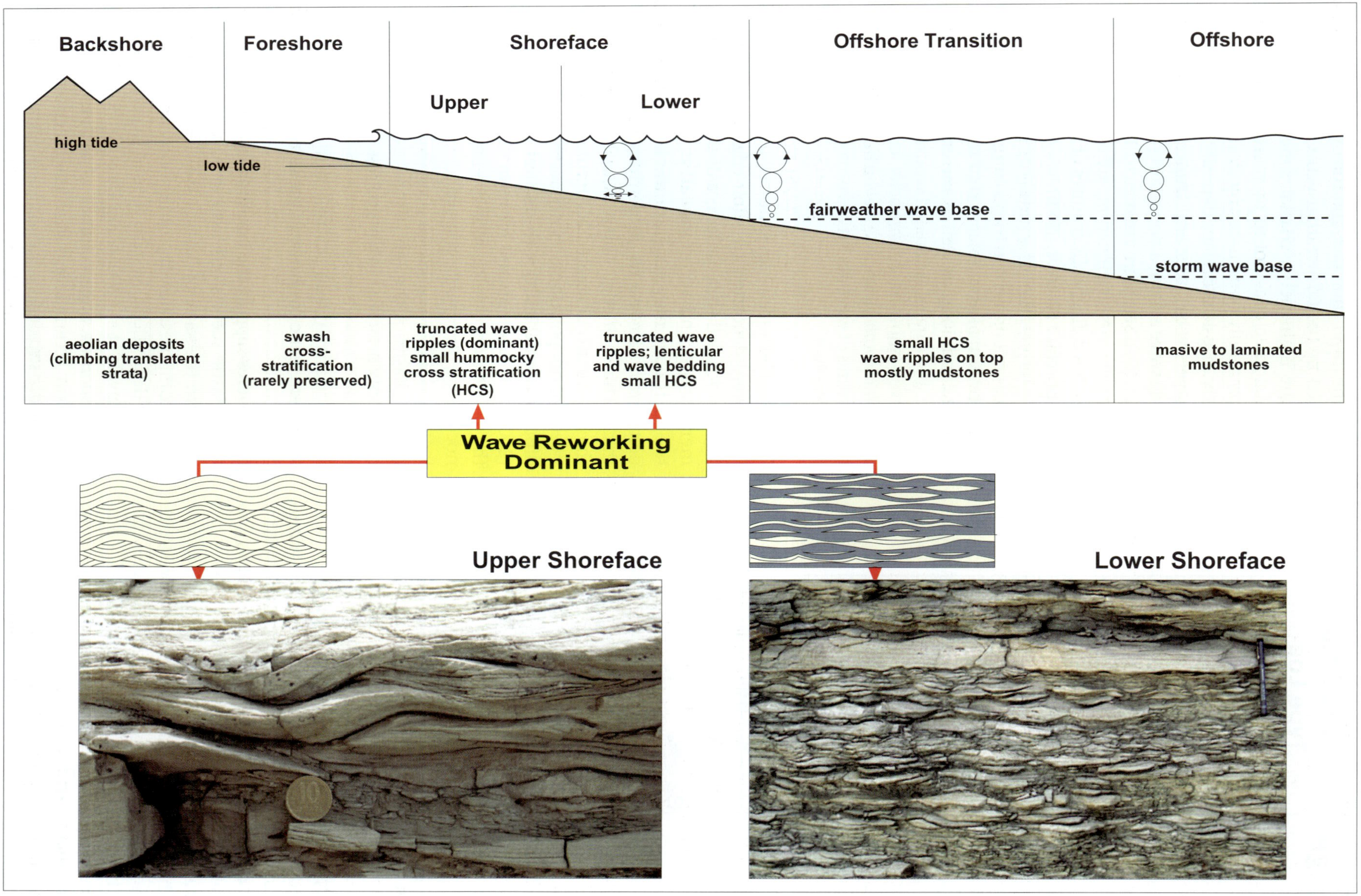

FIGURE 4. Depositional characteristics of shoreface deposits (C. Zavala, 2007, personal communication). Main sedimentary structures diagnostic of lower-to-upper shoreface environments. The dominant process in the shoreface is wave action, with truncated wave ripples being the most characteristic sedimentary structure in this environment.

Hyperpycnal Systems: New Depositional Models

A hyperpycnal system is the subaqueous extension of the fluvial system (Zavala et al., 2006c). It originates when excess in density causes a sediment-laden fluvial discharge to plunge below the receiving water body (Bates, 1953; Mulder and Alexander, 2001), mainly during flood conditions. During these periods, huge volumes of clastics are commonly transferred into the basin by a subaqueous extension of the fluvial system (in the sense of Schumm, 1981). The resulting deposits are termed "hyperpycnites" (Mulder et al., 2003) and constitute a land-generated turbulent flow deposit commonly associated with long-lived and quasi-steady flow discharges. These deposits have poorly known facies and facies associations. Although ancient hyperpycnal deposits are poorly documented in the literature, some advances have appeared in recent publications, referring to both ancient and modern successions (Mulder and Alexander, 2001; Mulder et al., 2003; Nakajima, 2006; Zavala et al., 2006c; Zavala et al., 2011). Recently, hyperpycnal flows are receiving more attention as efficient mechanisms for transporting enormous quantities of sands into a basin (Mutti et al., 1996; Mulder and Alexander, 2001; Mutti et al., 2003; Gamero et al., 2005; Pattison, 2005; Zavala et al., 2006c; Gamero et al., 2007, 2008; Zavala, 2008). Growing evidence emphasizes a close relationship between turbidity currents and rivers in flood in both modern (Nakajima, 2006) and ancient depositional settings (Mutti et al., 1996, 2003; Zavala et al., 2006b, c). Mutti et al. (2003) discussed the importance of flood-dominated fluvial-deltaic systems and their genetic relationship with hyperpycnal depositional models in shallow- and deep-water marine basins.

The occurrence of hyperpycnal flows in marine basins was initially received with skepticism (Nakajima, 2006), but it has been recognized at least since the published works by Mulder et al. and Mutti et al. in 2003 (Plink-Bjorklund and Steel, 2004; Gamero et al., 2005; Nakajima, 2006; Zavala et al., 2006c, 2007; Gamero et al., 2007). The origin of the long meandering submarine channels stemming from deltas, now widely attributed to sustained turbidity currents, was heavily debated a quarter of a century ago (Pickering and Hiscott, 1985).

Extensive work in outcrops and core studies (Mulder and Alexander, 2001; Gamero et al., 2005; Zavala et al., 2006a, b, c, 2007; Zavala, 2008) show that sedimentary facies related to quasi-steady hyperpycnal flows are very distinctive and could be easily differentiated from surgelike (resedimented) turbidite flows. A hyperpycnal flow is a land-derived relatively slow-moving turbulent flow carrying interstitial fresh water basinward (Hesse et al., 2004; Mansurbeg et al., 2006). According to Zavala et al. (2006c), the movement of hyperpycnal flows does not require steep slopes because the energy of the flow is maintained by the river discharge during a flood, and the distance reached by these flows will be more dependent on the duration of the flood event.

Zavala (2008) and Zavala et al. (2011) introduced a genetic facies tract for the analysis of sustained turbulent (noninertial) hyperpycnal flow deposits (Figure 5). This hyperpycnal facies tract is composed of three main genetically related facies groups termed B, S, and L, corresponding to bed-load, suspended-load, and lofting transport processes, respectively. This genetic facies model was followed in this study for the analysis of the Oilbird field.

DATA AND METHODOLOGY

A sedimentologic interpretation was conducted in the Pliocene B4 reservoir sand at the Oilbird field (Figure 2). The general stratigraphy of the main reservoir sandstones in the Oilbird field are shown in the type log of Figure 3. Integration of multiple data including open-hole logs, core descriptions, borehole images, mud logs, and biostratigraphic interpretations was used to develop a new depositional model. Table 1 summarizes the data available including the interval of image data for each well. A total of 1149 m (3770 ft) of image data together with 18 m (60 ft) of core data were used to develop the sedimentologic model of the B4 sand. Borehole electrical imagery, integrated with selective conventional core data, constitutes a very powerful and effective technique for use in detailed reservoir characterization and further field development.

The methodology used in this study included the following:

1) Detailed facies analysis of 18 m (60 ft) of core data included the identification of facies and facies associations. The different facies were described and interpreted on the basis of the sedimentary processes involved during their origin. Facies associations were interpreted based on the genetically oriented facies tract (Zavala, 2008; Zavala et al., 2011) proposed for the analysis of hyperpycnal systems.
2) Recognition of core-based facies on the images was based on a detailed core-to-image- and -to-log calibration. As a result, an image-based facies

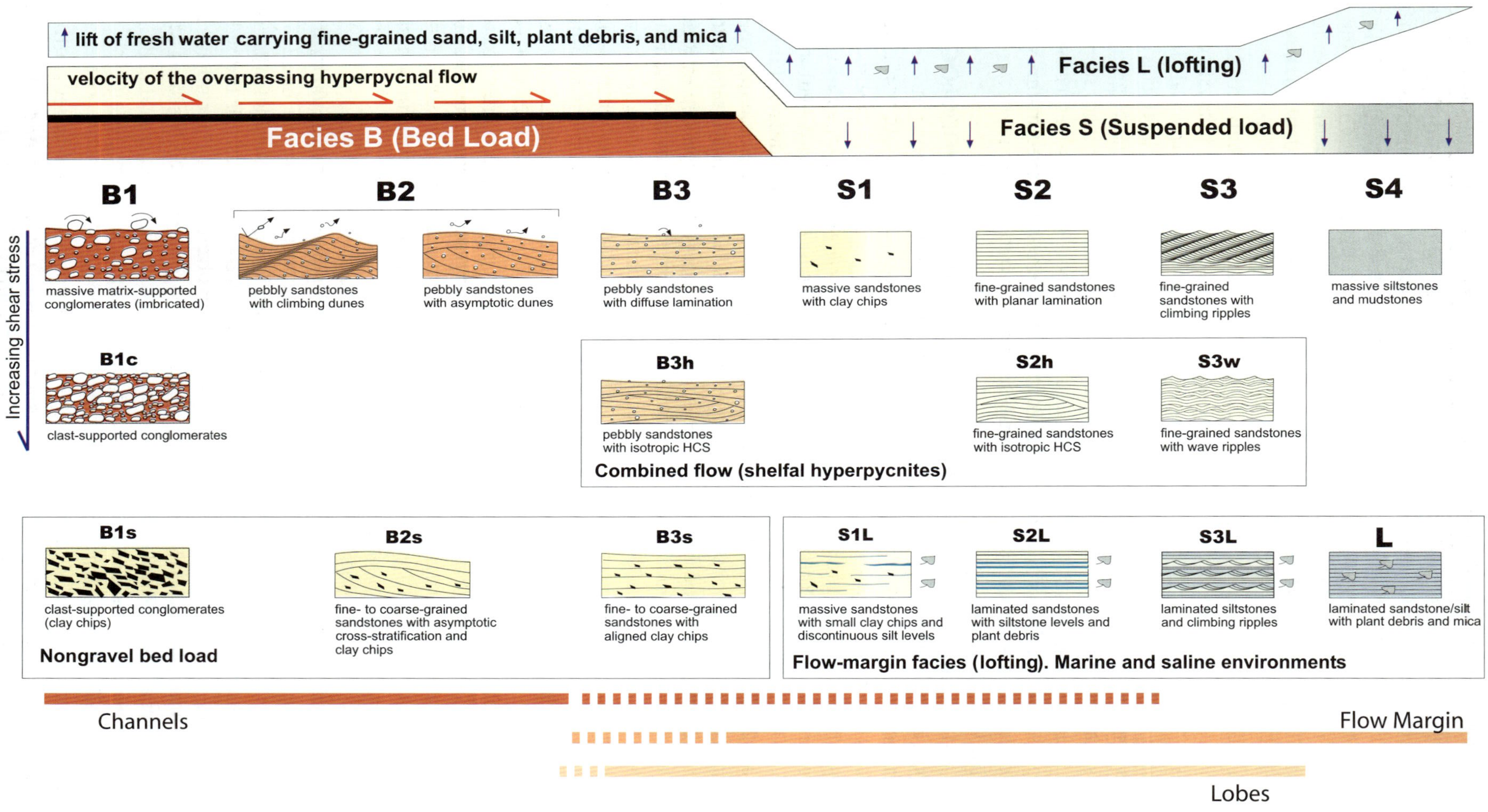

FIGURE 5. Genetic facies tract for the interpretation of hyperpycnal deposits (Zavala, 2008; Zavala et al., 2011). Type B facies relates to bed-load processes at the base of an overpassing long-lived turbulent flow. Type S facies originate from the sedimentation of sand-size suspended materials carried in the turbulent flow. Type L facies relates to fallout from lofting plumes. These facies are common in hyperpycnal flows entering a marine basin. HCS = hummocky cross-stratification.

Table 1. Summary of the available data for each well.

Well Name	*Open-Hole Logs*	*Mud Logs*	*Core Descriptions*	*Biostratigraphic Report*	*Borehole Images*	*Total Image (ft)*
Well 1	x			x		
Well 2	x	x			x	970
Well 3	x	x				
Well 4	x	x	x		x	853
Well 5	x	x			x	1740
Well 6	x	x				
Well 7	x	x			x	207

classification was used to interpret the remaining 1149 m (3770 ft) of image data.

3) Static and dynamic normalized borehole images were interpreted in conjunction with open-hole logs, such as gamma ray (GR), neutron-porosity, bulk density, and resistivity. Manual dip data were extracted directly from 1149 m (3770 ft) of borehole image data. The sedimentologic descriptions from borehole image data were integrated with mud-log results and complemented by biostratigraphic analysis from one well.
4) Borehole image data were used to determine sediment transport directions. It is important to highlight that borehole images provide paleocurrent determinations and sand orientations that support the sedimentologic interpretation obtained from the analysis of conventional nonoriented cores.
5) Stratigraphic cross sections were built to analyze the lateral variations of different facies associations. This requires recognition of facies associations from well logs and image data.
6) Depositional units were recognized based on the vertical stacking of facies associations and the GR log patterns.
7) Generation of facies maps. Facies maps were useful to analyze lateral facies changes through time for the different reservoir units throughout the field.

RESULTS

Core Data Analysis

The B4 reservoir sand comprises a variety of facies dominated by fine- to very fine-grained sandstones. In this study, a total of 12 facies were recognized on the basis of their texture, sedimentary structures, and origin. These facies are (1) clay chip conglomerate (B1s), (2) medium- to fine-grained sandstone with diffuse bedding (B3s), (3) medium- to fine-grained sandstone with asymptotic cross-stratification (B2s), (4) fine-grained massive sandstone (S1), (5) fine-grained sandstone with parallel laminations (S2), (6) fine-grained sandstone with climbing ripples (S3), (7) fine- to very fine-grained sandstone with hummocky cross-lamination (S2h), (8) fine- to very fine-grained sandstone with oscillatory ripples (S3w), (9) very fine-grained sandstones and siltstones with parallel laminations and abundant plant debris (S2L), (10) fine- to very fine-grained sandstones and siltstones with starved ripples (S3L), (11) thinly laminated siltstones with plant remains (L), and (12) very fine-grained sandstones and siltstones with water-escape structures and convolute bedding (Pd). Figure 6 summarizes the description and the origin of each recognized facies based on the main depositional processes involved during their accumulation. Main facies of the B4 reservoir were identified from an 18-m (60-ft)-long core acquired in well 4 (Figure 2). Core study was performed by core descriptions and continuous core photographs. Unfortunately, the core interval covers only 18 m (60 ft) of the lower part of the B4 reservoir sandstone. This core shows that the lower part of the B4 reservoir sandstone in well 4 is dominated by rhythmic, fine- to very fine-grained, parallel-laminated and massive sandstones interbedded with plant-derived carbonaceous material and thinly laminated sandstones and siltstones. As shown in Figure 6, the origin of these facies could be related to long-lived and land-derived hyperpycnal flows (Zavala et al., 2011).

Borehole Image Analysis

Core interpretation was calibrated with the borehole image data acquired in well 4, which allowed the identification of 12 image-based facies (Figure 6).

Facies	Core Photo	Image	Lithology	Sedimentary Structures	Origin
B1s		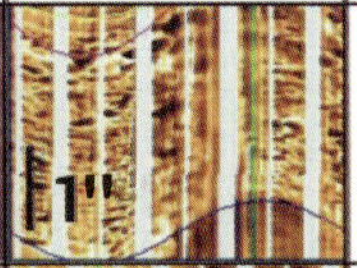	Clay chip conglomerates within a medium-grained sandstone matrix.	Clasts alignment	Bed-load transport from an overpassing quasi-steady turbulent flow, over a consolidated substrate.
B3s			Medium- to fine-grained sandstones.	Crude plane parallel stratification	Bed-load transport from an overpassing quasi-steady turbulent flow, over a consolidated substrate.
B2s	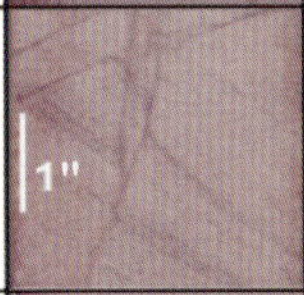		Medium- to fine-grained sandstones.	Asymptotic cross-stratification	Bed-load transport from an overpassing quasi-steady turbulent flow, over a consolidated substrate.
S1			Fine-grained sandstones with floating clay chips.	Massive	Traction plus fallout from an overpassing quasi-steady turbulent flow.
S2			Fine-grained sandstones.	Parallel lamination	Traction plus fallout from an overpassing quasi-steady turbulent flow.
S3			Fine- to very fine-grained sandstones.	Climbing ripples	Traction plus fallout from an overpassing quasi-steady turbulent flow.
S2h	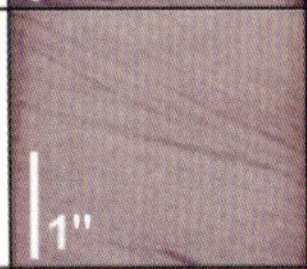		Fine- to very fine-grained sandstones.	Hummocky cross-stratification	Combined unidirectional flow and an oscillatory component flow.
S3w			Fine- to very fine-grained sandstones.	Oscillatory ripples	Traction plus fallout from an overpassing quasi-steady turbulent flow combined with an oscillatory component.
S2L		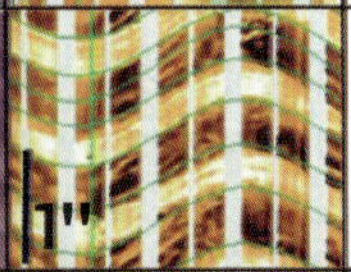	Fine- to very fine-grained sandstones interbedded with sandstones.	Parallel laminations interbedded with laminated siltstones and plant remains	Traction plus fallout from an overpassing quasi-steady turbulent flow plus precipitation from lofting plumes.
S3L			Fine- to very fine-grained sandstones interbedded with sandstones.	Ripple laminations interbedded with laminated siltstones and plant remains	Traction plus fallout from an overpassing quasi-steady turbulent flow plus precipitation from lofting plumes.
L			Light gray siltstones with plant remains.	Thinly laminated	Precipitation from lofting plumes.
Pd			Very fine-grained sandstones and laminated siltstones.	Water-escape structures and convolute bedding.	Water-escape structures is related to high sedimentation rates formed shortly after deposition.

FIGURE 6. Core- and image-derived facies scheme recognized from well data in the Oilbird field. It summarizes the description and the origin of each recognized facies based on the main depositional processes involved during their accumulation. HCS = hummocky cross-stratification.

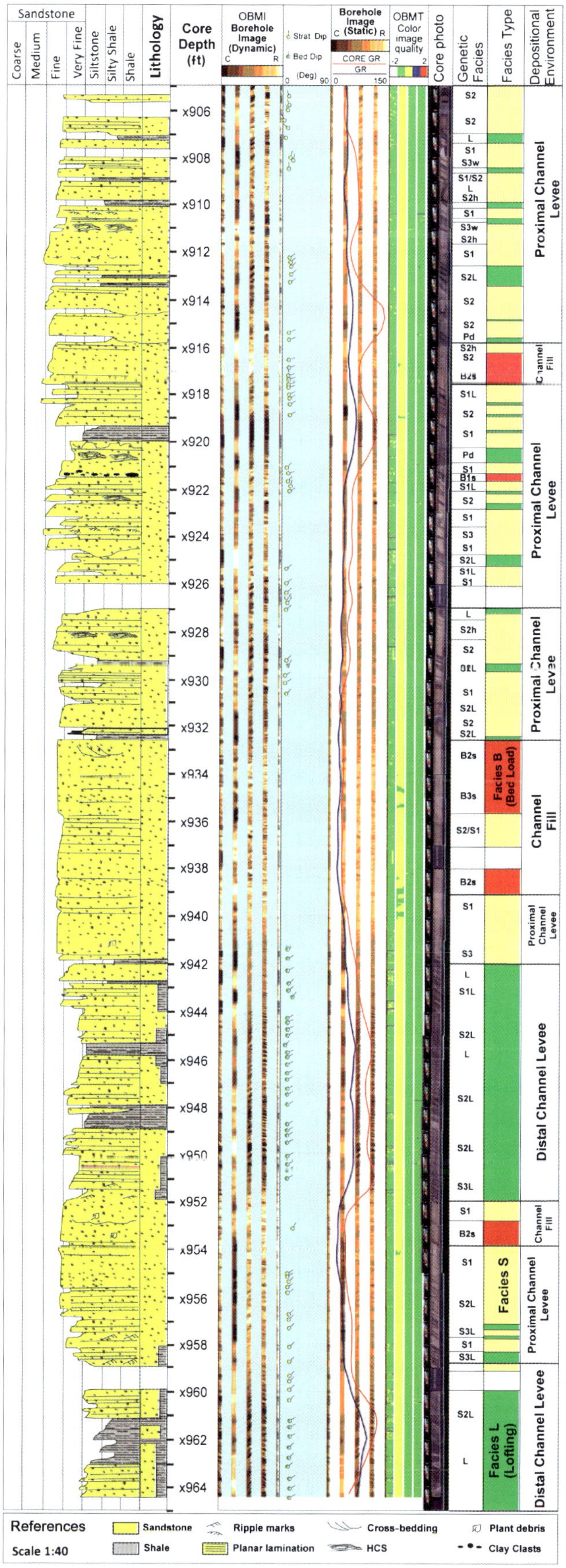

Figure 6 shows the image-based facies together with the core-based facies. These image-based facies were recognized on the basis of their texture, sedimentary structures, and origin. Sedimentary structures, such as cross-laminations, appear as sine-shaped bands of varying electrical contrast on dynamic normalized images. Dynamic normalization or image enhancement processing is used to bring out the details and is useful for detailed sedimentologic interactive analysis on the Image Examiner Workstation (Contreras and Gamero, 2000). The interpreter interactively fits a sine curve over each sedimentary structure to generate its orientation and assigns it to a specific sedimentologic category, such as planar lamination, cross-lamination, hummocky cross-stratification, and cross-bedding. The texture information in the wells with no core data was obtained from mud-log data analysis. Paleocurrent data were obtained by fitting a sine curve over the cross-bedding and cross-lamination structures providing orientation data. A structural dip removal was applied to determine the true paleocurrent orientation before being structurally altered.

This calibration permitted the identification of 12 image-based facies (Figure 6). Figure 7 shows the core-to-image calibration and the facies interpretation performed in well 4. The core-to-image calibration method is very important because it allows the extrapolation of the facies identified from the core throughout the 1149 m (3770 ft) of borehole image data without any additional core information. A new image-based facies scheme derived from image and core data was generated for the B4 sandstones in the Oilbird field based on the origin of the sedimentary structures. The interpreter's experience is important because of the possibility of recognizing a wide

FIGURE 7. Sedimentologic data sheet based on core and image calibration of well 4. The first track on the left shows a lithologic column generated from core photographs and core descriptions. The second track indicates the core depth, the third track the dynamic normalized borehole image, the fourth track the manual dip interpretation, and the fifth track the static normalized image. The sixth track shows the quality control of the image, in two colors: green indicates good quality; yellow indicates intermediate quality. The seventh track depicts the core photographs. The eighth, ninth, and tenth tracks represent the sedimentary facies, color-coded facies type, and depositional environment, respectively. Descriptions of the genetic facies are found in Figures 5 and 6. C = conductive; R = resistive; HCS = hummocky cross-stratification; GR = gamma ray; Strat. = stratigraphic; Deg = degree; OBMI = Oil-Base MicroImager (OBMI™) manufactured by Schlumberger.

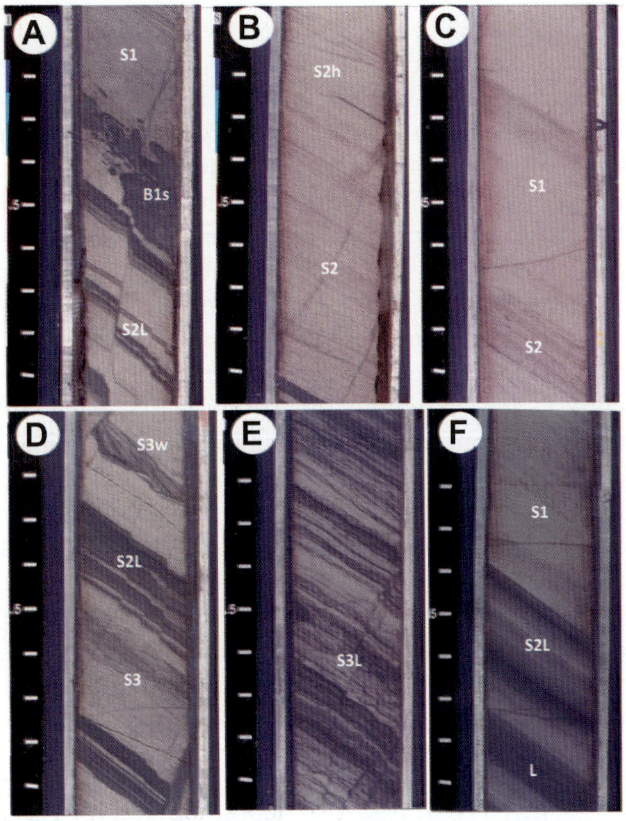

Figure 8. Core photographs showing examples of the main sedimentary facies recognized in the analyzed core. (A) Clay clast conglomerate within a massive to laminated, fine-grained sandstone, interbedded with organic-rich silty laminations. (B) Parallel-laminated fine-grained sandstones with hummocky cross-stratification (HCS) toward the top. (C) Massive fine-grained sandstones commonly associated with parallel-laminated fine-grained sandstones. (D) Oscillatory ripples (toward the top of the photograph) interbedded with thinly laminated carbonaceous rich siltstones and very fine-grained sandstones. Bioturbation is very scarce. (E) Heterolithic facies consisting of thinly laminated carbonaceous rich siltstones and very fine-grained sandstones interbedded with parallel- and ripple-laminated very fine-grained sandstones. (F) Thinly laminated carbonaceous-rich siltstones and very fine-grained sandstones overlain by massive fine-grained sandstones. No bioturbation is observed.

range of features in outcrop and core, therefore providing an accurate interpretation of borehole images (Contreras and Gamero, 2000).

High-resolution borehole electrical imagery is regarded as a powerful tool to improve sedimentologic models that, calibrated with core, can be useful for identifying facies and providing paleocurrent data that can support the sedimentologic interpretation derived from conventional core analysis (Gamero et al., 2000).

Biostratigraphic Data

A biostratigraphic analysis was performed in the B4 reservoir sandstone on the discovery well 1 (Figure 2B). This study reported inner-to-outer neritic (shelfal) conditions during deposition of the sandstones.

Depositional Model

A new depositional model was proposed for the B4 sandstone in the Oilbird field on the basis of integrating core data, borehole images, and cuttings descriptions. Sedimentologic evidences suggest that the origin of the B4 sandstones could be related to fluvial-derived long-lived hyperpycnal flow deposited in a shelfal setting. During the late Pliocene, the paleo-Orinoco River discharges generated hyperpycnal flows that extended into the Columbus Basin. As previously mentioned, the different facies types recognized in the B4 sandstone were related to sustained turbulent hyperpycnal flows carrying interstitial fresh water inherited from the direct fluvial discharge. The main facies types interpreted from core and borehole images (Figure 6) have been classified primarily into bed-load (B)-, suspended-load (S)-, and lofting (L)-derived transport mechanisms within a genetic facies tract developed for hyperpycnal flows (Zavala et al., 2011). Bed-load facies (B1s, B3s, B2s; Figures 6, 8) range from clay chip conglomerates with a medium-grained sandstone matrix to medium- to fine-grained sandstones. Sedimentary structures exhibit clast alignment (B1s; Figure 8A) and crude plane-parallel stratification/banding (B3s) to asymptotic cross-stratification (B2s). Bed-load facies are thought to have originated from bed-load transport from an overpassing quasi-steady turbulent flow, over a consolidated substrate. The B2s facies is related to the migration of straight or sinuous bed forms at the base of a long-lived turbulent flow carrying high suspended load.

Suspended-load facies (S1, S2, and S3; Figure 6) range from fine-grained sandstones with floating clay chips to fine- to very fine-grained sandstones. These facies exhibit sedimentary structures that range from massive (S1; Figure 8C) parallel lamination (S2; Figure 8B) to climbing ripples (S3; Figure 8D) and would have been deposited as a result of traction plus fallout from an overpassing quasi-steady turbulent flow. Experimental studies (Arnott and Hand, 1989; Sumner et al., 2008) have shown that massive sandstones will originate from a turbulent flow with a relatively high fallout rate (>0.44 mm/s), and parallel-laminated sandstones will originate from lower rates of sediment fallout (<0.44 mm/s) with equivalent flow velocities. Lamination in these massive sandstones is diffuse and appears as a recurrent feature within the massive intervals. The recurrence of massive and parallel-laminated sandstones without any break indicates fluctuating fallout rates from an overpassing sustained turbidite flow. Occasionally, we could see changes from parallel-laminated into climbing-ripple sandstones, which may suggest variations in flow conditions (Zavala et al., 2006c). These internally complex beds are called "composite beds" (Zavala et al., 2007) and typically display a vertical succession of different facies showing cyclical recurrence and transitional to sharp passages between them.

The S2h and S3w facies (Figure 6) are composed of fine- to very fine-grained sandstones. Main sedimentary structures are hummocky cross-stratification and oscillatory ripple laminations, respectively. The occurrence of hummocky cross-stratification (S2h; Figure 8B) and oscillatory ripples (S3w; Figure 8D) could be indicative of shallow-water conditions (shelfal hyperpycnites) in confined depocenters. The evidence of wave reworking at the top of major sandstone bodies is not always related to continuous wave-diffusion processes, but instead a consequence of the final reworking by the oscillating component of a density flow. In most cases, the oscillating component can be induced in the receiving basin by the disturbance provoked by the volume of the incoming flow in shallow-water bodies (Mutti et al., 1994).

Lofting-related facies (L, S2L, and S3L facies; Figure 6) are composed of fine- to very fine-grained sandstones interbedded with light gray siltstones with carbonaceous material. The associated occurrence of plant debris also favored the interpretation of a direct fluvial supply by rivers in flood (hyperpycnal systems). Main sedimentary structures are parallel and ripple laminations interbedded with thinly laminated siltstones with carbonaceous material. In general, these deposits show very little to no bioturbation. The S2L and S3L facies (Figure 8E, F) resulted from traction plus fallout from an overpassing quasi-steady turbulent flow combined with precipitation from lofting plumes. According to Zavala et al. (2006b), lofting facies constitute a diagnostic evidence of hyperpycnal sedimentation in a marine environment, resulting in vertical and lateral relationships between lofting and suspended-load facies, termed S2/L (laminated sandstones with abundant plant debris and micas) and S3/L (siltstone levels interbedded with sandstones having small climbing ripples). Lofting facies are related to the suspended fallout of fine-grained materials from lofting plumes. These plumes are related to the inverted buoyancy associated with the lift-up forces caused by the less dense interstitial fresh water contained in the hyperpycnal flow (Sparks et al., 1993; Hesse et al., 2004; Zavala et al., 2006b; Zavala et al., 2011).

The Pd facies (Figure 6) is composed of very fine-grained sandstones and laminated siltstones with water-escape structures and convolute bedding caused by high sedimentation rates.

The main facies types recognized in the B4 sandstone show a dominance of suspended-load processes (S and L facies types). Facies related to the fallout of suspended materials from turbulent flows predominate in the medium to distal part of the depositional system.

Facies Analysis

This study applied a genetic facies approach for performing facies analysis and the further development of a depositional model. Zavala et al. (2011) proposed a genetic facies approach that distinguishes facies on the basis of the main transport mechanisms involved during their deposition, resulting in a conceptual-based classification.

Facies associations were identified based on GR log trends integrated with the vertical stacking of facies interpreted from borehole images, core data, cuttings description, and biostratigraphic data within a conceptual-based channelized flow system derived from outcrop analysis. Figure 9 shows the lateral distribution of facies and facies associations within a channelized flow system (Zavala et al., 2006c). Six facies associations were identified from the image analysis of the B4 sandstone (Figure 10). These facies associations are discussed below.

Shelfal Facies Association

Description

Shelfal or background facies are composed of noncalcareous, dark gray, faintly to poorly laminated shales and siltstones (S4 facies).

This facies association appears in the images as poorly laminated to massive shales and siltstones having a relatively high GR response. These sediments might be moderately to highly bioturbated. Unfortunately, no highly bioturbated shelfal facies were identified in the core interval. Occasionally, images show interbedded deformed sandstone beds within the poorly layered siltstones and shales (Figure 10A).

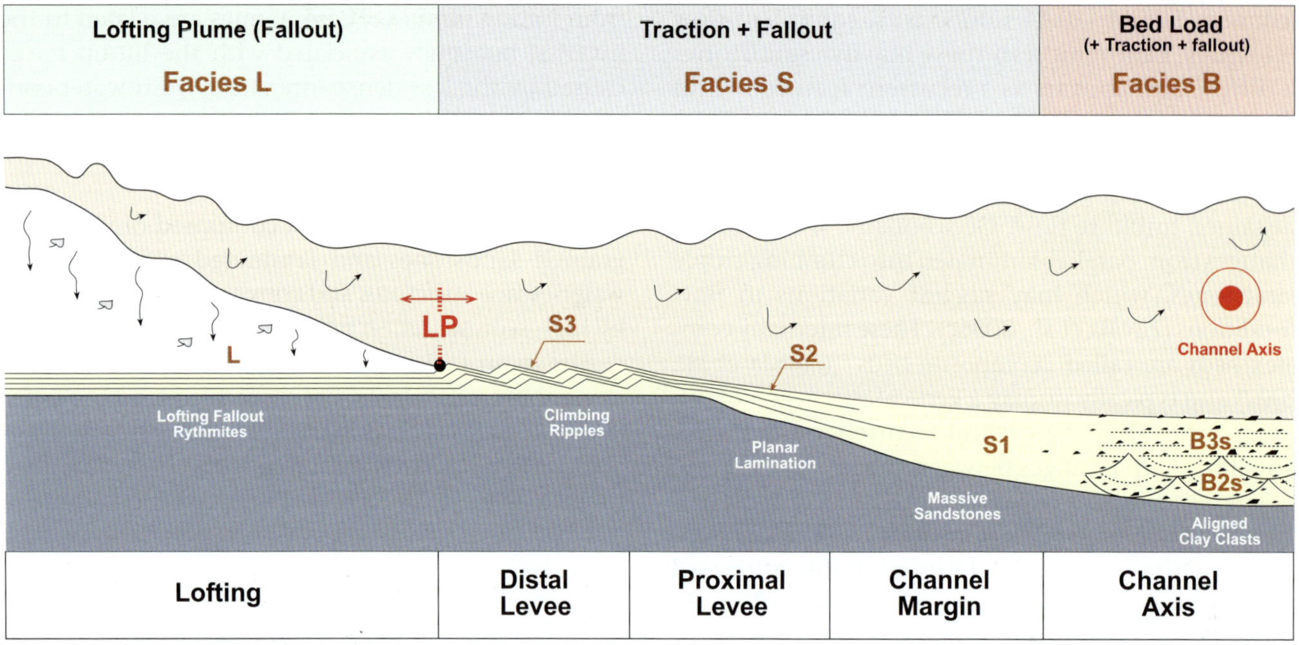

FIGURE 9. Lateral distribution of facies associations in a channelized flow system (adapted from Zavala et al., 2011). The channel-axis association is characterized by alternating bed-load facies B1s, B2s, and B3s; the channel-margin association is characterized by facies S1 and possible association of B3s, S2, S2L, S3w, and S2h; proximal levee facies are characterized by alternating S-type facies (S1, S3, S2h, and S3w) and lofting facies (S2L); distal levee facies are characterized by lofting facies (S2L, S3L) alternating with suspended-load facies (S3, S2). Lofting facies represent the most lateral facies of a channelized flow and are characterized by facies L, S2L, and S3L. LP = lofting point.

Interpretation

This facies association originated by the settling of mud and silt from suspension clouds related to delta plumes during periods of low clastic input into the basin. These low clastic input periods coincided with maximum flooding surfaces, as suggested by biostratigraphic analysis.

Lofting Facies Association

Description

The lofting facies association is composed of thinly laminated light gray siltstones with plant remains and lignite. In the analyzed cores, this facies association commonly does not show any evidence of bioturbation, with remarkably well-preserved laminae. This facies association appears in the images as highly and thinly laminated siltstones, with corresponding high GR (Figure 10B).

Interpretation

Hesse et al. (2004) were the first to recognize lofting-related facies associated with ice-rafted debris in Pleistocene deposits of the Labrador Sea. The recognition of lofting facies in marine environments is extremely important because it allows the diagnosis of a direct fluvial connection and a hyperpycnal origin for the associated deposits (Zavala et al., 2006b; Zavala, 2008). Lofting facies, in outcrops and cores, are composed of thinly laminated siltstones separated by thin intercalations of plant debris and micas, and their identification is very useful to predict the lateral occurrence of hyperpycnal channels in subsurface studies. This facies association is dominated by facies L commonly associated with facies S3L and S2L.

Distal Levee Facies Association

Description

The distal levee facies association is composed of stacked coarsening-upward successions (between 1 and 3 m [5 and 10 ft] thick), characterized from base to top by lofting facies (L, S3L), followed by 0.3- to 0.5-m (1- to 1.5-ft)-thick sharp-based, parallel-laminated, very fine-grained sandstones interbedded with thinly laminated argillaceous siltstones (S2L facies; Figure 10C). These facies appear in the images as highly and thinly laminated siltstones that grade

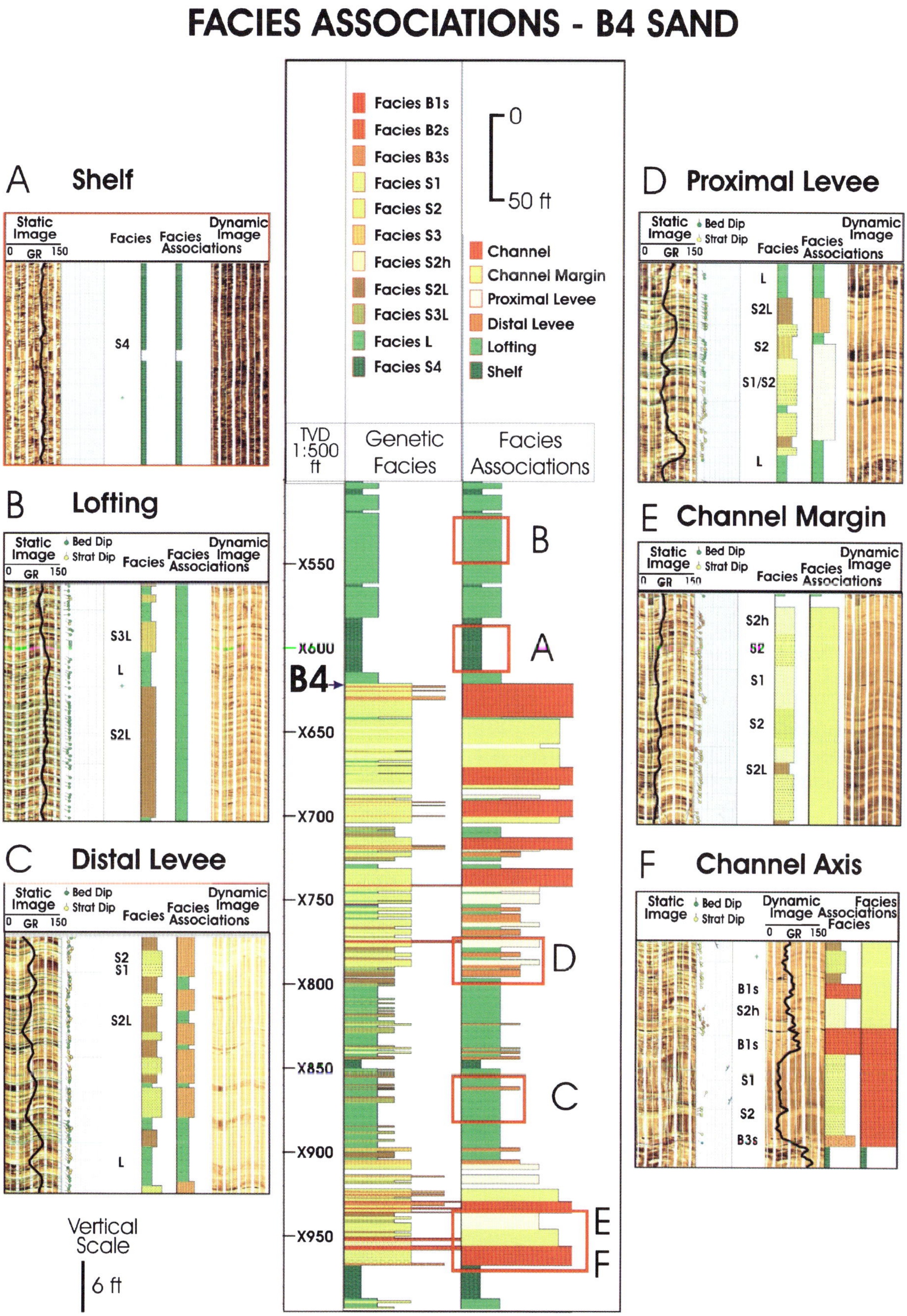

FIGURE 10. Summary of main characteristics of the six facies associations described and interpreted for the B4 sandstone in the Oilbird field. It shows an example of the main characteristics of each facies association, in terms of gamma ray (GR) log trends, static and dynamic normalized image responses, and vertical stacking of facies interpreted from borehole images. TVD = true vertical depth; Strat = stratigraphic.

upward into 0.3- to 0.5-m (1- to 1.5-ft)-thick parallel-laminated sandstones, composed of a small-scale coarsening-upward (funnel-shaped) GR pattern.

Interpretation

This facies association is characterized by traction plus fallout processes that alternate with pure fallout from lofting plumes. The complex internal characteristics of these deposits could have originated from fluctuations in velocity and concentration during a single hyperpycnal discharge. This facies association is interpreted as having accumulated in a distal levee of a channelized system.

Proximal Levee Facies Association

Description

The proximal levee facies association is composed of stacked coarsening- and thickening-upward successions (between 3 and 8 m [10 and 26 ft] thick), characterized from base to top by heterolithic or lofting facies (L, S2L, and S3L), consisting of sharp-based parallel or massive sandstones (up to 0.5 m [1.5 ft] thick) interbedded with thinly laminated shales and siltstones (up to 0.3 m [1 ft] thick), followed by 0.6- to 1-m (2- to 4-ft)-thick sharp-based ripple- to parallel-laminated sandstones (facies S2 and S3) (Figure 10D). This facies association appears in the images as a thinly horizontally layered interval that grades upward into 0.3- to 0.6-m (1- to 1.5-ft)-thick parallel and/or ripple-laminated sandstones interbedded with thinly horizontally layered beds, overlain by 0.6- to 1-m (2- to 4-ft)-thick parallel to massive sandstone beds. This facies association shows a coarsening-upward (funnel-shaped) GR pattern.

Interpretation

This facies association is mostly composed of facies types originated by traction plus fallout processes with minor intercalations of lofting-related deposits. These characteristics suggest a marginal (lateral) position with respect to the main flow, and these deposits are interpreted as having accumulated in proximal levee areas (thickening-upward successions) located proximal to a channelized system.

Channel-margin Facies Association

Description

This facies association is composed of stacked fining- and coarsening-upward successions (between 1 and 4 m [4 and 14 ft] thick), characterized by a vertical recurrence of parallel-laminated and massive sandstones (S1 and S2 facies) interbedded with thin (<0.3-m [1-ft]-thick) shale/siltstones/carbonaceous levels (S1/L). The recurrence of these facies creates thick (>15-m [50-ft]-thick) internally complex sandstone beds with no clear facies boundaries, defined as composite beds by Zavala et al. (2007) (Figure 10E). These facies appear in the images as interbedded, parallel-laminated, and massive sandstones interbedded with thin horizontal beds. These deposits may show stacked coarsening-upward and fining-upward GR patterns.

Interpretation

Composite beds typically display a vertical succession of different facies, showing cyclical recurrence and transitional to sharp passages between them. These cyclical and gradual changes are the result of near-continuous deposition from a fluctuating quasi-steady turbulent flow. This facies association is laterally related to amalgamated channel-fill facies. In this facies association, composite beds are almost entirely composed of suspended-load facies (S1, S2, and S3).

Channel-axis Facies Association

Description

The channel-axis facies association is composed of stacked thinning-upward successions (between 6 and 15 m [20 and 50 ft] thick), characterized from the base to the top by massive and stratified medium-grained sandstones interbedded with pebbly conglomerates and asymptotic cross-bedded sandstones and overlain by parallel-laminated sandstones interbedded with massive sandstones. This facies association shows composite beds with a predominance of bed-load facies (Figure 10F). This facies may appear over an erosive basal boundary, followed by parallel-bedded sandstones with interbedded aligned resistive clasts, cross-bedded sandstones, and parallel-laminated and massive sandstones interbedded with thin horizontal beds. This facies association commonly shows a blocky to fining-upward GR pattern.

Interpretation

This facies association is interpreted as channel-axis deposits. Channel axis is dominated by bed-load facies (B2s, B3s, and B1s). In this case, composite beds can be very thick (up to 15 m [50 ft] thick), depending on the duration of the hyperpycnal flow and the available accommodation space. These composite beds evolve laterally into lofting rhythmite

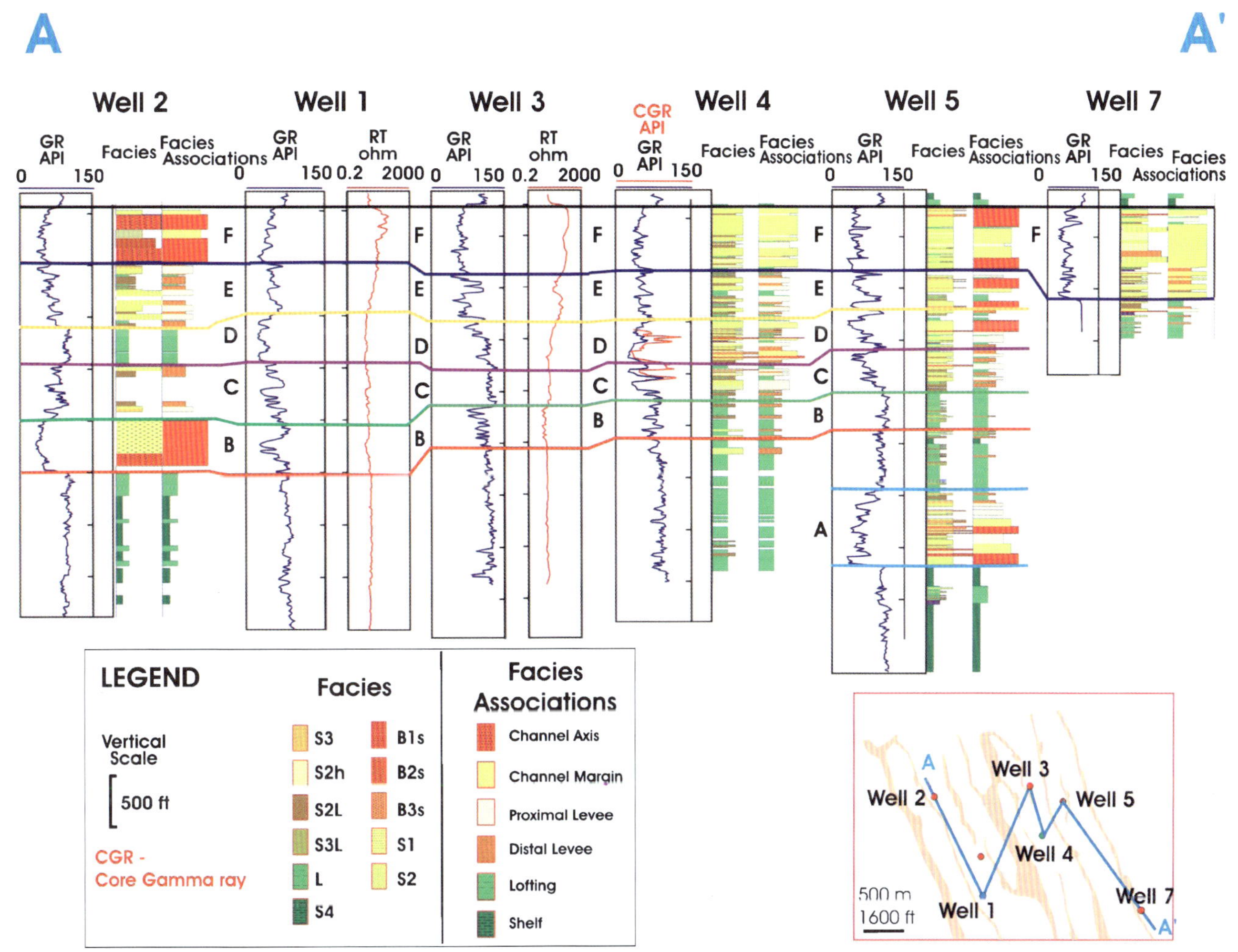

FIGURE 11. Regional detailed stratigraphic cross section showing lateral variation of sedimentary facies, facies associations, and depositional units of the B4 sandstone in the Oilbird field. The depositional units, named A, B, C, D, E, and F, were recognized based on the vertical stacking of facies associations and the gamma ray (GR) log patterns. GR = gamma ray in API units; RT = Resistivity in ohm m units; CGR = core gamma ray in API units. The API unit is a measure of radioactivity used for natural gamma-ray logs.

packages. Composite beds are coarser grained and better developed in the center of the channelized flow (Zavala et al., 2007).

Facies Maps

Based on the extrapolation of the image-derived facies associations and the GR log patterns, the B4 reservoir sand was further subdivided into six depositional units, denominated from base to top as A, B, C, D, E, and F. Figure 11 shows a cross section showing the main facies, facies associations, and depositional units interpreted on the B4 sandstone in the Oilbird field. Facies maps for the B4 reservoir have been developed and constrained by paleocurrent data derived from the analysis of image data. Figure 12 shows the facies maps developed for each depositional unit.

The depositional unit B4_A is present only in well 5, which records the filling of a depositional low in this area as a north-northwest- to south-southeast-oriented channel (Figure 12A). Unit B4_B is interpreted as being composed of channel-axis facies in well 2 and distal levee facies in wells 1 and 3. In wells 4 and 5, this unit is interpreted as lofting facies (Figure 12B). Unfortunately, no recorded paleocurrent data were available for this unit. Unit B4_C is interpreted as distal levee facies in wells 1, 2, and 4. Lofting facies in well 3 and proximal levee facies were interpreted in well 5 (Figure 12C). Again, no paleocurrent data were available. Unit B4_D is interpreted as channel margin in wells 1, 4, and 5, and proximal/

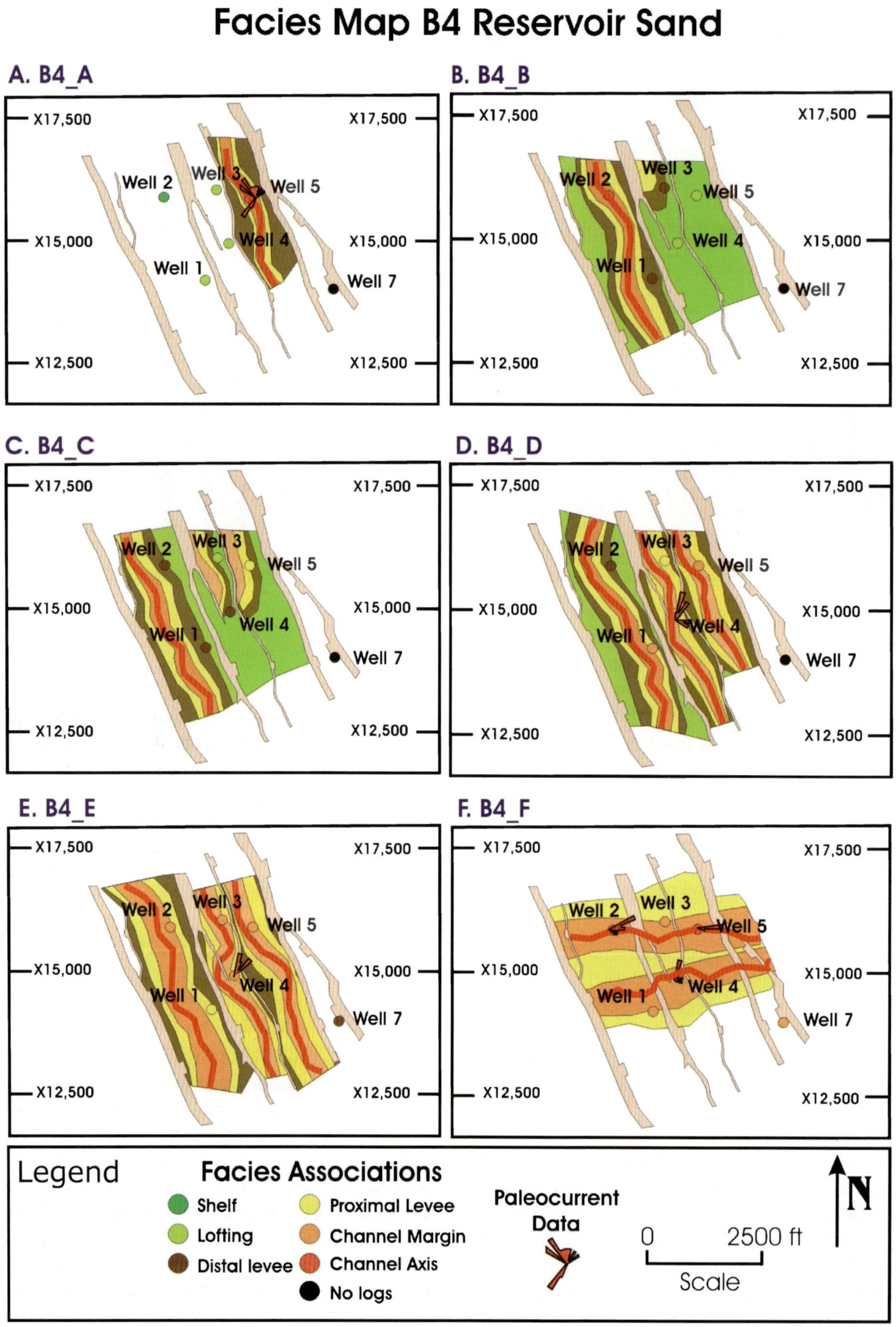

FIGURE 12. Facies maps of the B4 sandstones in the Oilbird field. Each map shows a schematic depositional environment interpretation based on the sedimentologic study performed on each well. Paleocurrent data were overlain on the well location.

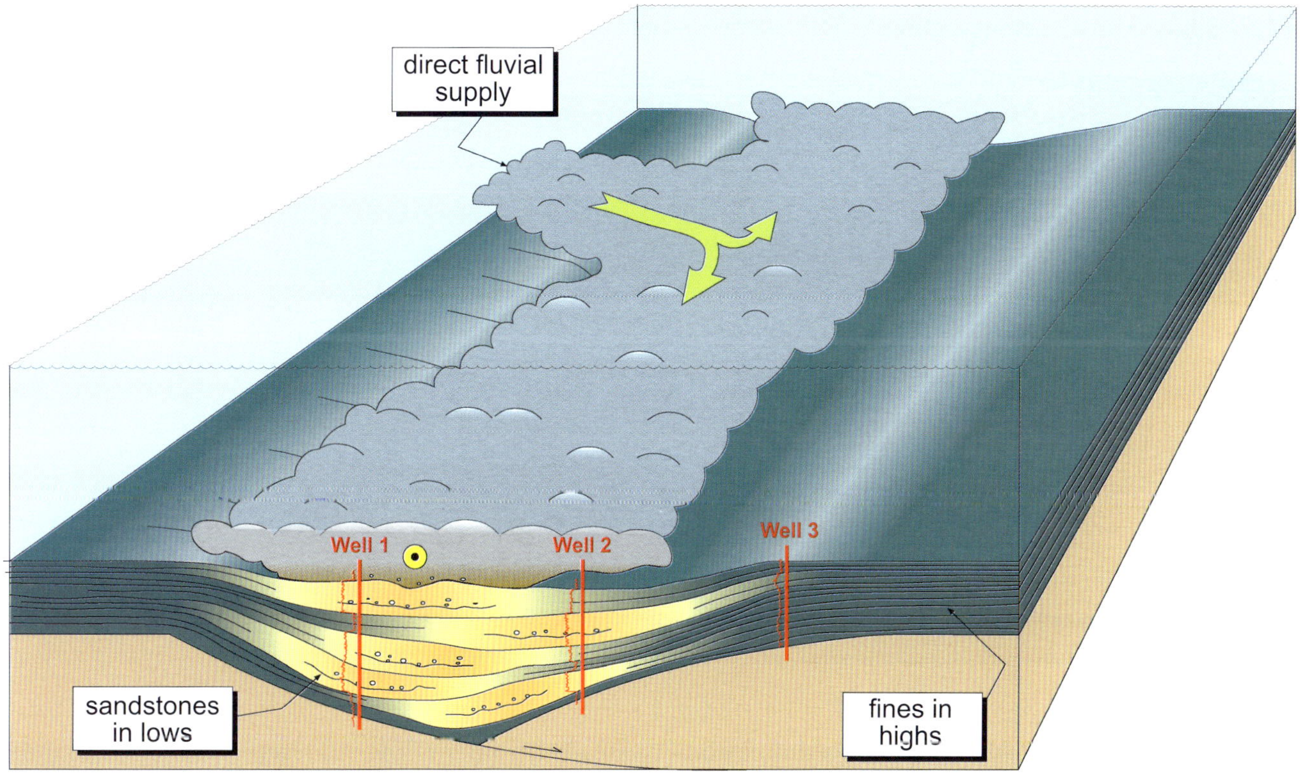

FIGURE 13. Schematic model of a fault-bounded depocenter filled by axial hyperpycnal flows.

distal levee facies in wells 3 and 2, with paleocurrent data indicating a northeast-north-northeast and east-southeast orientation (Figure 12D). Unit B4_E is interpreted as channel margin in wells 2, 3, 4, and 5, and proximal levee facies in wells 1 and 7. The interpreted channels show a northeast and east-northeast orientation based on paleocurrent data (Figure 12E). Finally, Unit B4_F is interpreted as channel axis in wells 2, 4, and 5, and channel-margin facies in wells 1, 3, and 7; with paleocurrent data indicating east, northeast, and southeast paleocurrent directions (Figure 12F).

The B4 reservoir sandstone shows a progradational pattern, reflecting the filling of a fault-controlled depocenter or minibasin. The paleoflow data indicate axial transport parallel to the fault system, indicating that these faults were active, progressively creating accommodation space (Unit D; Figure 12). The upper sandstone represents the final stage of sand accumulation, evidenced by a change in transport direction toward the east (Unit F; Figure 12). A schematic model of the axial transport of these channelized flows filling the lows (Figure 13) shows that lofting facies can be interpreted either as related to a topographic high or to a lateral position with respect to the channelized flow. Although all the sedimentologic interpretations were consistent with the observations made on the core and borehole images, a need to incorporate detailed mapping of these depositional units across the field using three-dimensional seismic data still exists. Unfortunately, seismic data in the Oilbird field are of poor quality and therefore not useful to support this model. However, based on seismic attribute analysis, a preferential north–south sand accumulation is seen (N. Lewis, 2007, personal communication), which may reflect the filling of these fault-bounded depocenters. In the literature, some examples that document axial transport of sediment gravity flows (Popescu et al., 2004; DeRuig and Hubbard, 2006) exist.

Implications in Reservoir Distribution

The hyperpycnal model introduces substantial changes for the prediction of the geometry and position of the sandstone accumulations, and understanding these changes will require detailed mapping of the associated paleotopography during deposition. Figure 14 is a conceptual model showing differences between a shelf-margin delta model and an axial hyperpycnal model. The axial filling of these fault-bounded depocenters (or minibasins) will control the geometry of the associated sand bodies with consequent changes in determining potential new plays for exploration. In the case of the line-source delta model (Figure 14A),

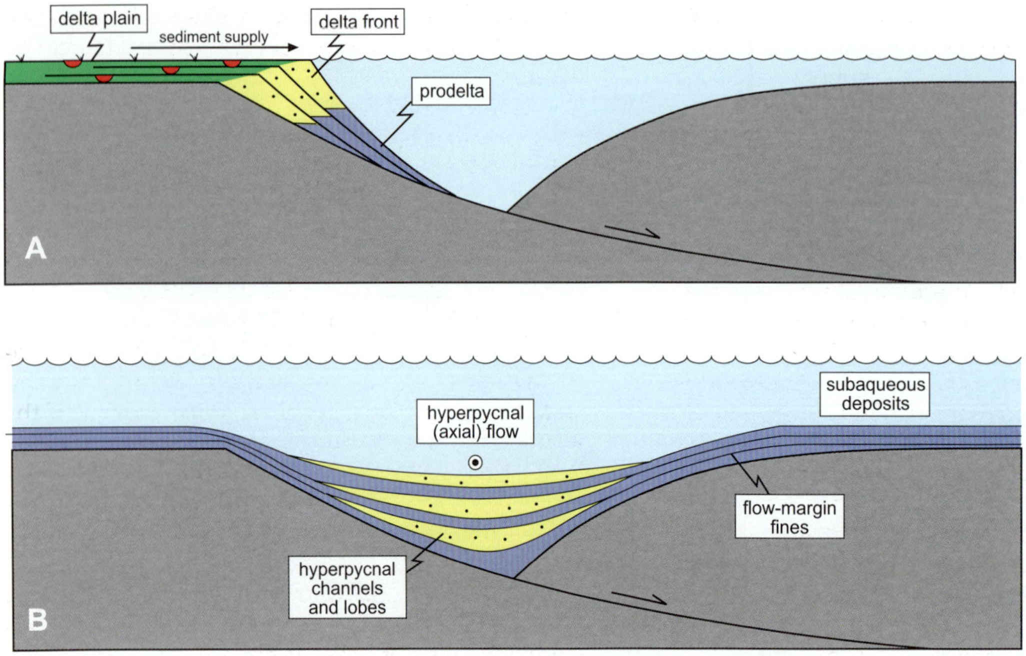

FIGURE 14. Schematic comparison between the conventional deltaic model for the Columbus Basin versus the new proposed model for the Oilbird field. (A) Line-source littoral delta model. (B) Hyperpycnal model.

it predicts the existence of sandstone accumulations on one side (one source point) of the receiving basin, whereas the lower parts of the lower paleotopography will be characterized by the accumulation of fine-grained deposits. However, the hyperpycnal model (Figure 14B) predicts the occurrence of sand accumulation in the lowest part of the paleolandscape, whereas the higher parts of the basin (margins) will be characterized by fine-grained sediments. Thus, the hyperpycnal model will represent a change in the prospectivity guide for new exploration plays. Understanding the evolution of the structures in the Oilbird field and its relationship with the creation of accommodation space will help predict the accumulation of sandstones and its distribution for future prospects.

SUMMARY AND CONCLUSIONS

Calibration of full-hole core data to borehole image log data has been critical to understand the depositional environment of the Oilbird field sandstones. Interpretation of these data has led to the identification of 12 distinct genetic facies. These genetic facies were deposited by hyperpycnal flows related to the subaqueous extension of the paleo-Orinoco fluvial system. Sedimentary facies related to hyperpycnal systems are very distinctive and could be easily differentiated from episodic surgelike classic turbidites. The presence of sandstone beds showing a recurrence of massive and parallel-laminated structures indicates sustained or near-continuous deposition from a quasi-steady turbulent flow. All reservoir sandstones show a progradational pattern, reflecting the filling of a fault-controlled depocenter. Paleoflow data indicate axial transport parallel to the fault system, indicating that these faults were active during deposition of these sandstones. The upper sandstones, located at the top of the progradational sequences, represent the final accumulation stage evidenced by a change of paleoflow from northwest-southeast toward the east. The hyperpycnal model introduces substantial changes for the prediction of the shape and location of sand accumulations. The axial filling of these fault-bounded depocenters results in different sand body geometries with substantial changes for the prediction of new exploration plays.

ACKNOWLEDGMENTS

This chapter presents the results of a sedimentologic study on the Oilbird field for EOG Resources Trinidad Limited. Mark Larsen and Maria Lourdes Diaz de Gamero provided reviews of the manuscript. We thank EOG Resources Trinidad Limited for giving its permission to publish the results of this study.

REFERENCES CITED

Arnott, R. W. C., and B. M. Hand, 1989, Bedforms, primary structures and grain fabric in the presence of suspended sediment rain: Journal of Sedimentary Petrology, v. 69, p. 1062–1069.

Babb, S., and P. Mann, 1999, Structural and sedimentary development of a Neogene transpressional plate boundary between the Caribbean and South American plates in Trinidad and the Gulf of Paria, *in* P. Mann, ed., Caribbean basins, sedimentary basins of the world: Elsevier Science B.V., v. 4, p. 495–557.

Bates, C., 1953, Rational theory of delta formation: AAPG Bulletin, v. 37, p. 2119–2162.

Bowman, A., 2003, Sequence stratigraphy and reservoir characterization in the Columbus Basin, Trinidad, Ph.D. dissertation, University of London, London, England, 530 p.

Bowman, A., and H. Johnson, 2006, Storm-dominated shelf-edge deltas in a high-accommodation setting: An outcrop example from the Columbus Basin, Trinidad, West Indies: AAPG/GSTT Hedberg Research Conference, Port of Spain, Trinidad West Indies, AAPG Search and Discovery Article 90057, 1 p., http://www.searchanddiscovery.net/abstracts/html/2006/hedberg_intl/abstracts/bowman.htm (accessed July 2010).

Contreras, C. C., and H. Gamero, 2000, High-resolution borehole images as powerful reservoir characterization tools: Transactions of the GSTT 2000 SPE Conference, Port of Spain, Trinidad 10–13 July, 2000, http://www.gstt.org/publications/transactions_of_gstt_2000_spe.htm (accessed July 2010).

DeRuig, M., and S. Hubbard, 2006, Seismic facies and reservoir characteristics of a deep-marine channel belt in the Molasse Foreland Basin, Puchkirchen Formation, Austria: AAPG Bulletin, v. 90, no. 5, p. 735–752, doi:10.1306/10210505018.

Diaz de Gamero, M. L., 1996, The changing course of the Orinoco River during the Neogene: A review: Palaeogeography, Palaeoclimatology, Palaeoecology, v. 123, p. 385–402, doi:10.1016/0031-0182(96)00115-0.

Gamero, H., C. C. Contreras, P. Pestman, and A. Mizobe, 2000, Borehole electrical images as a reservoir characterization tool in the Merecure Formation, Guarico 13 field, eastern Venezuela: VII Simposio Bolivariano, Exploracion Petrolera en las Cuencas Subandinas, Memoria, p. 620–641.

Gamero, H., C. Zavala, and C. Contreras, 2005, A reinterpretation of the Misoa facies types: Implications of a new depositional model, Maracaibo Basin, Venezuela (abs.): 2005 AAPG International Conference and Exhibition, Evolving Stratigraphic Techniques and Interpretation III, Paris, http://www.searchanddiscovery.net/documents/abstracts/2005intl_paris/gamero.htm (accessed July 2010).

Gamero, H., J. Reader, C. Izatt, C. Zavala, and C. Contreras, 2007, Herrera sandstones in the Southern Basin Area, Trinidad: Evidence of hyperpycnites deposited away from ancient Oficina delta systems in eastern Venezuela (abs.): AAPG Annual Meeting, Abstracts volume, p. 51, http://www.searchanddiscovery.net/abstracts/html/2007/annual/abstracts/lbGamero.htm (accessed July 2010).

Gamero, H., N. Lewis, R. Welsh, C. Zavala, and C. Contreras, 2008, Evidences of a shelfal hyperpycnal deposition in the Pliocene sandstones in the Oilbird field, SE Coast, Trinidad: Impact on reservoir distribution and field redevelopment, *in* J. J. Ponce and E. B. Olivero, conveners, Conference Proceedings, Research: Sediment transfer from shelf to deep water: Revisiting the delivery mechanisms, March 3–7, 2008, Ushuaia-Patagonia, Argentina, Conference Proceedings, 4 p., AAPG Search and Discovery Article 90079, http://www.searchanddiscovery.net/abstracts/html/2008/hedberg_argentina/extended/gamero/gamero.htm (accessed July 2010).

Gibson, R., K. E. Meisling, and J. C. Sydow, in press, Columbus Basin, offshore Trinidad: A detached pull-apart basin in a transpressional foreland setting, *in* A. Bally and D. Roberts, eds., Phanerozoic regional geology of the world: Columbus Basin, v. 3, Elsevier.

Hesse, R., H. Rashid, and S. Khodabakhsh, 2004, Fine-grained sediment lofting from meltwater-generated turbidity currents during Heinrich events: Geology, v. 32, p. 449–452, doi:10.1130/G20136.1.

Jackson, C., A. A. Zakaria, H. D. Johnson, F. Tongkul, and P. D. Crevello, 2009, Sedimentology, stratigraphic occurrence and origin of linked debrites in the west Crocker Formation (Oligo–Miocene), Sabah, NW Borneo: Marine and Petroleum Geology, v. 26, p. 1957–1973, doi:10.1016/j.marpetgeo.2009.02.019.

Kneller, B., and M. Branney, 1995, Sustained high-density turbidity currents and the deposition of thick massive sands: Sedimentology, v. 42, p. 607–616, doi:10.1111/j.1365-3091.1995.tb00395.x.

Lamb, M. P., and D. Mohrig, 2009, Do hyperpycnal-flow deposits record river-flood dynamics?: Geology, v. 37, p. 1067–1070, doi:10.1130/G30286A.1.

Lamb, M. P., P. M. Myrow, C. Lukens, K. Houck, and J. Strauss, 2008, Deposits from wave-influenced turbidity currents: Pennsylvanian Minturn Formation, Colorado: Journal of Sedimentary Research, v. 78, p. 480–498, doi:10.2110/jsr.2008.052.

Leonard, R., 1983, Geology and hydrocarbon accumulations, Columbus Basin, offshore Trinidad: AAPG Bulletin, v. 67, no. 7, p. 1081–1093.

Mansurbeg, H., M. A. K. El-ghali, S. Morad, and P. Plink-Björklund, 2006, The impact of meteoric water on the diagenetic alterations in deep-water, marine siliciclastic turbidites: Journal of Geochemical Exploration, v. 89, p. 254–258, doi:10.1016/j.gexplo.2006.02.001.

Mulder, T., and J. Alexander, 2001, The physical character of subaqueous sedimentary density flows and their deposits: Sedimentology, v. 48, p. 269–299, doi:10.1046/j.1365-3091.2001.00360.x.

Mulder, T., J. P. M. Syvitski, S. Migeon, J. C. Faugéres, and B. Savoye, 2003, Marine hyperpycnal flows: Initiation, behavior and related deposits: A review: Marine and Petroleum Geology, v. 20, p. 861–882, doi:10.1016/j.marpetgeo.2003.01.003.

Mutti, E., G. Davoli, and R. Tinterri, 1994, Flood-related gravity-flow deposits in fluvial and fluvio-deltaic depositional systems and their sequence-stratigraphic implications, *in* H. W. Posamentier and E. Mutti, eds.,

Second High-Resolution Sequence Stratigraphy Conference, Tremp, Abstract Book: Italy, Instituto di Geologia, Universita di Parma, p. 137–143.

Mutti, E., G. Davoli, R. Tinterri, and C. Zavala, 1996, The importance of ancient fluvio-deltaic systems dominated by catastrophic flooding in tectonically active basins: Memorie di Scienze Geologiche, Universita di Padova, v. 48, p. 233–291.

Mutti, E., R. Tinterri, G. Benevelli, D. Di Biase, and G. Cavanna, 2003, Deltaic, mixed and turbidite sedimentation of ancient foreland basins: Marine and Petroleum Geology, v. 20, p. 733–755, doi:10.1016/j.marpetgeo.2003.09.001.

Nakajima, T., 2006, Hyperpycnites deposited 700 km away from river mouths in the Central Japan Sea: Journal of Sedimentary Research, v. 76, p. 59–72.

Pattison, S., 2005, Storm-influenced prodelta turbidite complex in the Lower Kenilworth member at Hatch Mesa, Book Cliffs, Utah: Implications for shallow marine facies models: Journal of Sedimentary Research, v. 75, no. 4, p. 420–439, doi:10.2110/jsr.2005.033.

Pickering, K. T., and R. N. Hiscott, 1985, Contained (reflected) turbidity currents from the Middle Ordovician Cloridorme Formation, Quebec, Canada: An alternative to the antidune hypothesis: Sedimentology, v. 32, p. 373–394, doi:10.1111/j.1365-3091.1985.tb00518.x.

Pindell, J. L., R. Higgs, and J. Dewet, 1998, Cenozoic palinspastic reconstruction, paleogeographic evolution and hydrocarbon setting of the northern margin of South America, *in* Paleogeographic evolution and nonglacial eustasy, Northern South America: SEPM Special Publication 58, p. 45–85.

Plink-Björklund, P., and R. J. Steel, 2004, Initiation of turbidite currents: Outcrop evidence for Eocene hyperpycnal flow turbidites: Sedimentary Geology, v. 165, p. 29–52, doi:10.1016/j.sedgeo.2003.10.013.

Popescu, I., G. Lericolais, N. Panin, and A. Normand, 2004, The Danube submarine canyon (Black Sea): Morphology and sedimentary processes: Marine Geology, v. 206, no. 1–4, p. 249–265, doi:10.1016/j.margeo.2004.03.003.

Robertson, P., and K. Burke, 1989, Evolution of the southern Caribbean plate boundary, vicinity of Trinidad and Tobago: AAPG Bulletin, v. 73, no. 4, p. 490–509.

Schumm, S. A., 1981, The evolution and response of the fluvial system, sedimentologic implications, *in* F. G. Ethridge and R. M. Flores, eds., Recent and ancient nonmarine depositional environments: Models for exploration: SEPM Special Publication 31, p. 19–29.

Sparks, R. S. J., R. T. Bonnecaze, H. E. Huppert, J. R. Lister, M. A. Hallworth, J. Phillips, and H. Mader, 1993, Sediment-laden gravity currents with reversing buoyancy: Earth and Planetary Science Letters, v. 114, p. 243–257, doi:10.1016/0012-821X(93)90028-8.

Sumner, E. J., L. A. Amy, and P. J. Talling, 2008, Deposit structure and processes of sand deposition from decelerating sediment suspensions: Journal of Sedimentary Research, v. 78, p. 529–547, doi:10.2110/jsr.2008.062.

Walker, R., 1996, Facies modeling and sequence stratigraphy: Journal of Sedimentary Petroleum, v. 60, p. 777–786.

Wood, L., 2000, Chronostratigraphy and tectonostratigraphy of the Columbus Basin, Eastern offshore Trinidad: AAPG Bulletin, v. 84, no. 12, p. 1905–1928.

Wood, L., and C. Roberts, 2001, Opportunities in a world-class hydrocarbon basin: Trinidad and Tobago's eastern offshore marine province: Houston Geological Society Bulletin, v. 43, no. 10, p. 37–45.

Zavala, C., 2008, Toward a genetic facies tract for the analysis of hyperpycnal deposits, *in* J. J. Ponce and E. B. Olivero, conveners, Conference Proceedings, Research Sediment transfer from shelf to deep water: Revisiting the delivery mechanisms: Ushuaia-Patagonia, Argentina, Conference Proceedings, AAPG Search and Discovery Article 50075, 2 p., http://www.searchanddiscovery.net/abstracts/html/2008/hedberg_argentina/extended/zavala/zavala.htm (accessed July 2010).

Zavala, C., M. Arcuri, and H. Gamero, 2006a, Toward a genetic model for the analysis of hyperpycnal systems (abs.): Geological Society of America Abstracts with Programs, v. 38, no. 7, p. 541, http://gsa.confex.com/gsa/2006AM/finalprogram/abstracts_110453.htm (accessed July 2010).

Zavala, C., H. Gamero, and M. Arcuri, 2006b, Lofting rhythmites: A diagnostic feature for the recognition of hyperpycnal deposits (abs.): Geological Society of America Abstracts with Programs, v. 38, no. 7, p. 541, http://gsa.confex.com/gsa/2006AM/finalprogram/abstracts_110667.htm (accessed July 2010).

Zavala, C., J. Ponce, D. Drittanti, M. Arcuri, H. Freije, and M. Asensio, 2006c, Ancient lacustrine hyperpycnites: A depositional model from a case study in the Rayoso Formation (Cretaceous) of west-central Argentina: Journal of Sedimentary Research, v. 76, p. 41–59, doi:10.2110/jsr.2006.12.

Zavala, C., M. Arcuri, H. Gamero Díaz, and C. Contreras, 2007, The composite bed: A new distinctive feature of hyperpycnal deposition (abs.): AAPG Annual Convention & Exhibition, v. 16, p. 157, http://www.searchanddiscovery.net/abstracts/html/2007/annual/abstracts/lbZavala.htm (accessed July 2010).

Zavala, C., M. Arcuri, M. Di Meglio, H. Gamero Diaz, and C. Contreras, 2011, A genetic facies tract for the analysis of sustained hyperpycnal flow deposits, *in* R. M. Slatt and C. Zabala, eds., Sediment transfer from shelf to deep water—Revisiting the delivery system: AAPG Studies in Geology 61, p. 31–51.